Historical Perspectives on Erklären and Verstehen

Uljana Feest
Editor

# Historical Perspectives on Erklären and Verstehen

Springer

*Editor*
Uljana Feest
Technische Universität Berlin
Germany
Uljana.Feest@TU-Berlin.DE

ISBN 978-94-007-3134-9 e-ISBN 978-90-481-3540-0
DOI 10.1007/978-90-481-3540-0
Springer Dordrecht Heidelberg London New York


Softcover reprint of the hardcover 1st edition 2009


Printed on acid-free paper

Springer is part of Springer Science+Business Media (www.springer.com)

# Contents

# Chapter 1
# Historical Perspectives on Erklären and Verstehen: Introduction

**Uljana Feest**

## 1.1 Methodological Preliminaries

The conceptual dichotomy of *Erklären* and *Verstehen* (explaining vs. understanding) has a revealing dual status. On the one hand, it has something of an antiquated air to it, as we loosely associate its origins with the work of Wilhelm Dilthey and other nineteenth-century German philosophers who are not widely read any more, at least not within contemporary Anglo-American history and philosophy of the human sciences. At the same time, however, remnants of the dichotomy still come up in various guises and in various areas of contemporary philosophy and philosophy of science. One example is the long-standing debate over the logical status of action explanations ("reasons vs. causes") in philosophy of mind (Davidson 1980), and associated issues of "teleological explanations" and the explanatory status of laws of nature in the philosophy of the human sciences (Dray 1957; Hempel 1965; von Wright 1971). Another is the question of whether the subject matter of the social sciences requires a special type of interpretative, hermeneutic, or perhaps even empathetic, "access" (Collingwood 1946; Winch 1964; Taylor [1971] 1985). More recently, there has been renewed interest in the question of how to explicate our capacity to interpret another person's actions (see the recent suggestion that the "theory–theory" vs. "simulation theory" distinction is similar to some aspects of the Erklären/Verstehen distinction) (Kögler and Stueber 2000). And within mainstream analytical philosophy of the social sciences, one of the central topics has long been the question of whether social facts/events can be reduced to the explanations of the actions of individuals (e.g., Kincaid 1997), raising questions about the units at which explanatory and/or interpretive efforts ought to be directed. ("individualism" vs. "holism"). This question can be traced back to early twentieth-century debates in economics and other emerging social sciences (Udehn 2001).

U. Feest (✉)
Technische Universität Berlin, Institut für Philosophie,
Straße des 17. Juni 135, 10623 Berlin
email: uljana.feest@tu-berlin.de

U. Feest (ed.), *Historical Perspectives on Erklären and Verstehen*, Archimedes, Vol. 21,
DOI 10.1007/978-90-481-3540-0_1, © Springer Science+Business Media B.V. 2010

While this list of topics is far from complete, it suggests that contemporary debates in the philosophies of mind, action, and science touch on a variety of aspects of the earlier dichotomy between explanation and understanding. Keeping this in mind is important for at least two reasons. First, it reminds us that from a philosophical perspective the Erklären/Verstehen distinction is richer and more multi-faceted than might appear at first sight, suggesting that perhaps some systematic insights might be gained from investigating the historical precursors and the contexts in which they originated. Second, it raises the question of whether the categories employed by the current philosophical literature may provide us with analytical tools that we can use when entering the thicket of nineteenth and early twentieth-century issues and debates. The essays in this volume are selected to juxtapose precisely these two questions, i.e., (1) what (if any) novel philosophical insights can be gained from analyses of the varied and diverse previous debates about aspects of the dichotomy between explanation and understanding, and (2) what (if any) historical insights can be gained by means of analytical tools taken from current philosophical discussions?

The second question, in particular, raises an important problem for conducting this sort of historical/philosophical enterprise: the worry that by using our contemporary philosophical categories, we are in danger of anachronistically reading issues into the historical debates that did not actually concern the historical actors. We might then assume that our current categories form some kind of teleological endpoint of a historical trajectory. However, if our interest in historical questions is motivated at least in part by a desire to learn something about current debates, then surely we need a language that makes past events and discussions comprehensible and relevant to our present concerns. These are legitimate worries. Ironically, it was precisely these types of questions that were behind some of the nineteenth-century writings in the philosophy of history, which found one articulation in terms of the dichotomy between *Erklären* and *Verstehen*. The problem at issue is how to gain knowledge about a subject matter both intimately tied to our own sense of who we are and what we value, and at the same time remote from our current lives and practices. Or, as the issue was put as part of debates about *historicism*, how can we appreciate the extent to which previous categories may not have been informed by our own values and interests without thereby relativizing those very values and interests, and how can we get interpretive access to the past without assuming that our current categories are at least partially valid? (e.g., Wittkau 1992; see also Jacques Bos's contribution in this volume). In a similar vein, a philosopher who looks at aspects of historical debates about explanation and understanding may worry that she must refrain from any judgment of what is the *correct* way of thinking about this dichotomy, whereas a historian may worry that such philosophical prejudices might distort the historical narrative.

Several contributions to this volume address these questions, either explicitly or implicitly. Beginning with worries more likely to be formulated by historians, we can in particular identify methodological questions that concern the use of *sources*. This concern has two aspects. First, can we do justice to the history of a philosophical category – or any category, for that matter – only by looking at historical sources that explicitly addresses this category (for example, in our case, nineteenth or early twentieth-century literature about the distinction between explanation and understanding)? Second, can we do justice to a particular writer's notion of that category by taking at face value what he writes about his own motivation for adopting it? In response to both types of questions, there is by now a consensus amongst many historians of science and of philosophy that (a) intellectual history – like other kinds of history – has to be careful not to uncritically adopt actors' categories, and (b) more generally, even the actors' own thinking about a particular issue has to be contextualized vis-à-vis their other intellectual commitments and interests, as well as the complex conditions that make the totality of their commitments possible. Such conditions include cognitive as well as practical, institutional, and cultural factors. The articles in this volume respond to these challenges in several ways. For example, one author (Christopher Pincock) seeks to read some of the nineteenth-century philosophical writings about *Erklären* and *Verstehen* as standing for a more fundamental problem, which he terms the problem of the "unity of experience". In turn, other authors contextualize aspects of the Erklären/Verstehen dichotomy in relation to debates about educational reforms in nineteenth-century Germany (Denise Phillips), controversies about the relationship between science and religion in the wake of the rise of Darwinism (Bernhard Kleeberg), aspects of the material culture of the Austrio-Hungarian empire (Katherine Arens), and a growing appreciation – throughout the nineteenth century – of the notion of *individuality*, both with respect to persons and with respect to historical events (Jacques Bos).

The philosophical concern with the question of whether an appreciation of the contingent nature of intellectual history forces us to regard our current philosophical positions as unfounded or arbitrary is confronted especially clearly by Christopher Pincock, who tries to strike a middle ground between writing a history of arguments for whatever positions we currently hold and writing a history that makes our current positions entirely contingent on their historical developments. He concludes with a plea for a type of intellectual history that aims at identifying important philosophical problems and keeping a wide range of solutions – both past and present – on the table. In a similar vein, Warren Schmaus's comparative analysis of French and German debates in the nineteenth and early twentieth centuries highlights the fact that there is no reason to suppose that a shared set of problems would automatically lead to similar philosophical solutions, thereby also expanding the range of philosophical options available to us. Thomas Uebel's narrative, in turn, shows that even if we restrict our attention to an analysis of developments that took place *within* one (e.g., the logical empiricist) tradition, we may find that the dynamics of the emerging story provide us with a much more fine-grained picture than standard textbook accounts might suggest.

The dichotomy between *Erklären* and *Verstehen* is located somewhere between science and philosophy, in that it expresses philosophical thoughts about the epistemological or methodological foundations of the human sciences. Moreover, it emerged at a time when the line between philosophy and the sciences was less clear-cut and more contested than it is today. Hence, it seems especially clear that if we aim at an analysis of the historical and systematic status of this dichotomy, a combination of approaches from the history of science, the philosophy of science, and the history of the philosophy of science is called for. It is for this reason that the current volume presents a range of articles by authors from different disciplines, where this interdisciplinarity plays out both in terms of the *academic training* of the contributors (bringing together philosophers, historians, sociologists, and literary scholars), and in terms of their *subject matters* (social science, psychology, history, theology, philosophy, literature, and intellectual culture). The volume therefore attempts not only to offer different disciplinary perspectives on the history of the *Erklären/Verstehen* dichotomy, but also to overcome a narrow focus on disciplinary histories. In this vein, several contributions offer insights into the writings of well-known figures (such as Wilhelm Dilthey or Max Weber) by relating these writings in novel ways to other academic and/or scientific developments. For example, Michael Heidelberger argues that Max Weber's conception of an understanding sociology (which is often characterized as coming out of an engagement with the work of Heinrich Rickert) was in fact stimulated by the philosopher/psychologist and theorist of probability Johannes von Kries, while Daniel Ŝuber argues that Wilhelm Dilthey's work had a much bigger impact on twentieth-century sociology than is commonly assumed, and Safia Azzouni describes the ways in which Dilthey's theory of poetics helped shape early twentieth-century popular science writings.

Finally, the volume takes a *comparative* perspective, insofar as a number of contributions compare and contrast the issues discussed, and concepts used, in debates that took place in Germany, Austria, France, Britain, and the USA. In taking this perspective, the volume seeks to highlight commonalities and divergences in the approaches adopted by writers in different countries and national traditions. The contributions of Warren Schmaus and Philipp Müller, for example, bring out the specifics of the French debates, with Schmaus focusing more on differences between French and German debates, and Müller revealing some unexpected commonalities between the thought of Wilhelm Dilthey and Hippolyte Taine. In a similar vein, David Leary traces similarities and dissimilarities between William James and Wilhelm Dilthey with respect to their notions of understanding and explanation, while Roger Smith provides a detailed analysis of the issues that dominated British debates about the relationship between different areas of learning.

## 1.2 Overview of the Papers

In the German context, the distinction between *Erklären* and *Verstehen* is usually seen as closely linked to the aim of securing an epistemological basis for the distinction between the *Naturwissenschaften* and *Geisteswissenschaften* (the natural and the human sciences). Denise Phillips (University of Tennessee at Knoxville) contextualizes the latter dichotomy by relating it to debates that started in the 1830s, regarding the notion of *Bildung*, i.e., the question of what constitutes the notion of knowledge and education that was essential to the self-fashioning of the educated middle class in the nineteenth century. Central to these debates, Phillips argues, were the questions of what kinds of personality traits were crucial to being a good scientist, and what kind of training was required to assist the development of such traits. She thereby brings out the close relationship between notions of *Bildung* (the German term "Bildung" has a double meaning of "education" and "molding") and philosophical writings about scientific methodology. Using as a case study a specific debate in Dresden, Phillips points out that what was at stake were two models of knowledge, one that emphasized the study of *texts,* and one that emphasized the study of *nature*. While the former was traditionally more highly valued and deeply rooted in the German tradition of a humanist secondary education, Phillips points to the complex set of circumstances in which the study of nature not only began to become part of the university curriculum (raising the question of whether high school students were adequately prepared), but also began to be organized in local associations for the study of nature, whose members increasingly protested their marginalization by traditional educational culture.

Phillips's contribution provides an important context for understanding the emerging distinction between *Natur-* and *Geisteswissenschaften*. It makes clear that the underlying notions of both the natural *and* the human sciences were in a state of flux, and underwent some changes in the course of the nineteenth century. For example, while we find one of Phillips's protagonists referring to the natural sciences as based in *sense experience* and the human sciences as allowing for a *text-based* understanding of the past, Wilhelm Dilthey would later distinguish between the two in terms of a science based in *hypotheses* (natural sciences) vs. one based in *lived experience* (human sciences) (see Feest 2007). This points to an important shift that took place, namely that parts of the human sciences themselves "went empirical". In his contribution, Bernhard Kleeberg (Universität Konstanz) analyzes a particular aspect of this shift, namely the question of how German protestant theologians in the nineteenth century responded to the fact that the new science of evolution was producing an empirically based alternative narrative of the history of the earth and the place of human beings on it. As a consequence, theologians felt themselves in danger of losing their interpretive authority concerning issues of ethics and meaning. As Kleeberg points out, we can make out an overall development, in the course of which theology became *one humanity among others*, even though many theologians still wanted to resist this conclusion, emphasizing the distinct status of religious experience as the basis of theology. The central thesis of Kleeberg's paper is that in the course of this development, the very notions of *meaning* and *interpretation* underwent some changes as theologians developed a sophisticated hermeneutic methodology, which, in many cases, sought to show that science and religion could coexist peacefully. As Kleeberg shows, theologians adopted different strategies in their attempts to reconcile science and religion. Some turned to neo-vitalism in support of the idea that both material and non-material factors contribute to evolution, while others took the position that while science provided *explanations* of events and phenomena in natural history, it still had to rely on theology to provide an *understanding* of the symbolism of nature or to interpret the moral narrative inherent in the biblical account of creation.

Like Kleeberg and Phillips, Safia Azzouni (Max Planck Institute for the History of Science) also addresses the issue of how notions of *Erklären*, *Verstehen*, and related ideas played out in the cultural and educational context of turn-of-the-century Germany. More specifically, she describes the way in which the literary genre of *science popularization* constituted an important new model for the presentation and distribution of knowledge. This model, she argues, borrowed significant elements from both the natural and the human sciences, in that it attempted to convey scientific explanations by means of literary techniques, thereby re-creating in the reader a lived experience of the scientific facts in question. As a point of departure, Azzouni takes Wilhem Dilthey's work, *Poetik*, of 1887, arguing that the ideas expressed there were a major influence on Dilthey's one-time student, the high-school teacher and popular science writer Kurd Laßwitz. While Dilthey had in other places emphasized the gulf between explanation and understanding as constituting an important epistemological difference between the natural and the human sciences, his *Poetik* posited the notion of lived experience both as an important *explanatory concept* (i.e., one that explains the poet's creative potential) and as something that enables us to *understand* that which the poet conveys to us. As Azzouni shows, Laßwitz took this to mean that the poet is ideally suited to evoke in his audience the lived experience necessary for an understanding of scientific ideas. She then provides an illustration of how he applied this idea in his own popular science writings, seeming thereby to bridge the gap between the human and natural sciences.

As laid out in the contribution by Philipp Müller (Humboldt Universität), the French philosopher Hippolyte Taine, like Dilthey and Laßwitz, thought that artistic expression was an important vehicle of psychological insights. Müller reminds us that Dilthey used Taine, along with Buckle and Mill, as a prime example of the kind of positivistic philosophy of history that he rejected. However, Müller argues that in fact Taine's idea of founding history in psychology was closer to Dilthey's conception of the human sciences than is commonly assumed. While Taine did appeal to the role of psychological laws in the explanations of historical events, he in fact viewed the structure of the human mind as resulting from the interplay between mental and historical forces, and, furthermore reflected on the cultural and historical context of the science of psychology itself. Müller places Taine's philosophy of history in its historical context, showing how Taine's views about the necessity to naturalize the human mind have to be read as criticisms of spiritualists' writings, such as Victor Cousin's "méthode psychologique", and that his own outlook on psychology was the result of close intellectual contacts with contemporary novelists and literary critics who emphasized the historical situatedness of human thought. Taine then sought to investigate this situatedness by means of historical studies of artistic products.

The comparison between Taine and Dilthey with respect to the status of psychological laws raises the more general question of the relationship between French and German approaches to the issues we might summarize under the heading of *Erklären* and *Verstehen*. This task is taken on by Warren Schmaus (Illinois Institute of Technology). Schmaus argues that while reflecting on differences in the subject matters of the natural and human sciences, French philosophers did not conclude that any significant methodological distinctions should be drawn. In this, he suggests, they differed both from Dilthey, who derived the distinction between *Erklären* and *Verstehen* from the different ways in which we gain epistemic access to mind and nature, and from Windelband and Rickert, who derived the distinction between nomothetic and idiographic method from the different scientific goals of gaining knowledge about laws of nature vs. individual events. Beginning with an account of Dilthey's early thoughts on descriptive psychology and his notion of lived experience, Schmaus shows that there were some parallels in the work of various French philosophers, such as the idea that we have direct, unmediated experience of our inner lives. He argues, however, that few in France thought that this implied a distinct foundation for the human sciences. Similarly, while some French scholars distinguished between the nomothetic and idiographic in ways similar to Windelband, they did not argue that this entailed a *methodological* distinction, since they believed that all scientific knowledge was inductive and hypothetical. Schmaus's paper concludes with a comparison between (the later) Dilthey's method of hermeneutical interpretation and Durkheim's notion of an interpretive social science, pointing to the fact that while Durkheim disagreed with some of Dilthey's most fundamental assumptions (i.e., the idea of grounding the social sciences in psychology), Durkheim nonetheless formulated his social science as an interpretive endeavor.

In his contribution, "Instead of Erklären and Verstehen: William James on Human Understanding", David Leary (University of Richmond) also offers a comparative perspective, by analyzing some aspects of William James's philosophy of psychology. He argues that while James in fact shared with Wilhelm Dilthey some views about the nature of psychology, James never became involved in anything like the Erklären/Verstehen debate. As Leary explains, James seems not to have found such a distinction useful. For James, the question of whether we take an explanatory or a descriptive stance was ultimately a matter of preference vis-á-vis an open-ended, changing world. Leary contextualizes this analysis of James by providing an overview of James's views on human understanding, highlighting some significant biographical factors that may have contributed to James's views, such as his extensive reading of Goethe and Shakespeare and his association with the metaphysical club around Chauncey Wright. As Leary shows, James's views about explanation comes out particularly clearly in his response to the philosopher-psychologist Ladd, who had argued (in a review of James's book *Principles of Psychology*) that James's approach was not able to provide truly scientific explanations. James's reply suggests that he did think that psychology aimed at providing explanations (by which he meant ultimate descriptions of how things fit in), but that he thought of the actually existing explanatory statements as a provisional body of propositions. Leary concludes with a brief discussion of why Dilthey and James might have been content with such differing conceptions of the nature of science.

The contribution by Katherine Arens (University of Texas at Austin) draws attention to the fact that the notion of *understanding* cannot be reserved for those who posited it as a distinguishing methodology of the *human sciences*, to be contrasted with the *explanatory* methodology of the *natural sciences*. Positivism, especially in its turn-of-the-century instantiation, explicitly rejected the notion of natural science as providing explanations that appealed to hypothetical constructs. For positivists, such as Ernst Mach, the task of philosophy of science was to provide an analysis of the ways in which our *understanding of both the natural and social world is based in our phenomenal experience*. Arens argues that Mach's way of thinking was in fact representative of a particular cognitive style that she attributes to features of the Austrio-Hungarian empire (such as its school system). This style, which emphasized historical situatedness as foundational to knowledge, was not, according to Arens, easily reconciled with the Erklären/Verstehen dichotomy, at least in the sense in which these notions were employed in the context of German philosophy. Arens examines the work of three otherwise quite diverse theorists (Ernst Mach, Karl Menger, and Alois Riegl), and argues that they shared in common the idea that scientific knowledge is firmly grounded in specific phenomenal experiences, which in turn are tied to specific material practices. Arens refers to this way of thinking about scientific understanding as "materialist phenomenological" and argues that its emergence has to be placed in the context of Viennese culture and history, in which all three thinkers developed their views.

In two complementary papers about British philosophy and the human science in the nineteenth century, both Roger Smith and Christopher Pincock argue that nothing like the Erklären/Verstehen dichotomy existed in the English language context. In his paper, "British thought on the relations between the natural sciences and the humanities, c. 1870–1910", Roger Smith (Russian Academy of Sciences) presents a detailed and comprehensive overview of the development and disciplinary formation of what we might today refer to as the human sciences, arguing that (a) the primary concern of English-language writers in the philosophy of scientific knowledge was *naturalism*, and (b) both proponents and critics of naturalism shared a commit to having their scholarly work provide a *moral foundation* for society. These concerns are traced back to Mill's 1843 *Logic of the Moral Sciences*, whose naturalistic outlook, however, was not widely shared until the 1860s. Smith then provides an overview of the development of different humanities disciplines (social theory, philosophy, history, psychology) in the decades following the 1870s, providing both intellectual and institutional contexts for each, and pointing to factors that may have been responsible for the different forms the debates took in Britain, as compared to debates in Germany at the same time. As Smith remarks, it is probably no coincidence that in the 1950s and 1960s, British critics of a positivist philosophy of the social sciences turned to the older German debates for inspiration. There was no prior British tradition to embrace.

In "Accounting for the Unity of Experience in Dilthey, Rickert, Bradley and Ward", Christopher Pincock (Purdue University) approaches the difference between German and English work by asking what was the *common philosophical problem* to which the German-language Erklären/Verstehen dichotomy proposed to provide an answer, as distinct from answers we find in the English-language context. According to Pincock, the problem may briefly be summarized as that of showing *how experience relates to scientific knowledge*. He refers to this as the "problem of the unity of experience" and shows that it can be demonstrated especially clearly in John Stuart Mill's phenomenalist philosophy of science. Pincock presents a reading of two German authors (Dilthey, Rickert) and two British authors (Bradley, Ward), according to which they each attempted to tackle the problem of experience in different ways. Dilthey and Rickert, Pincock argues, disagreed fundamentally in several respects, but each end up with a conception of the unity of experience that has as a consequence the distinction between different types of sciences (human vs. natural sciences) and both invoke a notion of understanding as supporting the distinction. In contrast, Pincock argues, Bradley and Ward (while also disagreeing widely) each proposed solutions that did not invoke a distinction between understanding and explaining.

By focusing his attention on the unity of experience rather than Erklären/Verstehen divide, Pincock is also able to draw attention to interesting *similarities* between Dilthey's and Ward's conceptions of psychology.

As already mentioned, one approach in Germany that attempted to delineate the human sciences from the natural sciences came from proponents of the Southwest school of Neo-Kantians, in particular Wilhelm Windelband and Heinrich Rickert. According to this school, the relevant epistemological distinction was not one between explanation and understanding, but rather between the explanations of individual events and the search for general laws (Windelband used the well-known terminology of "idiographic" vs. "nomothetic" sciences for this contrast). In the view of Jacques Bos (University of Amsterdam) the big category underlying many discussions of the human and natural sciences during the nineteenth century was that of individuality. Bos argues that the notion of historical events as *unique* was central to the nineteenth-century notion of *historicism*. While the term "historicism" has different connotations, Bos uses it to refer to a movement that rejected Hegelian and Romantic ideas about the *telos* of historical developments, instead turning to discussions of empirical methods in the investigation of past events. Bos contends that the depiction of historical events as unique was linked to an insistence on the uniqueness of individual human beings and their central role in shaping historical events. To substantiate his claims, Bos analyzes the work of three important nineteenth-century historicists – Wilhelm von Humboldt, Leopold von Ranke, and Johann Gustav Droysen. He argues that each rejected both Hegelian and positivist philosophies of history, because they implied a disavowal of the idea of individuality. Nonetheless, Bos emphasizes, while agreeing on the importance of individuality, each also championed his own understanding of the concept.

Given the notion of individuality, the question arises whether, and in what way, *individual agency* can be appealed to when accounting for particular historical or social phenomena, i.e., at what level of analysis should explanatory and/or interpretive efforts be pitched? This question, commonly known as the individualism/holism debate, was famously addressed in the so-called "Methodenstreit" between the Austrian school of economics (in particular, Carl Menger) and the German school of national economy (in particular, Gustav Schmoller). In her contribution, Filomena de Sousa (Technical University of Lisbon) addresses this debate, arguing that it neither began nor ended with Menger and Schmoller. De Sousa traces the origins of this debate back to the nineteenth-century historical school of German economics, which (following the literature in economics) she terms "historicist". In contrast to the type of historicism presented in Jacques Bos's contribution – where the term is used to describe a particular historiographical approach – de Sousa's focus is on a type of economic theory that viewed economic phenomena as bound to particular historical time periods. De Sousa argues that this historical approach to economic theory, which often made appeal to concepts such as "Volksgeist", was inherently *holistic* in its orientation, an approach at odds with the methodological *individualism* of the Austrian school. The debate touches importantly on the question of the relationship between *Erklären* and *Verstehen*, since the notion that individual preferences can explain individual economic behavior and, ultimately, the economy as a whole, presupposes that it is possible to access the preferences that individuals in fact have. In other words, it presupposes that there is some sense in which economists can "understand" what goes on in the minds of individual agents. De Sousa argues, however, that there was no consensus amongst members of the Austrian school on this issue. She suggests that the methodological disagreement between the individualist theoretical approach of the Austrians and the more holistic empirical approach of the Germans also must be seen against the background of conflicting political ideologies, i.e., that of the social reformism of the German national economists vs. the political and economic liberalism of the Austrians.

Max Weber, whose historical sociology places him in a critical position both with respect to the German and the Austrian schools, is commonly credited with having developed a unique notion of *Verstehen* as a necessary precondition for *explanations* in the human sciences. According to Weber, explanations of particular historical or social events require some presupposition to the effect that these events are the results of a means-end rationality on the parts of social agents. Our understanding of such an idealized means-end rationality then serves as a norm ("ideal type"), drawing our attention to the necessity of providing explanations where this norm appears to be violated. While Weber's methodological framework is commonly regarded as having been significantly influenced by Heinrich Rickert's analysis of *values*, the contributions of Michael Heidelberger (Universität Tübingen) and Daniel Ŝuber (Universität Konstanz) call this assumption into question, though in rather different ways. In his contribution, Michael Heidelberger examines the impact on Max Weber of Johannes von Kries's work on the relationship between statistical laws and the attribution of an adequate cause. Heidelberger argues that once we appreciate the central status of von Kries's work in Weber's thinking, we have to recognize that Weber's notion of an "understanding sociology" in fact bears a surprisingly close resemblance to notions of causal explanation that we ordinarily associate with the natural sciences. Contrary to Heidelberger, the contribution by Daniel Ŝuber argues that the impact of Dilthey on the proponents of classical sociology (Max Weber, Georg Simmel, Karl Mannheim) has been much neglected in favor of emphasizing the neo-Kantian elements in sociology. Ŝuber contends that the simplified accounts found in many disciplinary histories of sociology are due to the lack of an adequate understanding of the philosophical positions being debated at the turn of the century. Ŝuber's own methodology is informed by Mannheim's theory of structure, which aims at displaying the logical structure of prior systematizations in a given field of research. He argues that such an analysis reveals that Dilthey's ontology was a holistic one (as opposed to Rickert's dualistic ontology) and that subsequent work in sociology can be shown to bear some marks of Dilthey's holism.

The Erklären/Verstehen dichotomy regained some currency within English-language debates by the 1940s and 1950s, following the publication in 1942 of Carl Gustav Hempel's "The Function of General Laws in History". In his contribution, "Opposition to Verstehen in Orthodox Logical Empiricism", Thomas Uebel (University of Manchester) situates Hempel's work within the broader historical trajectory of logical empiricists' changing views about the unity of science. Uebel argues that it is necessary to distinguish between not only different stages of logical empiricism, but also different aspects of what was considered problematic about *understanding*. As Uebel shows, the early logical empiricists' rejection of *Verstehen* should be seen as (a) a special case of the rejection of intuition as a validational method, and (b) an expression of their logical behaviorism and verificationism. With the relaxation of the latter dogmas by the mid-1930s, logical empiricists began to wonder whether an understanding of mental states could be appealed to in explanations and what status should be accorded to understanding in the context of *validating* such explanations. After recounting the relevant arguments, Uebel turns to a 1952 paper by Hempel, in which Hempel (importantly drawing on Max Weber's ideal-type method) analyzed at length the relationship between explanation and understanding in both the natural and the human sciences. This analysis, Uebel suggests, leads more or less directly into current discussions in the philosophy of the social sciences.

## 1.3 Concluding Remarks and Acknowledgements

The articles in this volume are based on papers that were presented at the conference *Historical Perspectives on Erklären and Verstehen – an Interdisciplinary Workshop*, which took place at the Max Planck Institute for the History of Science in Berlin in June, 2006. The aim of the conference was to bring together scholars from various fields to reflect about both the histories and the current status of the dichotomy between explanation and understanding. Like the workshop, this volume tries to address not only the question of how to make sense of a particular set of philosophical concepts (*Erklären* and *Verstehen*) by paying attention to the contexts of their emergence, but also how to use these concepts as analytical tools with the aim of gaining some insights into a particular complex of intellectual, social, cultural, scientific and institutional changes that took place around the emerging human sciences from the mid-nineteenth to early twentieth centuries. Clearly, a collected volume such as this one cannot hope to be comprehensive and I am painfully aware of some gaps. For example, a fuller historical treatment of the dichotomy between explanation and understanding would also examine aspects of the histories of psychiatry and psychotherapy, such as Sigmund Freud's psychoanalytic method or Karl Jaspers's understanding psychology. Moreover, while this volume provides some analyses that compare aspects of the German debate with those of other national traditions (Austria, France, Britain, the United States), it also leaves out national contexts, such as Russia or Italy, that would have provided material for further analytical reflection.

This collection of papers should therefore be viewed as opening up a field of analysis and discussion, rather than as providing any definitive answers. Nonetheless, I think it is possible to highlight some findings and questions. Most prominently, it appears that the distinction between *Erklären* and *Verstehen*, as underwriting two separate methodological approaches for the natural and the human sciences, had a distinctly German flavor to it. That is to say, while the question of differences and commonalities between the emerging human sciences and other scientific endeavors was also debated in other places, the dividing line between explanation and understanding was apparently not drawn anywhere else as sharply as in the work of (for example) Wilhelm Dilthey. At the same time, however, even within the German (and German language) context, there were other ways of demarcating the natural and the human sciences (a particularly prominent example being the distinction between nomothetic and idiographic methods, coming out of Rickert and Windelband's school of Neo-Kantianism). Moreover, as several contributions to this volume remind us, the continuous branching off of the human sciences into separate disciplines with their own methodological concerns helped to give rise to more general worries about the disintegration of knowledge. These worries, in turn, spurred some to initiate the *Unity of Science* movement, which would turn out to be especially influential for the development of philosophy of science in North America. We may therefore ask (as Warrren Schmaus does at the end of his contribution) what set of circumstances led to the specifically "German" model of thinking about *Verstehen* (and its critiques!) as a distinguishing feature of the human sciences, given the many similarities between German and other (e.g., French) approaches to the study of human minds and societies. While there is surely no *one* right answer to this question, the analyses provided in this volume contribute several pieces of the puzzle, by pointing to educational, religious and political factors at work in mid nineteenth and early twentieth-century Germany. Elaborating further on these analyses, both in comparison with other countries and with respect to the complex interrelations between these and other factors, strikes me as opening up promising topics for future research.

In the same vein, however, we may also remark on the fact that there was certainly no consensus on the significance of the distinction between *Erklären* and *Verstehen* within Germany, let alone across the other national and cultural contexts. Moreover, as Roger Smith emphasizes in his contribution, it is not clear that there was no *one* set of shared questions and concerns associated with this conceptual dichotomy. This observation, too, is brought out clearly in a number of articles in this volume, raising the question of why, in spite of this vagueness, the dichotomy between *Erklären* and *Verstehen* has had such tenacity in twentieth-century philosophical discourse. This question has two aspects, one historical and one philosophical, both of which go beyond the scope of what is being presented in this volume, and both of which are well worth investigating further. The historical question is how the dichotomy came to travel from the contexts of its origin to be represented in various fields of contemporary philosophy of mind, action, and science. Important events, in this respect were surely the renewed debates about the philosophy of history in the 1940s and 1950s (this is touched upon in Thomas Uebel's contribution to this volumes) as well as Georg Henrik von Wright's philosophy of action explanation (see Kusch 2003, for a recent reappraisal), but more research into this question is definitely called for. Apart from such specific questions, however, I would like to suggest that it is precisely the heterogeneity of questions and concerns bundled together under the rubric of the *Erklären/Verstehen* dichotomy that accounts for some of its continuing popularity. This then brings to the fore the philosophical question of what is the current relevance of the distinction. In response to this question, too, I would like to suggest that it is the heterogeneity of usages and philosophical topics behind the conceptual pair of *Erklären* and *Verstehen* that makes it hard to dismiss it *tout court*. Careful and historically informed analysis of the different ways in which notions of *explanation* and *understanding* figure in current philosophical debates remains central for systematic philosophical work. Conversely, if used with the requisite historical sensibility these notions can provide analytical tools for intellectual history.

I would like to thank Hans Jörg Rheinberger and the Max Planck Institute for the History of Science for making this project possible, the contributors to this volume for stimulating discussions, John Carson for helpful comments on this introduction, and – last but certainly not least – my assistant, Christine Gross, for the painstaking work she put into proof-reading and assembling the bibliographies for the articles in this volume.

## Bibliography

Collingwood RG (1946) The Idea of History. Oxford University Press, Oxford
Davidson D (1980) Actions, reasons, and causes. In: Essays on Actions and Events. Clarendon, Oxford, pp 3–19 (first published 1963)
Dray W (1957) Laws and Explanation in History. Oxford University Press, London
Feest U (2007) 'Hypotheses, everywhere only hypotheses!' On some contexts of Dilthey's critique of explanatory psychology. Stud Hist Philos Biol Biomed Sci 38(1):43–62
Hempel CG (1965) The function of general laws in history. In: Aspects of Scientific Explanation and Other Essays. The Free Press, Glencoe, IL (first published in 1942)
Kincaid H (1997) Individualism and the unity of science. Essays on Reduction, Explanation, and the Special Sciences. Rowman & Littlfield, Lanham MD
Kusch M (2003) Explanation and understanding: The debate over von Wright's philosophy of action revisited. Poznani Stud Philos Sci Human 80:327–353
Kögler HH, Stueber K (2000) Introduction: Empathy, simulation, and interpretation in the philosophy of social science. In: Kögler H, Stueber K (eds) Empathy and Agency. The Problem of Understanding in the Human Sciences. Westview Press, Boulder, CO, pp 1–61
Taylor C ([1971] 1985) Interpretation and the sciences of man. In: Philosophy and the human Sciences: Philosophical Papers 2. Cambridge University Press, Cambridge, pp 15–57 (first published in 1971: Rev Metaphys 3–51)
Udehn L (2001) Methodological Individualism: Background, History and Meaning. Routledge, London
von Wright G-H (1971) Explanation and understanding. Cornell University Press, Ithaca, NY
Winch P (1964) Understanding a primitive society. Am Philos Quart 1(4):307–324
Wittkau A (1992) Historismus. Zur geschichte des begriffs und des problems. Vandenhoeck & Ruprecht, Göttingen

# Chapter 2
# Epistemological Distinctions and Cultural Politics: Educational Reform and the *Naturwissenschaft/Geisteswissenschaft* Distinction in Nineteenth-Century Germany

**Denise Phillips**

The *Erklären/Verstehen* debates of the late nineteenth century were, as several essays in this collection point out, peculiar to German-speaking Europe. Even when French or British scholars dealt with similar philosophical arguments, an epistemological distinction between two kinds of knowledge – one about human beings and texts, the other about the natural world – never took on the ubiquity or importance that it held in the German context (see Smith and Schmaus, in this volume). Why was this the case? What about the German intellectual scene made the issue of *methodological* differences between the sciences such an important theme in the second half of the nineteenth century?

Any answer to this question would have to include a history of the two other categories used to organize these late nineteenth-century debates – the terms *Naturwissenschaft* and *Geisteswissenschaft*. This binary division of the sciences, still common in German academic discourse today, evolved out of a far more complex set of early nineteenth-century classificatory schemata (Diemer 1968; Flint 1905, 57f.). Each of these two concepts had early modern precedents, but historians generally agree that this distinction, as a stable feature of German intellectual culture, appeared only around the middle of the nineteenth century (Wise 1983; Veit-Brause 1999; Reill 1994, 2005). In its starkest form, this bifurcation has been described, borrowing a term from C. P. Snow, as the rise of "two cultures", one scientific and one scholarly (MacClean 1988; Hörz 1997, 10f.,15).

As a cause for this new divide, historians have often pointed to the emergence of a more reductionist and mechanistic form of science, typified by the work of figures like Hermann von Helmholtz and Emil DuBois-Reymond (Wise 1983; Veit-Brause 1999). Yet this point of origin is both too specific and chronologically too late. Discussions of the methodological differences between the natural sciences and their humanistic cousins were actually already widespread in the 1840s, well before physicalist physiology was a significant force in German scientific life. Speeches like Helmholtz's [1862] 1876 often-cited 1862 "*Ueber das Verhältniss*

D. Phillips (✉)
University of Tennessee, Department of History, 915 Volunteer Boulevard, 6th Floor, Dunford Hall, Knoxville, TN 37996-4065
e-mail: aphill13@utk.edu

U. Feest (ed.), *Historical Perspectives on Erklären and Verstehen*, Archimedes, Vol. 21, DOI 10.1007/978-90-481-3540-0_2, 

*der Naturwissenschaften zur Gesammtheit der Wissenschaft*" (Helmholtz [1862] 1876), were not opening salvos, but interventions in a debate that was by that point several decades old.

Helmholtz himself recognized this fact; he blamed Hegel for the bad blood that had existed for so long between the *Natur-* and the *Geisteswissenschaften* (Helmholtz [1862] 1876, 6–11). The real historical force behind this split, however, was something broader than the intellectual aftershocks of a single philosopher's work. This divide within German *Wissenschaft* first emerged in the context of a lengthy dispute over educational policy that, beginning in the 1820s, set German *Naturforscher* against many of their colleagues in the humanistic disciplines. These debates dealt, among other things, with the place of the natural sciences in the curriculum of the *Gymnasium*, the German secondary schools that prepared students for the universities, and this seemingly quotidian issue was in fact of serious importance for the public standing of the natural sciences. The school debates raised the question of how the natural sciences fit into the dominant cultural ideal of *Bildung*, or self-cultivation, an ideal closely allied with the *Gymnasium* as an institution (on *Bildung*, see, Turner 1971; Vierhaus 1972; Koselleck 1990; Olesko 1991, 21–60; Kaschuba 1993; Bollenbeck 1994). Could the natural sciences be used to cultivate a form of selfhood that was properly *gebildet*, balanced and harmonious in aesthetic, moral and intellectual terms? For German *Naturforscher*, the answer to this question needed to be yes.

Much more than narrow bickering about the number of classroom hours assigned a given subject, the school debates encouraged a thorough public vetting of the supposed intellectual and moral benefits of different kinds of knowledge, and it was in this context that a new, binary system for classifying knowledge first became linked with the question of methodological difference. At the heart of these debates was the same basic issue at stake in the later *Methodenstreit* – namely, whether or not there was a difference between the kind of knowledge one can have about nature and about human beings. Advocates of a curriculum based around the study of Greek and Latin claimed that a philological education provided students with a universal set of intellectual skills that could be readily applied later in life to any other field, including the natural sciences. *Naturforscher* and many educational reformers, in contrast, argued that knowledge about nature required a unique set of skills, different from the one employed with texts, and as a result, a narrow classical education left future medical doctors and scientists woefully unprepared for their later studies. The conflict between these two camps had two major results. It popularized the idea that *Wissenschaft* was divided into two basic branches, which already in the 1840s began to take on their later labels, the *Naturwissenschaften* and the *Geisteswissenschaften.* It also linked this distinction to epistemological questions; these two branches of *Wissenschaft*, it became common to argue, were not just different in the topics they covered, but in their forms of knowledge, as well.

"*Erklären*" and "*Verstehen*", the key terms of the later nineteenth-century debates, were not as central to the *Vormärz* discussions, although the words were used occasionally and some of the contrasts drawn between the two branches of *Wissenschaft* were broadly similar to the arguments of later thinkers. My primary interest in what follows, however, is not to analyze these earlier exchanges as philosophical precursors to later, more famous debates. I hope, instead, to suggest one reason why the question of methodological difference seemed so important to German intellectuals. Through the debates over school reform and *Bildung*, ideas about moral and intellectual character formation were closely tied to ideas about the structure of knowledge. In order to defend the cultural value of their disciplines, nineteenth-century German *Naturforscher* needed to answer an epistemic question – what kind of knowledge does your field produce, and what sorts of procedures does it employ? After this question was answered, others followed as a matter of course – how does your kind of knowledge shape its practitioners? What kind of person does it produce? And would that person possesses the necessary intellectual breadth, order and balance to count as truly "*gebildet*"?

## 2.1 Knowledge and Bildung in the Gymnasium Debates

The dominant educational power in Germany throughout the nineteenth century was neohumanism, a movement that had begun at the end of the preceding century as a kind of cultural revolt, a critique of contemporary life that looked to ancient Greece and Rome as a model for the aesthetic, philosophical and moral regeneration of the present. This loose and diverse movement, woven out of many strands of late eighteenth-century intellectual life, went on to play a central role in the educational policy of the nineteenth century, spreading out from its North German origins to become a powerful force in school policy and in educated culture more broadly (O'Boyle 1968; Jäger 1987; Marchand 1996, 4–35). By the 1830s, neohumanism was well-established as a bureaucratic ally of Germany's dynastic states, many of whom were trying to standardize the curricula taught in their secondary schools. State officials hoped to bring greater uniformity to an educational landscape characterized by a great deal of local variation, and in their efforts, the rhetoric and pedagogical goals of neohumanism played a leading role. Neohumanist pedagogues captured key bureaucratic posts and successfully preserved the study of Latin and Greek as the centerpiece of preparatory schooling, breathing new life into a curriculum that had come under fire during the Enlightenment (Jäger 1987). In the process, they assured that knowledge of Latin and Greek would continue to be a defining hallmark of the well-educated man (O'Boyle 1968).

Because of the breadth of neohumanist influence, the period's debates about educational reform were pan-German in scope, sharing a common set of themes and categories despite local and regional differences in school policy. The new school plans followed two basic models across the different German states, and both were heavily weighted towards the study of classical languages. Prussia and most of the middle German states adopted curricular guidelines that included a smattering of "modern" subjects (the natural sciences, for example, and the study of modern languages) alongside Greek and Latin. Bavaria and Austria, in contrast, approved plans that focused almost entirely on classics. In both cases, the natural sciences played only a minor role in the curriculum, and the natural historical fields in particular were especially weakly represented (Jeismann 1987; Schubring 1987; Olesko 1991, 21–60; Bonnekuh 1992).

For many defenders of the classical curriculum, limiting the time devoted to natural science made sound pedagogical sense. Natural science, orthodox neohumanists claimed, did not possess the necessary intellectual qualities to fulfill the *Gymnasium's* central purpose – the task of *Bildung*, or self-cultivation. For example, Christian Grossmann, an educational official in Leipzig, argued that students would never develop the necessary moral and mental character studying a misshapen combination of classical languages and natural science. It could only be to their detriment to spend their formative years wandering between classical texts and natural historical objects, going from "Solon's … laws to baobabs, quartzes and pebbles … from reason to unreason, from the ideals of humanity to the beasts, from the high and the eternal to the changeable, the ordinary, and the trivial." In fact, Grossmann claimed, only a man with a solid early training in the universals of classical language and culture would have the mental and moral capacity necessary for later greatness in the natural sciences. A philological education was the best preparation for future intellectual achievement, regardless of the specific field one hoped to enter. Grossmann cited Herder to the effect that it was "the intensity of the spiritual powers" fostered by their humanist education that had made Francis Bacon, Kepler, Newton, Leibniz, Haller, Euler, Linné, and Buffon the great men – and the great natural researchers – that they were (Grossmann [1834] 1847, 180f.).

The preceding passages capture most of the neohumanists' favorite arguments against the pedagogical benefits of the natural sciences. A philological curriculum supposedly trained students in a set of universally applicable intellectual skills, skills that would allow them to master other subjects with ease later on in their lives. Indeed, a philological education was the *best* possible kind of intellectual preparation for the future university student. Boys whose future lay in the world of *Wissenschaft* needed training in abstract, ordered thought, a kind of education otherwise known as "formal *Bildung*". They needed to master "strict formal relationships" and "strict external lawfulness" (as *Gymnasium* director Friedrich Lindemann explained), and grammar was the subject most suited to teach students these habits of disciplined, law-governed thought (Grossmann [1834] 1847, 26). Learning Latin and Greek initiated students into a set of elegant and internally consistent rules, especially since these languages surpassed modern tongues in the purity and logic of their grammar. (Marchand 1996, 31).

The classical curriculum was also supposed to mold the student's character, not just his intellect. The major sin neohumanist pedagogy sought to avoid was "one-sidedness" *[Einseitigkeit]*, a lack of intellectual and moral harmony in the personality of the young student. The best way to protect against one-sidedness was for students to spend their time studying subjects that were themselves internally consistent and harmonious, qualities embodied to perfection, so the argument ran, in Greek and Roman culture and in the Greek and Roman languages (Marchand 1996, 28). The subjects students mastered shaped the structure of their souls, and the natural sciences, according to neohumanist critics, were too utilitarian, too base in their subject matter, and too internally diverse to offer the right moral resources to young students. Even in cases where natural science might merit inclusion as a minor subject, it could never serve as a cornerstone of a *Gymnasium* education. It did not lend itself to so-called "formal" *Bildung*, the training in abstract, internally consistent forms of knowledge that was the main goal of a *Gymnasium* education (e.g., Thiersch 1830; Raschig 1847).

The opposite of formal knowledge was knowledge that was merely "positive" – knowledge that dealt not with law-governed abstractions, but with specific empirical content. Another often-invoked opposition was between the spiritual and the material, with the classical curriculum aligned with the former. The *Gymnasium* was devoted to training the *Geist*, and subjects that dealt only with base matter were supposedly ill-suited to that purpose. While material or positive knowledge might be valuable, it was out of step with the *Gymnasium*'s primary mission. The so-called *Realien*, disciplines like modern history, the natural sciences, and modern languages, were first and foremost kinds of positive knowledge, and these subjects, according to neohumanist orthodoxy, had their proper place in the *Realschule* or *Bürgerschule*, secondary schools intended for students who would not continue on to the university (Thiersch 1830; Grossmann [1834] 1847; Raschig 1847).

While the study of languages held pride of place, many neohumanists believed that there was another body of knowledge that met the criteria for generality and internal consistency they required in a *Gymnasium* subject – mathematics. Along with the classical languages, mathematics was also a respectable medium for formal *Bildung*, and for many neohumanists, mathematics served as a sufficient proxy for the natural sciences as a whole. Math was "the grammar of natural phenomenon" (Lindemann 1834, 22), and like grammar, it provided " a system of laws." What grammar offered for the realm of the human spirit, mathematics provided for nature, and the pedagogical benefits of these two subjects were assumed to be similar (Snell 1833; Lindemann 1834).

Even mathematics, however, was not safe from criticism. Given its association with mechanical, practical tasks, it was not always considered morally elevated enough to be a means of formal *Bildung*. When asked in 1843 about plans to expand the mathematical curriculum in Weimar's *Gymnasium*, for example, the majority of the philosophical faculty of Jena concluded that the plan was a bad idea. *Realschulen* and technical institutes were the proper place for mathematics, and mathematics, in their minds, was already overvalued in the *Gymnasium*. As evidence for this assertion, they pointed to the fact that where teachers had once written qualitative evaluations of their students, they now recorded numerical grades. Under such a grading system, the professors complained, the student saw "his whole personality reduced to a few heartless numbers" (Philosophische Fakultät [1843] 1992, 185f.).

A critique of numerical grading might seem, at first glance, an odd answer to a question about the recommended content of school curriculum, but there was a certain logic in the Jena faculty's response. The link between these two issues – how students were evaluated and what they were taught – lay in the question of moral character. How does someone trained to place high value on mathematics approach the world? Reductively, the professors worried, and without the necessary attention to the individual human personality.

Neohumanists' criticisms of natural science did not go unanswered; not surprisingly, they provoked a sustained and often angry response. The announcement of the Bavarian school plan in 1829, for example, happened to coincide with the annual meeting of the *Gesellschaft deutscher Naturforscher und Aerzte* in Heidelberg, and it caused a universal uproar at the meeting. The school plan, crafted by the philologist Friedrich Thiersch, contained no natural science at all, a fact that inspired outrage among the *Naturforscher* assembled in Heidelberg. Lorenz Oken reported that everywhere he went at the meeting, people were complaining about the new Bavarian curriculum (Oken 1829).

Men as intellectually distant from one another as Lorenz Oken and Justus Liebig, though they agreed on little else, saw eye-to-eye on this one issue, and found, in the neohumanists' pedagogical agenda, a common enemy. Both of these men – Oken, the aging hero of Romantic science, and Liebig, the vocal critic of *Naturphilosophie* – complained about the overweening ambition of philologists who sought to promote their own form of *Bildung* as the golden standard by which all other kinds of knowledge should be measured. Oken had been attacking classical educational ideals since the early nineteenth century (Oken 1809, 13f.), and in the *Vormärz* he used his journal *Isis* to complain about marginal role of natural science in the schools (Oken 1829). Similarly, Liebig's famous 1840 speech on chemistry in Prussia, long seen as defining document for the "new" science that followed the Romantic era (Heuser and Zott 1992), was structured through and through by the school debates. The speech has been read primarily as an attack on *Naturphilosophie* (which it was), but it was also an attack on neohumanist pedagogy. Liebig contrasted the characteristics of the *Naturforscher* with those of the *Philolog*, arguing that the natural sciences would train a "newer, more powerful generation" than the one raised on philological learning; he ended the speech with an extended attack on the neohumanist model of the *Gymnasium* (Liebig 1840, 11–21, 43–47).

But how, precisely, did the issue of *method* fit into the debates over school curriculum and *Bildung*? The first step towards answering that question is to ask what mid-nineteenth-century Germans included under discussions of *Methodologie*. *Methodologie* was, in fact, a recognized branch of philosophy in this period, one covered most frequently in propaedeutic works that provided an introductory overview of a given field like medicine or theology, or that covered *Wissenschaft* as a whole. Carl Schmid described his 1810 work *Allgemeine Encyklopädie und Methodologie der Wissenschaften* as a contribution to "*die Wissenschaft der Wissenschaften*." This "science of the sciences" had two branches. *Encyklopädie* was its "objective side"; that is, it described the objects of scientific inquiry. *Methodologie* was its "subjective side", or "the *Wissenschaft* of the study of the *Wissenschaften*." In other words, methodology described how scientific knowledge was produced ("subjective" in this context referred to the knowing subject, the person who possessed or made knowledge) (Schmid 1810, iv). *Encyklopädie* dealt with the "what" (what external object does this science explore?); *Methodologie* with the "how" (how does one, as a knowing subject, create knowledge about this object?). In addition to general works like Schmid's, which covered all of the sciences, similar books were common in university disciplines like medicine, and lectures on "*Encyklopädie und Methodologie*" were also offered as university courses, both within the professional faculties and as a general introduction to university study (e.g., Universität Halle 1829–1840; Wagner 1838; Heusinger 1839).

In their treatment of method, works of *Encyklopädie und Methodologie* took a strongly anthropological approach. *Methodologie* was a science of personas as well as a science of processes. In addition to epistemological procedures, these works discussed the personality traits and specific skills needed in a given field. For example, one of the rare treatments of the natural sciences in this genre, Gustav Suchow's obscure *Systematische Encyklopädie und Methodologie der theoretischen Naturwissenschaft* included under its discussion of *Methodologie* two sections: one on "general methodology" and one on "Requirements for Natural Scientific Study". The latter included things such as a healthy body and good sense organs, a gift for observation and a good memory (Suchow 1839).

*Methodologie*, as a matter of course, dealt with personal qualities, the concrete competencies and character traits necessary to practice a certain science or profession. What kind of person was best suited for a certain kind of *Wissenschaft*? And how, in turn, did the practice of natural science, philology, or history shape the individual personality? These were all questions that *Methodologie* sought to answer. Discussions of method and discussions of *Bildung*, in other words, covered overlapping intellectual ground, and the issues raised in the school debates blended easily into questions of method. Indeed, proper *method* had been the rallying cry of neohumanists since the late eighteenth century. When philologists argued for the universality of a classical education, they often did so by citing the universality of the "method" this education imparted; it provided a set of basic intellectual capacities that could be used to advantage in any branch of *Wissenschaft* (Turner 1983, 460–462). This claim, as the primary one *Naturforscher* needed to counter, gave the representatives of the natural sciences increased incentive to emphasize the particularity of their own forms of knowledge, the unique intellectual benefits offered by the study of nature.

## 2.2 History, Natural Science and Method in the School Debates

Questions about the structure of knowledge, then, were central to the education debates, with the issue of methodology never far from center stage. By the mid-1840s, s widespread discussion had emerged about the differences between the natural and the philological sciences. One of the clearest formulations of this supposed difference came from writings of a classical philologist, Hermann Köchly, a liberal young schoolteacher who published a number of works on the *Gymnasium* question. Köchly took a particularly strong stance on the differences between the natural and the historical sciences, and as a result, the responses his proposals inspired offer a useful survey of the various intellectual options on the table in *Vormärz* debates over the classification of knowledge. Köchly's proposed reforms, which took as their starting point an assumed methodological difference between the natural and the historical sciences, were both controversial and widely discussed. By the late 1840s, the young Saxon schoolteacher had a national reputation, and in 1848, he was elected the provisional chair of the first meeting of the *Allgemeine Deutsche Lehrerversammlung*. After the revolutions of 1848–1849, his reputation as a school reformer helped win him the chair of classical philology at the University of Zurich (Böckel 1904, 53, 91–102).

In opposition to many of his colleagues, Köchly argued that there were two distinct forms of education that could prepare students for future work in the world of *Wissenschaft*; consequently, there ought to be two separates types of secondary schools. The traditional *Gymnasium* could train students for later careers in the historical sciences, which included philology, history, theology and law, while the *Realschule* would prepare students for university study in the natural sciences and medicine. Both kinds of schools, contrary to contemporary practice, should be accorded equal prestige and status (Köchly 1845, 4).

In the early 1840s, Köchly had been part of the same Dresden literary circle as the liberal Young Hegelian Arnold Ruge (Böckel 1904, 29–31), and Köchly relied on Hegel to justify his plan for a new, dual-track school system. While *Wissenschaft* itself was unified, Köchly argued, it had two primary objects, "*Natur*" and "*Geist*". Nature was characterized by "its cyclically recurring states"; the realm of spirit, in contrast, by "progressive development" (Köchly 1845, 4; Köchly 1846, 47; on Hegel's original use of these categories, Pinkard 2002, 266–304). Johann Gustav Droysen, traditionally seen as a seminal figure in the history of historical methodology, later used this same distinction in his famous essay "*Natur und Geschichte*" (Wise and Norton 1983), and this parallel usage was one of several congruencies between the careers of these two historians. Droysen also had ties to Young Hegelian circles in the 1830s and 1840s, and, during the same period, he criticized classical philology's excessive focus on language at the expense of historical development (MacClean 1982, 352f.) – another central tenet of Köchly's writings, as we shall shortly see.

Like Droysen, Köchly went beyond an initial topical distinction to argue that the two branches of *Wissenschaft* were also *methodologically* distinct. The natural sciences had "a strictly observational and demonstrative method" that was unique to them (Köchly 1847c, 25), and extensive historical training would be wasted on future *Naturforscher*. Philological training provided them with the wrong kind of "*Gymnastik des Geistes*"; *Naturforscher* needed a bodily and sensory education that textual study could not provide, since the study of nature involved learning to move from concrete sensory experience to the abstractions of natural law. Future historians, theologians and lawyers, in contrast, needed to learn to bring the abstractions presented to them in texts into dynamic relationship with their own concrete inner lives. The highest calling of the historical sciences was to provide a vibrant understanding of the past (Köchly 1846, 59–62).

Köchly's ideas about historical method were at the heart of his criticism of the contemporary *Gymnasium*. Standard teaching methods, he claimed, failed to fulfill the *Gymnasium's* stated aims of *altclassische Bildung*. Instead of imbuing students with the spirit of ancient Greece and Rome, as early neohumanists had intended, the classical high schools simply provided language training. Education in the classics had lost its way in linguistic minutiae, and it needed to be thoroughly reformed to provide students with a deep *historical* understanding of the ancient world. Language study should be a means to an end, not an end in itself. Properly handled, the classical languages provided the conduit through which students came to know the great cultures of Greece and Rome, but if Greek and Latin were taught with dead, philological precision rather than living historical understanding, they were dull tools for training young minds. (This criticism was not an uncommon one; see [Grafton 1983; La Vopa 1990]). Köchly's historical approach to teaching the classics was well-received in reform circles (Mager 1846; Fuchs 1846, 317; "Geschichte und Verhandlungen" 1847, 50). He presented his proposals as a defense of the philological tradition, a way of saving philology from itself by countering the current "piling on of so-called *Realien*" and remedying the "lack of respect, not to say the general derision, in which the public opinion of today holds philology and philologists" (Köchly 1845, vi).

Köchly's reform plans presented, in a particularly stark form, a distinction that was already common in much that had been previously written about school reform. In describing the internal divisions of *Wissenschaft*, earlier pedagogical writers had often used the opposition between *Geist* and *Natur* or between an inner, spiritual and an outer, material world to analyze the different components of the school curriculum (e.g., Snell 1833, 19; Lindemann 1834, 21f., 41–43). Köchly's suggestion that the two halves of *Wissenschaft* needed to be taught in two separate schools gave institutional flesh and blood to an older philosophical division; the differences between the two kinds of *Wissenschaft* were so great, his reform plan assumed, that they required two separate kinds of *Bildung* from very early in life.

Although Köchly's use of this distinction was not unique, he assigned it a practical significance that went beyond the norm, and reviews of his early books acknowledged the relative novelty of his suggestions. Commentators saw Köchly's work as part of a more general innovation, not only in school policy, but also in the categorization of knowledge. Karl Mager's review of Köchly's first book, for example, grouped it with two others, both by small-town school directors, who made similar arguments about the internal differences between the *Wissenschaften*. Mager opposed Köchly's plan for two separate kinds of secondary schools (he felt that *Naturforscher* also needed a measure of erudition), but he accepted the epistemological division Köchly had proposed and considered the use of such categories a helpful recent development. "It is nice," Mager wrote, "that the distinction between the *Naturwissenschaften* and the ethical *Wissenschaften* is beginning to be common." (Mager 1846, 58–62, 66) Other reviews mentioned Köchly's new system of classification as noteworthy, and argued with him over the appropriate labels for the two branches ("Ueber Gymnasialreform" 1845 1239f.; "Schul- und Unterrichtswesen" 1846, 60; A.A 1847, 2).

Despite its roots in Idealist philosophy, then, this binary division of *Wissenschaft* was clearly considered relatively new in the 1840s. The label "*Naturwissenschaft*" was used fairly universally, but there was more disagreement about what to call the disciplines on the other side of the divide. Köchly himself preferred the term "historical sciences" because it emphasized the "unique method of the *Geisteswissenschaften*", which he considered the most important source of their distinctiveness. The precise label used, however, was something that Köchly considered inessential – the division was clear, whatever one chose to call the two fields. At one point he listed six different possible sets of terms that might be used to describe the two main branches of *Wissenschaft* (*Geisteswissenschaften* and *Naturwissenschaften*, the ethical and the physical sciences, the humanistic and the realistic, the historical and the exact, the spiritual and the sensual, the traditional and the experimental). Whatever labels one used, the fields of history, law and theology belonged on one side of a divide; the sciences that dealt with nature lay on the other (Köchly 1847a, 113, 129f.).

Not everyone agreed with Köchly's specific proposals, but, given the dividing lines produced by the school debates, the basic distinctions he used were beginning to seem increasingly like common sense to educated Germans by the mid-1840s. Köchly's relative nonchalance about terminology makes it clear that there was much more than just a philosophical distinction at stake. The categories *Naturwissenschaft* and *Geisteswissenschaft* (or their various equivalents) functioned as placeholders for the interests of different groups of scholars, teachers and students; they were practical tools to address issues of status and identity in a key social institution.

Among *Naturforscher*, the most intense discussions of Köchly's proposals took place in the schoolteacher's home city of Dresden, where, in 1846, he helped found a society for school reform, the *Gymnasialverein*. Köchly's own proposals offered a starting point for the *Gymnasialverein*'s work. He summarized the issue before the society as follows: the question was whether "depending on whether they deal with the development of the *Geist* itself, or the objects of external nature, the sciences break apart into historical or ethical sciences on the one hand and natural sciences on the other." Whether it made sense to have two different forms of *Gymnasien* to correspond to these two different branches of *Wissenschaft* was the next question, Köchly thought, that followed from this conclusion ("Geschichte und Verhandlungen" 1847, 59f.).

Several Dresden *Naturforscher* joined the *Gymnasialverein*, and all of them held strong, if varying, views on Köchly's proposal. The three most prominent and vocal natural scientific members of the society were Hermann Richter, a professor at the local medical academy; Ludwig Reichenbach, the director of Dresden's botanical garden; and Emil Roßmäßler, a forestry professor from nearby Tharandt. None of these men were among the most eminent of German researchers, but all went on to positions of some national importance in the second half of the nineteenth century. Reichenbach was later elected head of the national scientific association the *Leopoldina* ("Nekrolog" 1879), Roßmäßler was one of the foremost scientific popularizers of the nineteenth century (Daum 1998, 203–209; Daum 2002), and Richter was a leading figure in the national movement for medical reform (Richter 1964).

The positions these men took on the question of school reform had important similarities, but also equally important differences, both in tone and content.

Richter was perhaps the most acerbic and combative member of the *Gymnasialverein*, and also a strident defender of Köchly's proposed split. Richter had been interested in epistemological issues from early on in his career; his dissertation had examined the question of certainty in medical thought, a problem he tackled in conjunction with a careful study of Francis Bacon (he recommended the creation of *Codex empiricus* in which medical propositions could be collated with relevant observations) (Grosse 1896, 14; Böckel 1904, 51). At the Dresden medical academy he was an aggressive spokesman for a reformed, "rational medicine". An admirer of the Younger Viennese School and the methods of physiologist Johannes Müller, he antagonized some of his older colleagues, men he derided as aging medical Romantics (Richter 1964, 7, 25–30).

Richter agreed with Köchly that there was a clear methodological distinction between the two branches of *Wissenschaft*, but he had less than flattering things to say about Köchly's own field of classical philology. "The natural sciences have a completely different teaching and research method," he stated, "and it stands in relation to the humanistic method as oil does to water" (Richter 1847a, viii; Richter 1848, 105). The natural sciences allowed the student to learn to follow the logic of, as Richter put it, "things which can speak for themselves", while the traditional humanist curriculum only taught them to follow authority. Studying the classics cultivated, both literally and figuratively, a kind of blindness. The natural sciences taught students to see the world clearly, and hence provided, Richter thought, the perfect foundation for a new liberal political order. An education in the natural sciences would "give every student the ability to look around with a trained eye and an independently thinking spirit at all of the living relations around him", and give him the courage to sweep away all that was dead and sickly in the current political system (Richter 1847b, 55).

Fellow society member Ludwig Reichenbach defended natural science from a very different political and philosophical perspective; he held many of the Romantic commitments that were anathema to Richter. The main scientific accomplishment of Reichenbach's career had been a system of botanical classification similar to those proposed by other *Naturphilosophen* of his generation (Jardine 1996), and the botany professor spoke of Oken and Schelling with approval in the *Gymnasialverein*'s debates. Reichenbach also drew very different conclusions than did Richter about the meaning of science for social and political order. As a close confidant of the Saxon King Friedrich August II, Reichenbach believed the natural sciences would restore faith in traditional monarchy. He had argued in the early 1840s that the study of nature would create "devoted, peace-loving citizens", convinced of the beneficent rights of "the strong and the powerful" to rule, and he repeated this argument in his speeches in the *Gymnasialverein* (Reichenbach 1843, 88; Reichenbach 1847a, 13).

Reichenbach and Richter also disagreed about the relationship of natural science to the humanist tradition. Reichenbach, in addition to promoting the virtues of natural science, also praised the traditional classical curriculum, complaining about the excessive pull of "*Realismus*", neohumanism's pedagogical enemy, in current public discussions (Reichenbach 1847a, 16; "Geschichte und Verhandlungen" 1847, 61–67). Richter, in contrast, considered the traditional curriculum more or less useless ("Geschichte und Verhandlungen" 1847, 53–57). The forestry professor Emil Roßmäßler, a childhood schoolmate of Hermann Richter's, fell somewhere in the middle of the other two men. He shared Richter's political sympathies, but, like Reichenbach, was more conciliatory in his discussion of the humanist tradition. Roßmäßler's main aim was to defend natural science as an activity that could be pursued in a humanist spirit, a subject that provided sound *Bildung* in the best humanist tradition ("Geschichte und Verhandlungen" 1847, 70–76).

In arguing for the pedagogical value of natural science, however, all three *Naturforscher* shared several common reference points. Richter, Roßmäßler and Reichenbach all claimed that natural science developed unique mental capacities that could be created no other way. Richter was most explicit in using the concept of method to make a case for the particularity of natural science, but all three of the society's *Naturforscher* defended the epistemic distinctiveness of science in some form. The society's *Naturforscher* all held that the peculiar benefit of a natural scientific education was a sound training of the senses; the end result of this training was the ability to perceive universal laws. For Richter, the scientific method began with clear "*sinnliche Anschauung*", or sensory perception. From there, the *Naturforscher* continued on through a process of induction to the formulation first of sound scientific descriptions, and then of explanatory laws (Richter 1848, 101f.). Similarly, Reichenbach characterized science as a particular kind of "practical logic", grounded in solid sensory training, that led to the discovery of laws. "The spirit learns what law is through nature," he stated, citing earlier author Karl Snell with approval (Reichenbach 1847a, 9).

All three men also insisted that *Naturwissenschaft* was the sort of unified body of knowledge that offered general, formal *Bildung*. They defended an "organically unified *Naturwissenschaft*" that included natural history (and here they intentionally used the word *Naturwissenschaft* in the singular, not the plural, form) in an attempt to forestall competing proposals for reform that included physics and mathematics as *Gymnasium* subjects, but left out the natural historical sciences, the fields in which their own scientific interests lay (Reichenbach 1847b, 71; Richter 1847b, 1848, 102; "Geschichte und Verhandlungen" 1847, 70–76). Despite their political and philosophical differences, both Richter and Reichenbach found in Alexander von Humboldt a worthy embodiment of their scientific principles, praising Humboldt's work *Kosmos* as an example of the organically unified natural science they defended (Richter 1847a, vi; Reichenbach 1847b, 69–88).

In discussing natural science's unique qualities, several members of the *Gymnasialverein* made mention of natural science's "exact methods" (e.g., Köchly 1847b, 24; Richter 1847c, 174). What did they mean by this term? For the *Verein* members, "exact" did not just mean quantitative. For Richter, properly scientific expressions came in qualitative forms, too, in "word, line and number" (Richter 1848, 101f.). Indeed, for people interested in arguing for the equal status for the natural historical sciences, keeping mathematics and its allies in the physical sciences in their proper place was an important concern. Reichenbach hoped to dispel "the misunderstanding that *Naturlehre* [physics] was the entirety of *Naturwissenschaft*", and Richter wrote that *Naturwissenschaft* must be taught as "a great organically coherent whole" (Reichenbach 1847b, 71; Richter 1848, 102). "The exact sciences", then, included botany, zoology and geology. In addition, the term "method", when used in the *Gymnasialverein*'s discussions, covered broad ground. The members slid back and forth between discussions of pedagogical methods, methods of knowledge production, and common personality traits – a generous use of the term that was in keeping with the standard topics covered by "*Methodologie*" in the period.

The Dresden *Gymnasialverein* kept stenographic records of their debates, so one can observe at particularly close range the tensions that curricular questions sparked between representatives of philology and natural science. Even among this group of reformers, all of whom were critical of the neohumanist curriculum in its standard form, the sense of an epistemological divide was clear. In the second meeting of the *Gymnasialverein,* Richter gave a brief, polemical speech summarizing his views and roundly criticizing the classical curriculum. The main question, according to Richter, was to observe "the difference between the intellectual [*geistigen*] Bildung that takes place through sensory perception and that which takes place through word and text." The rest of his comments came close to dismissing the latter kind of *Bildung* completely. Not only were the methods of natural science and humanism completely different, humanist training, he implied, might be not only worthless, but actively harmful ("Geschichte und Verhandlungen" 1847, 55).

Richter's claims left many of his fellow society members unnerved. Köchly commented that he felt a bit like the sorcerer's apprentice, someone who had called forth forces he could no longer control; another society member said that Richter had only confirmed his fear that the *Gymnasien* were truly on the defensive against the merciless onslaught of the *Realien*, subjects like natural science and modern languages. When Köchly pressed Richter on whether or not he had really meant to imply that all branches of *Wissenschaft* that dealt with "*Wort und Schrift*" were without value, Richter answered that of course he had not meant to question the legitimacy of the historical sciences, or of law, theology or philosophy; he had only been speaking about educational methods ("Geschichte und Verhandlungen" 1847, 58–60). (Judging from his other writings, this distinction was disingenuous. Richter later argued that what Köchly's field of classics really needed to do was to adopt the "*naturwissenschaftliche Methode*" and bring "*sinnliche Anschauung*" to bear on historical questions) (Richter 1847b, 41f.). Reichenbach and Roßmäßler gave more conciliatory speeches at later meetings, and in the end, the society was able to agree on a compromise curriculum – a unified *Gymnasium* that prepared students for future work in both branches of *Wissenschaft* by including two different courses of study within a single school (Richter et al. 1848).

Not everyone in the *Gymnasialverein* agreed that the methods of the natural and the historical sciences were unique; some members of the society argued that all *Wissenschaften* shared a single method. The day after Richter's controversial speech in the *Gymnasialverein,* fellow society member Eduard Calinich sat down to write the preface for his recently completed book, *Philosophische Propädeutic für Gymansien, Realschulen und höhere Bildungsanstalten,* which, like others of its genre, was an overview of philosophy for a secondary school audience (Calinich chose to cover

*Seelenlehre, Denklehre,* and *Kunstlehre* in the work). The reference to philosophy in his title, he emphasized in the preface, was to philosophy as a set of empirically grounded *Geisteswissenschaften* (as distinct from philosophy as "purely speculative philosophy"). In his opening paragraph, Calinich penned several sentences aimed directly at Richter's speech from the previous day. He complained about those who believed that one could only gain "true and positive knowledge [*ein wahres Wissen und positive Kenntnisse*]" from the natural sciences. The *Geisteswissenschaften* did not merely "sway in the air" without any grounding. Both the *Geisteswissenschaften* and the *Naturwissenschaften* shared a common method, the method of "the *Wissenschaften* in general" – "sensory understanding, experiment and observation" (Calinich 1847, iv, 3).

## 2.3 Positive Knowledge, Realism, and Natural Science at Mid-century

Calinich's criticisms of Richter, in particular his use of the term "positive knowledge", raise a number of broader questions. General histories of nineteenth-century Germany have presented the "rise of natural science" as a defining feature of German culture in the 1840s,1850s and 1860s (Nipperdey 1983, 484–498; Sheehan 1989, 802–820). Natural science's new confidence has sometimes been attributed to an emergent positivism or realism (e.g., von Engelhardt 1979, 159–220; Lenoir 1997, 131–178), with the triumph of a positivist natural science blamed for breaking apart the older, idealist-inspired unity of *Wissenschaft*. But how valuable are these two labels – "positivism" or "realism" – in capturing what was novel in discussions of natural science at mid-century? At first glance, neither term adequately describes the range of positions adopted by the *Naturforscher* examined above, either in Dresden or in the national scene more broadly. For one thing, the heritage of German Idealism and *Naturphilosophie* was still evident in a variety of ways, in forms too diverse to be described in terms of simple acceptance or rejection. If some *Naturforscher*, men like Justus Liebig or Hermann Richter, raged against speculative philosophy and the foibles of aging romantics, those same aging romantics (say, Lorenz Oken or Ludwig Reichenbach) were still prominent public figures, and they defended the particularity of natural science with their own set of intellectual resources.

Both "positive knowledge" and "realism" were important categories in the school debates, but one needs to be careful in analyzing what these concepts meant in this particular context. In the terminology of the period, both the defenders of the human and the natural sciences understood themselves to be dealing, at least in part, with "positive" knowledge. Ernst Calinich's angry reaction to the Hermann Richter's attempt to appropriate this privilege solely for the natural sciences was typical (MacClean 1988, 474f.). More importantly, positive knowledge could be defined in contradistinction to speculative philosophy, but it was also used to mean the opposite of "formal" kinds of knowledge like grammar or mathematics.

For *Naturforscher* and historians who defended the value of positive knowledge, the main target was the excessive formalism of the classical neohumanist position, the idea that "formal *Bildung*" should not be diluted by the addition of too much detailed empirical content. Defending the value of positive knowledge, however, did not necessarily mean rejecting the ideal of formal *Bildung* (on the continued commitment to the ideal of *Bildung* among *Naturforscher* later in the century, see von Engelhardt 1991, 106–116; Cahan 1994; Daum 1998, 52–57). More commonly, authors assigned value to formal *Bildung* as well, and the approbatory use of the term "positive knowledge" revealed little about what other epistemological or metaphysical positions a writer might hold.

Talking about a simple shift from idealism to realism at mid-century is similarly problematic, because, like defenses of positive knowledge, defenses of realism were rarely one-sided. The term "*Realismus*", as it appeared most often in public discussion, referred to a specific pedagogical tradition (a curriculum centered around the *Realien*, the modern languages, the natural sciences, modern history, etc), and the values that tradition embodied. It did not denote a coherent set of philosophical commitments. Just as support for positive knowledge did not entail a complete rejection of formal knowledge, "*Realismus*" did not usually receive unqualified support at the expense of the spiritual or the ideal (von Engelhardt 1991, 106–116; Daum 1998, 52–57). Both sets of terms could, and usually did, function as binaries that stood in a productive relationship with one another. Both poles – the positive and the formal, the real and the ideal, the material and spiritual – were considered necessary. To present these categories as mutually exclusive choices would be to miss the dialectical way in which they were typically employed.

Yet the cumulative result of the school debates was a decided shift in the kind of rhetoric *Naturforscher* employed when speaking about their own forms of knowledge. The debates added a new emotional force to defenses of natural science, and they also made assertions of natural science's epistemic particularity more common. Given the diversity of figures who came forward to defend natural science in the face of neohumanist criticism, however, there is little to suggest that this new stridency was the side-effect of some general shift in philosophical first principles. In making the case for natural science's epistemic distinctiveness, many *Naturforscher* did not necessarily break significantly with arguments that had been common earlier in the nineteenth century. For example, the prominence of *Anschauung*, or sensory apprehension, in the Dresden society's debates reflected a consensus among *Naturforscher*, and one that was not new to the 1840s. In the first third of the nineteenth century, there was widespread agreement among *Naturforscher* of otherwise varying commitments that "*Anschaulichkeit*" or intuitive clarity, was the best criteria for judging the soundness of a scientific theory (Caneva 1978; Heidelberger 1979, 8f.). More generally, the idea that refined sensory perception was the hallmark of the *Naturforscher* (and by extension the medical doctor) was widespread in introductory textbooks, both in the natural sciences proper and in medicine (e.g., Ficinus 1828, 1f.; Choulant 1829, 116–119; Wagner 1838, viii). Padagogical writers also regularly praised natural science as a "school for the senses." (e.g., Freese 1845, 57f.; Beger 1845, 44,117). The consensus around the concept of *Anschauung* was beginning to shift by the mid-nineteenth century (developments like Ohm's mathematical treatment of electricity and the new strategies of measurement taught in Franz Neumann's Königsberg seminar were beginning to push other views to the fore, at least in the physical sciences) (Caneva 1978; Heidelberger 1979; Olesko 1991), but these older arguments could still work perfectly well to bolster claims about the unique qualities of the natural sciences in the face of neohumanist criticism.

What German *Naturforscher* shared at mid-century was not a new set of fundamental philosophical commitments but a broadly common position in the debate over *Bildung*. Beyond the basic unifying effects of this common cause, however, diversity remained the rule. A new emphasis on the uniqueness of "the natural scientific method" did not necessarily signify any strong philosophical agreement on what exactly that method was, a fact that remains true if one looks ahead to the second half of the nineteenth century. In the numerous memorials for Alexander von Humboldt held after his death, for example, "the natural scientific method" appeared in many of the speeches, though in inconsistent form, with the unique properties accorded to science differing according to the disciplinary allegiances of the speaker. The ethnologist Adolph Bastian, speaking in 1869 to a joint meeting of all of Berlin's natural scientific societies, said that the essence of the natural scientific method was rigorous, disciplined comparison, and he considered Humboldt's work the perfect epitome of its use. For the physiologist and physicist Emil Du Bois-Reymond, in contrast, real science involved the mathematization of phenomenon along the lines of Newtonian physics, a project Humboldt's work embodied only very imperfectly (Bastian 1869; DuBois-Reymond E [1883] 1912).

## 2.4 Conclusion

Though originating in the *Vormärz*, the school debates continued for the rest of the nineteenth century, forming one of the perennial themes of German cultural politics through the end of the *Kaiserreich* (Daum 1998, 51–57). Indeed, as Norton Wise has noted, a number of key texts in later versions of the *Naturwissenschaft/Geisteswissenschaft* debate were written with this context in mind. Helmholtz, Mach and Ostwald, for example, all wrote on the education debate (Wise 1983, 28f.). Some of Emil Du Bois-Reymond's most strident rhetoric came in reference to *Gymnasium* reform ("Conic sections! No more Greek composition!") (Cited in Lenoir 1997, 173f.). At the inaugural celebration for the University of Tübingen's natural scientific faculty (the first of its kind in Germany), Hugo von Mohl explained the occasion's significance in terms of the debate over *Bildung*. "In the creation of a new natural scientific faculty," he said, "one sees a break with the medieval idea that *Bildung* is only to be found in the humanistic studies" (von Mohl [1863] 1963, 208).

While the struggles to define *Bildung* and reshape the schools can hardly explain all the permutations of later debates over scientific methods, the school debates did provide a basic cultural scaffolding that made methodological questions a recurring issue for German academics. The German educational tradition, as embodied in the preparatory institution of the *Gymnasium*, linked epistemology to broader questions of moral and intellectual self-development. Because the individual personality supposedly reflected back the characteristics of the knowledge that formed it, claims about method (in the broad terms in which mid-century German intellectuals understood this term) were a necessary accompaniment to claims about the social value of knowledge. The ongoing argument over the nature of *Bildung*, along with the specific fissures the school debates produced, helps to explain why debates about the comparative methods of the different sciences loomed so large in German intellectual life.

The preceding pages have focused primarily on the emergence of a new division within the world of German *Wissenschaft*. In closing, it is worthwhile to dwell for a moment on the many interconnections and commonalities the school debates also revealed. While the divisions between *Naturforscher* and their neohumanist colleagues were real and often bitter, the "two cultures" metaphor so often used as a shorthand to describe this tension in nineteenth-century intellectual life has limited analytical value. Differences in disciplinary allegiance hardly rose the level of anything so all-encompassing as a "culture", and a commitment to the fundamental unity of *Wissenschaft* remained an important feature of German academic life throughout the rest of the century (Daston 1999).

If relations between *Naturforscher* and their scholarly colleagues sometimes seemed strained, countless other kinds of connections kept university-educated *Naturforscher* part of the same cultural world as their humanistic peers. To love natural science more was not necessarily to love Greece less, and many university-educated *Naturforscher* continued to share in the culture of erudition so important to the educated middle classes throughout the period under discussion. Ludwig Reichenbach's eulogy made special mention of his love of the classical tradition ("Nekrolog" 1879). Indeed, *Naturforscher* continued to use many of the same intellectual categories as their neohumanist educational rivals (terms like "formal *Bildung*", for example), a fact that had important implications for the development of the debate.

For German intellectuals, furthermore, academic disciplines were only one cause among many that might inspire intense allegiance. Hermann Richter and Hermann Köchly, despite their differing intellectual commitments, were in obvious ways a part of the same cultural and political milieu. In the early 1840s, both had belonged to the same liberal political and literary circles in Dresden, and both were leaders in the local *Turnverein* [Gymnastic Society], an important vehicle for liberal sentiment. Building on this earlier activism, both would go on to play prominent roles in the Revolution of 1848–1849 in Saxony. When the revolution failed, Köchly was forced to flee Dresden to avoid arrest, and Richter spent several years in prison (Böckel 1904, 45, 71–135; Richter 1964, 12–18). This shared caesura in their careers offers a valuable reminder of the many things that joined these erstwhile friends and opponents. Their disagreements were debates within a common culture, not shots fired across a wide and unbridgeable divide.

# References

Ueber Gymnasialreform (1845) Blätter für literarische Unterhaltung 2:1234–1236
Geschichte und Verhandlungen (1847) Des dresdener gymnasialvereins von michaelis bis weihnachten 1846. Vermischte Blätter zur Gymnasialreform 1:29–92
Nekrolog [H. G. Ludwig Reichenbach] (1879) Sitzungsberichte der naturwissenschaftlichen gesellschaft Isis 1–8
"Schul- und Unterrichtswesen" (1846) Leipziger repertorium der deutschen und ausländischen literatur 4:54–62
AA (1847) Zur Verständigung über das gymnasialwesen. Arnold, Dresden
Bastian A (1869) Alexander von Humboldt. Festrede bei der von den naturwissenschaftlichen vereinen berlins veranstalteten Humboldt-Feier. Wiegandt & Hemp, Berlin
Beger A (1845) Die idee des realgymnasiums. J. C. Hinrich, Leipzig
Böckel E (1904) Hermann Köchly. Ein bild seines lebens und seiner persönlichkeit. Heidelberg, Carl Winter
Bollenbeck G (1994) Bildung und kultur: glanz und elend eines deutschen deutungsmuster. Insel, Frankfurt am Main
Bonnekuh W (1992) Naturwissenschaft als Unterrichtsfach: Stellenwert und Didaktik des naturwissenschaftlichen Unterrichts zwischen 1800 und 1900. Peter Lang, Frankfurt am Main
Cahan D (1994) Helmholtz and the civilizing power of science. In: Cahan D (ed) Hermann von Helmholtz and the Foundations of Nineteenth-Century Science. University of California Press, Berkeley, pp 559–601
Calinich EAE (1847) Philosophische propädeutik für gymnasien, realschulen und höhere bildungsanstalten. Adler und Dietze, Dresden
Caneva K (1978) From galvinism to electrodynamics. Hist Stud Phys Sci 9:63–160
Choulant L (1829) Anleitung zum studium der medicin. Voß, Leipzig
Daston L (1999) Die Akademien und die Einheit der Wissenschaften: die Disziplinierung der Disziplinen. In: Kocka J et al. (eds) Die königlich preußische akademie der wissenschaften zu berlin im kaiserreich. Akademie, Berlin, pp 61–84
Daum A (1998) Wissenschaftspopularisierung im 19. Jahrhundert. R. Oldenbourg, München
Daum A (2002) Science, politics, and religion: Humboldtian thinking and the transformation of civil society in Germany, 1830–1870. Osiris 17:107–140
Diemer A (1968) Die Differenzierung der Wissenschaften in die Natur- und die Geisteswissenschaften und die Begründung der Geisteswissenschaften als Wissenschaft. In: Diemer A (ed) Beiträge zur entwicklung der wissenschaftstheorie im 19. Jahrhundert. Anton Hain, Meisenheim am Glan, pp 174–223
DuBois-Reymond E (1912) Die Humboldt Denkmäler an der Berliner Universität [1883]. In: DuBois-Reymond E (author), DuBois-Reymond E (ed) Reden von emil dubois-reymond. Veit, Leipzig, pp 249–284
von Engelhardt D (1991) Der Bildungsbegriff der Naturwissenschaft des 19. Jahrhundert. In: Koselleck R (ed) Bildungsbürgertum im 19. Jahrhundert. Klett-Cotta, Stuttgart, pp 106–116
Ficinus H (1828) Physik allgemein faßlich dargestellt. P. G. Hilscher, Dresden
Flint R (1905) Philosophy as Scientia Scientiarum: A History of Classifications of the Sciences. William Blackwell, Edinburgh
Freese E (1845) Das deutsche gymnasium, nach den bedürfnissen der Gegenwart. Arnold, Dresden
Fuchs A (1846) Schriften zur Pädagogik. Pädagogische Revue 12(220–226):306–328
Grafton A (1983) Polyhistor into Philolog: Notes on the transformation of german classical scholarship, 1780–1850. Hist Univ 3:159–192
Grosse J (1896) Hermann Eberhard Richter, der Gründer des deutschen Aerztevereinsbundes. Otto Wigand, Leipzig
Grossmann ([1834] 1847) Separatvotum des Superintendenten Dr. Grossmann aus Leipzig vom 13. Juli 1834. In: Richter H, Ludwig Reichenbach HG (eds) Der naturwissenschaftliche unterricht auf gymnasien. Arnold, Dresden, pp 173–187

Heidelberger M (1979) Some patterns of change in the Baconian sciences of early 19th-century Germany. In: Jahnke HN, Otte M (eds) Epistemological and Social Problems in the Sciences in the Early Nineteenth Century. D. Reidel, Dordrecht, pp 3–18

Heuser R, Zott E (1992) Die streitbaren gelehrten: Justus von liebig und die preußischen universitäten. ERS, Berlin

Heusinger CF (1839) Grundriss der encyklopädie und methodologie der natur- und heilkunde. Chr. Fr. Bärecke, Eisenach

Hörz H (1997) Brückenschlag zwischen zwei Kulturen: Helmholtz in der korrespondenz mit geisteswissenschaftlern und künstlern. Basilisken-Presse, Marburg

Jäger G (1987) Philologisch-historische Fächer. In: Jeismann K-E, Lundgreen P (eds) Handbuch der deutschen bildungsgeschichte, vol. 3, 1800–1870. C.H. Beck, München pp 191–204

Jardine N (1996) Naturphilosophie and the Kingdoms of Nature. In: Jardine N et al. (eds) Cultures of Natural History. Cambridge University Press, Cambridge, pp 230–245

Jeismann K-E (1987) Das höhere Knabenschulwesen. In: Jeismann K-E, Lundgreen P (eds) Handbuch der deutschen bildungsgeschichte, vol 3, 1800–1870. C.H. Beck, München, pp 152–180

Kaschuba W (1993) German Bürgerlichkeit after 1800: Culture as Symbolic Practice. In: Kocka J, Mitchell A (eds) Bourgeois Society in Nineteenth-Century Europe. Berg, Oxford, pp 392–422

Köchly H (1845) Ueber das Princip des gymnasialunterrichtes der Gegenwart. Arnold, Dresden

Köchly H (1846) Zur gymnasialreform. Arnold, Dresden

Köchly H (1847a) Polemik und Kritik. Vermischte Blätter zur Gymnasialreform 1:93–145

Köchly H (1847b) Erwiderung des Herausgebers an Herrn Rector M. Raschig. Vermischte Blätter zur Gymnasialreform 1:11–16

Köchly H (1847c) Zwei und siebzig Sätze zur Gymnasialreform. Vermischte Blätter zur Gymnasialreform 1:17–28

Koselleck R (1990) Einleitung – Zur anthropologischen und semantischen Struktur der Bildung. In: Koselleck R (ed) Bildungsbürgertum im 19. Jahrhundert. Klett-Cotta, Stuttgart, pp 11–46

La Vopa A (1990) Specialists against specialization: Hellenism as professional ideology in German classical studies. In: Cocks G, Jarausch KH (eds) German Professions, 1800–1950. Oxford University Press, Oxford, pp 30–45

Lenoir T (1997) Instituting Science: The Cultural Production of Scientific Disciplines. Stanford University Press, Stanford

Liebig J (1840) Ueber das studium der naturwissenschaften und ueber den zustand der chemie in preußen. Friedrich Vieweg und Sohn, Braunschweig

Lindemann F (1834) Die wichtigsten mängel des gelehrtenschulwesens im königreiche sachsen. Birr und Nauwerck, Zittau und Leipzig

MacClean M (1982) Johann Gustav Droysen and the development of historical hermeneutics. Hist Theor 21(3):347–365

MacClean M (1988) History in a two-cultures world: The case of the German historians. J Hist Ideas 49(3):473–494

Mager K (1846) Schriften zur Pädagogik. Pädagogische Revue 12:204–223; 13:41–69

Marchand SL (1996) Down from Olympus: Archaeology and Philhellenism in Germany, 1750–1970. Princeton University Press, Princeton, NJ

Nipperdey T (1983) Deutsche Geschichte, 1800–1866. C.H. Beck, München

O'Boyle L (1968) Klassische Bildung und soziale Struktur in Deutschland zwischen 1800 und 1848. Historische Zeitschrift 207:584–608

Oken L (1809) Ueber den werth der naturgeschichte, besonders für die bildung der deutschen. Friedrich Frommann, Jena

Oken L (1829) Für die Aufnahme der Naturwissenschaften in den allgemeinen Unterricht. Isis: 1225–1237

Olesko K (1991) Physics as a Calling: Discipline and Practice in the Königsberg Seminar for Physics. Cornell University Press, Ithaca

Philosophische Fakultät J ([1843] 1992) Gutachen, den mathematischen Lehrplan bei dem Großherzoglichen Gymnasium zu Weimar betreffend. In: Eccarius W (ed) Über eine auseinandersetzung um den mathematikunterricht an thüringischen gymnasien aus dem Jahre 1843. In: Demidov SS et al. (eds) Amphora: festschrift für hans wussing. Birhäuser, Basel, pp 183–187
Pinkard T (2002) German Philosophy 1760–1860: The Legacy of Idealism. Cambridge University Press, Cambridge
Raschig M (1847) Sendschreiben des Herrn Rektor M. Raschig in Zwickau an Dr. H. Köchly. Vermischte Blätter zur Gymnasialreform 1:1–10
Reichenbach HGL (1843) Blicke in das leben der thierwelt. Verglichen mit dem leben des menschen. Arnold, Dresden
Reichenbach HGL (1847a) Denkschrift der Gesellschaft für Natur- u. Heilkunde in Dresden. In: Richter HG, Reichenbach HGL (eds) Der naturwissenschaftliche unterricht auf gymnasien. Arnold, Dresden, pp. 1–16
Reichenbach HGL (1847b) Das Gymnasium und die Naturkunde. In: Richter H, Reichenbach HGL (eds) Der naturwissenschaftliche unterricht auf gymnasien. Arnold, Dresden, pp 67–88
Reill PH (1994) Science and the construction of the cultural sciences in late enlightenment Germany: The case of Wilhelm von Humboldt. Hist Theor 33:345–366
Reill PH (2005) Vitalizing Nature in the Enlightenment. University of California Press, Berkeley
Richter H (1847a) Vorrede. In: Richter H, Reichenbach L HG (eds) Der naturwissenschaftliche unterricht auf gymnasien. Arnold, Dresden, pp iii–viii
Richter H (1847b) Denkschrift der Gesellschaft Isis. In: Reichenbach L, Richter HE (eds) Der naturwissenschaftliche unterricht auf gymnasien. Arnold, Dresden and Leipzig, pp 17–66
Richter H (1847c) Die Vorbildung der Aerzte auf Gymnasien. Vermischte Blätter zur Gymnasialreform 1:169–185
Richter H (1848) Allgemeiner Bericht über Naturwissenschaft. Vermischte Blätter zur Gymnasialreform 2(3):99–109
Richter G-M (1964), *Hermann Eberhard Richter: Leben und Werk eines grossen Dresdner Arztes.* Ph.D. dissertation. Medizinische Akademie Carl Gustav Carus, Dresden
Richter H, Köchly H, Herz H (1848) Gesammtbericht. Vermischte Blätter zur Gymnasialreform 2(3):233–268
Schmid C (1810) Allgemeine encyklopädie und methodologie der wissenschaften. Akademische Buchhandlung, Jena
Schubring G (1987) Mathematische-naturwissenschaftliche Fächer. In: Jeismann K-E, Lundgreen P (eds) Handbuch der deutschen Bildungsgeschichte, vol 3, 1800–1870. C.H. Beck, München, pp 508–551
Sheehan JJ (1989) German History, 1770–1866. Clarendon, Oxford
Snell K (1833) Skizze einer philosophischen begründung des gymnasial-unterrichts. Karl Wagner, Dresden
Suchow G (1839) Systematische encyklopädie und methodologie der theoretischen naturwissenschaft. C.A. Schwertschke und Sohn, Halle
Thiersch F (1830) Ueber den angeblichen jesuitismus und obscurantismus des bayerischen schulplanes. J.G. Cotta, Stuttgart
Turner RS (1971) The growth of professorial research in Prussia, 1818 to 1848. Hist Stud Phys Sci 3:137–182
Turner RS (1983) Historicism, *Kritik*, and the Prussian Professoriate, 1790 to 1840. In: Bollack M, Wismann H (eds) Philologie und hermeneutik im 19. Jahrhundert, vol II. Vandenhoeck & Ruprecht, Göttingen, pp 450–477
Universität Halle (1829–1840) Verzeichniß der auf der vereinigten hallischen und wittenbergischen friedrichs-universität. zu Haltenden Vorlesungen, Halle
Veit-Brause I (1999) Scientists and the Cultural Politics of Academic Discourse in Late Nineteenth-Century Germany. Max-Planck-Institut für Wissenschaftsgeschichte, Berlin
Vierhaus R (1972) Bildung. In: Brunner O et al. (eds) Geschichtliche grundbegriffe. Ernst Klett, Stuttgart, pp 508–551

von Engelhardt D (1979) Historisches bewußtsein in der naturwissenschaft. Karl Alber, Freiburg

Wagner R (1838) Grundriß der encyklopädie und methodologie der medizinischen wissenschaften. J.J. Palm und Ernst Enke, Erlangen

Wise, MN (1983) On the relation of physical science to history in late nineteenth-century Germany. In: Graham L et al. (eds) Functions and Uses of Disciplinary Histories. D. Reidel, Dordrecht, pp 3–34

Helmholtz H ([1862] 1876) Ueber das Verhältniss der Naturwissenschaften zur Gesammtheit der Wissenschaft. In: Populäre wissenschaftliche vorträge. Friedrich Vieweg und Sohn, Braunschweig, pp 2–29

von Mohl H ([1863] 1963) Rede gehalten bei der Eröffnung der naturwissenschaftlichen Facultät der Universität Tübingen [1863]. In: von Engelhardt WF, Decker-Hauff H (eds) Quellen zur gründungsgechichte der naturwissenschaftlichen fakultät in tübingen. JCB Mohr, Tübingen, pp 185–208

# Chapter 3
# Vestiges of the Book of Nature: Religious Experience and Hermeneutic Practices in Protestant German Theology, ca. 1900

**Bernhard Kleeberg**

## 3.1 Introduction

His new "scientific point of view," he wrote in 1859, built on the study of "sensual things." "Evidence of experimental science" would ultimately lead to a new form of "consilience" that demonstrated "objective necessity" instead of "phantasmagoric ideas or arbitrary abstract reasoning." Natural scientists, philosophers and theologians had to acknowledge that there was but one universal ontology and epistemology, and that a new approach had set out to deal with any question thus arising.

Reading these passages, we might suppose that we face yet another mid-century positivistic materialist. Astonishingly, they stem from the "Theologia Naturalis," wherein Otto Zöckler, one of the leading conservative Lutherans of the second half of the nineteenth century introduced his concept of "Biblical Physics" (Zöckler 1860, III–V, 243). According to Biblical Physics, scientific explanations formed an integral part of biblical and natural hermeneutics, providing detailed knowledge about natural things. Science served theology, since it helped to extend its epistemological frontiers into formerly unknown areas, directing the attention of theologians towards new discoveries relevant to understand the true meaning of nature and scripture. Like Biblical Physics, other strong versions of natural hermeneutics since early Christianity had read natural phenomena as signs of direct divine interference, carrying a specific meaning that had to be understood. Weaker versions of natural hermeneutics, inspired by romantic *Naturphilosophie* and following Friedrich Schleiermacher's account of nature, referred to the meaning of nature as a whole, which in its beauty, harmony, and order represented the divine plan of creation and universal development – *providentia generalis*. Around 1900, both these natural theological attempts to understand God from the book of nature were equally endangered. The primary cause of concern was not the sciences explaining natural phenomena, but the claim of scientific naturalism to be able to provide answers even to ultimate questions about the meaning of nature.

B. Kleeberg (✉)
EXC 16, Department of History and Sociology, University of Konstanz,
78457 Konstanz, Germany
e-mail: bernhard.kleeberg@uni-konstanz.de

U. Feest (ed.), *Historical Perspectives on Erklären and Verstehen*, Archimedes, Vol. 21,
DOI 10.1007/978-90-481-3540-0_3, © Springer Science+Business Media B.V. 2010

A successful apologetic strategy in the struggle between science and religion was to rely on the distinction between two epistemic realms: being and meaning. If science explained the structures and mechanisms of natural things, and religion provided their (ultimate) meaning, the belief in a higher teleological force guiding nature could be maintained and the crucial doctrine of man as the *imago dei* be reassured. Similarly, biblical hermeneutics – classical paradigm of the art of interpretation – could further on be understood as a rational practice to gain theological knowledge. In contrast to Catholic theology, Protestant theologians followed the Lutheran principle *scriptura sui ipsius interpres*, according to which "neither the postulates of a world view alien to the essence of the testimonies of faith, like for instance the newer 'scientific religion,' nor the ecclesiastical judgments about the characteristics of the canon can guide the formation and application of the hermeneutical method, but only the contents of the bible itself in its mutual relations" (Heinrici 1899, 723). Still even Protestant theologians significantly differed in what they conceived as the epistemological relations between scientific and theological approaches: some considered their epistemic fields as corresponding, some as complementary, some as independent or incompatible. The range of positions within Protestant theology was linked to different contemporary concepts about the status of scientific theories, the kind of evidence necessary to establish facts (e.g., empirical, experimental, or: from faith) and about the epistemological foundations of the *Wissenschaften* – that is, the sciences *and* the humanities – in general (e.g., belief, trust, correspondence, coherence). Adding to this, the status and scientific objects of theological knowledge itself were under discussion. Concentrating on scripture, on questions of morality and faith, theologians less and less regarded nature as a central object of theological investigation.

Yet the interpretation of the bible from faith had been challenged as well. Historical criticism and secular hermeneutics since the late eighteenth century argued that the bible be interpreted as a *historical text*, contesting the fourfold sense of the scriptures (its literal, allegorical, tropological, and anagogical meaning). By 1900, most theologians accepted the historical relativity of scriptural knowledge about the empirical world: The sciences provided answers to genealogical and functional questions, and due to scientific progress these answers changed over time. Thus the bible at the time it was written could only refer to the phenomena known to its authors, and its references to the world could not be taken literally. Instead, biblical hermeneutics now tended to propose metaphorical, mainly tropological readings of scripture. Hermeneutics served as a methodology that developed rules to guide the interpretation, and as the art to revive the newly understood material (Heinrici 1899, 719). Instead of logical abstraction, these interpretations rested on the idea of an individual experience of faith, meaning and morality, and on the conviction that beings with similar experiences were able to empathically understand and thus correctly interpret its contemporary and historical expressions (see Dilthey [1900] 1974, 336; Dilthey [1910] 1981, 263–265).[1] Focusing on debates about

[1]Concerning the choice of most of theologians discussed in this paper I follow Hübner 1966, Gregory 1992, and Rohls 2007. All translations are mine, except where indicated otherwise. I am grateful to Robert Fülpesi and Viola Huang for bibliographical assistance.

Darwinism, this paper gives an account of hermeneutical practices and concepts of "meaning" in a time that saw theology become a modern *Geisteswissenschaft*, even though it accentuated the necessity of religious experience as its basis.

## 3.2 Secularizations

In the course of the nineteenth century the political, economic and cultural influence of the German churches steadily declined. The churches had been severely weakened with the confiscation of a vast amount of their property during the early nineteenth century process of secularization. In addition to this, since the 1820s Prussia had tried to further its influence on the local Protestant denominations by creating the "Union of Prussian Regional Churches," prompting a reaction of the Reformed and Lutheran churches which ultimately led to a denominational splitting of German Protestantism: when other German states made similar efforts, more and more so-called 'free churches' were founded in an attempt to establish independent and self-sustaining communities. These developments intensified in the 1860s and 1870s, when after the revolutions of 1848/49 most local rulers had made themselves head of the church, resulting in an immense heterogeneity of Protestant German theology connected to different theological schools and local environments (see Froböß 1903, 4; Rohls 1997, 602). Yet in spite of their great diversity, theologians considered themselves as members of one common academic discipline that was of key importance to the educational system of the German states. As a *Wissenschaft*, Protestant theology had developed a sophisticated hermeneutic methodology that contributed to the tendencies of scientific rationalization and ensured its disciplinary relevance even though it had lost its central role with the restructuring of German universities (see Howard 2006, 130–211). Thus secularization in the broader sense of disenchantment – a growing hegemony of immanent over transcendent explanations of man and nature – did not per se pose a threat to the disciplinary integrity of theology.

Still with the explanatory role ascribed to divine purpose diminishing, the subject area of theological interpretation seemed to be cut down bit by bit. The same year Zöckler finished his "Theologia Naturalis," Charles Darwin's "On the Origin of Species," (Darwin [1859] 1860) was published, providing a causal mechanism to explain the realm of organisms. This was crucial for apologetics, since while mechanistic explanations had by and large substituted arguments from divine interference in the inorganic world, difficulties to explain organic forms and the relation between the organic whole and its parts secured teleological interpretations in the realm of life. Here, arguments from divine interference remained a somewhat plausible alternative. Yet with Darwin, arguments against natural theology and teleology that had gained strength during the so called "materialism struggle" of the 1850s were reinforced. When Darwin's theory was introduced to Germany by zoologist Ernst Haeckel and others, the theological audience was already struggling to fight off the secular relativization or even thorough dismissal of the scientific, philosophical and social importance of the bible, searching for arguments to underpin its conception of reality and the moral and religious consequences it implied.

These threats to religious sovereignty shaped theological reactions to scientific accounts of man's place in, and the ultimate meaning of nature. In the narrower sense, it affected *biblical* hermeneutics: the meaning of scripture and its references to reality. But there were also developments within theology itself that challenged the traditional role and interpretation of the bible, above all the emergence of the historical–critical method and its consolidation as a central part of biblical hermeneutics since 1800. When Kant, Herder and others had pointed out the importance of source criticism, the old doctrine of the inspiration of the bible could no longer be easily upheld. The contents of the bible became conceived as largely historical, the hermeneutic question about the ultimate meaning of scripture was substituted for the historical one about how it had originated as a text. As scripture nevertheless was still held to contain the word of God, the dogmatic postulate and the historical axiom stood side by side and had to be reconciled. The solution that tradition had suggested for this problem was the principle of accommodation, understood as God's pedagogical attempt to ease reception of the divine message by accommodating it to the specific needs and apprehensions of the audience. This principle was modified in a crucial way, when the possibility of different authors of the bible was introduced, culturally conditioned under different circumstances. Regarding the biblical account of nature, this meant that incorrect depictions of reality might have been given, either due to the historical stage of knowledge, or as a result of the biblical authors following the principle of accommodation. Consequently, this concept led some exegetes to a thoroughly historical interpretation of scripture, as Georg Heinrici, professor for New Testament in Leipzig, bemoaned in an article on biblical hermeneutics in Heinrici (1899, 737f.).

It was the liberal Protestant theologian David Friedrich Strauss, who became the epitome of this new kind of 'unfaithful' theology (see Kuhn 1907). Assuming that the biblical texts directly corresponded to the historical past and that biblical assertions about the physical world had to be considered on the same ontological and epistemological level as scientific statements, Strauss substituted divine meaning with (natural) historical explanations. Stressing the historicity of the person 'Jesus' and the hermeneutical importance of distinguishing between an ideal and a literal meaning of the bible, Strauss ([1835/1836] 1895) in his "Leben Jesu, kritisch bearbeitet" introduced his 'mythical approach' as continuation of the allegorical interpretation: the bible presents a mythology, "a kind of narration, covering primordial Christian ideas" (Gregory 1992, 67–111; Rohls 1997, 604). A speculative demythologization was to eliminate the mythological contents from the gospels in order to get to their real truth content. This truth he conceived as the unity of God and Man, not rooted in the historical Jesus, but in mankind as such, Jesus being only part of the *form*, not the *content* of the idea of the divine man. Strauss (1840/41) advanced this approach in his "Christliche Glaubenslehre", introducing a "speculative Christology of mankind," a post-Christian religion of humanity that allowed for a reconciliation with Darwinism.[2] The development of Strauss' ideas is an excellent example for the secularization of biblical hermeneutics, dismissing all but historical–critical approaches, that is, the historization of scripture led to an interpretation of the bible as a text containing a mythological narration just like *any* other religious mythology. Though his writings alienated him from Protestant orthodoxy and led to him losing his job as professor of dogmatics in Zurich, after Strauss' "Life of Jesus" all theological schools from the mid nineteenth century onwards had to deal with the question of the historical development of religion and of the bible in particular.

## 3.3 Meaning Comprises Explanation: Conservatives

Provoked by the polemical position of Haeckel and his followers, who had described their opponents as relicts of the evolutionary past, intellectually inferior to "dogs, horses, and elephants" (Haeckel [1866] 1988, 436), many theologians did not follow Strauss' *Speculative Christology*, but openly denied the possibility of reconciling Darwinism and Christianity. Hübner (1966), Gregory (1992), and Rohls (2007) have shown that a common argument held it that science had illegitimately transgressed the boundaries of its epistemological realm. The influential neo-Lutheran Christoph Ernst Luthardt, one of the main protagonists of the 'Erweckungsbewegung,' argued that Darwinism misconceived the qualitative difference between man and animal, the huge gap that divided them: reason (Luthardt [1864] 1897, 85). Science and religion could only live side by side, if science did not claim competence beyond the realm of material phenomena, whilst theology as the comprising discipline rightly exerted its authority on scientific questions. The natural sciences had to refrain from interfering with religious issues, to which they had been encouraged by the 'Kulturkampf.'[3] Furthermore, whilst the transmutation of species or even their common descent could not be empirically confirmed, creation and the constancy of species were warranted by scripture and Christian tradition, and even backed up by observation (Luthard [1880] 1908, 172; Hübner 1966, 39–41). A slightly more cautious view was held by the conservative apologetic and biblical realist Robert Benjamin Kübel, for whom the aims of the biblical and scientific accounts of nature differed. While science explained *how* matter developed, theology asked *why* something had come into being and thus provided meaning (Kübel 1883, 245). Kübel distinguished between the scientific and the theological content of the bible, but did not argue for them relying on different epistemologies – the biblical

[2] Strauss' ([1872]1873) "Der alte und der neue Glaube" can be regarded as the most elaborate adaptation of theology to evolutionism, and it is not surprising that he referred to Ernst Haeckel's Monistic religion (see Kleeberg 2005, 2007a) when outlining his new approach.

[3] Between 1871 and 1887 Prussia and the German Reich under Otto von Bismarck tried to minimize the political and educational influence of Catholicism, which affected the Protestant side as well: The "Kulturkampf" was not only directed against Rome, but against certain religious principles common to Christianity, Luthardt ([1880] 1967) remarked; see Rohls (2007, 124).

account of creation causally explained the origins of nature just like science explained natural development. Hence the direct creation of man was undisputable. If the human *Geist* was a vital power deriving from God, man's first appearance could never be explained by immanent natural development. Man had been created in a "specific act of God," "God's immediate, personal interference ad hoc" (Kübel 1883, 252; Hübner 1966, 42f.; Rohls 2007, 124f.). Similarly, conservative theologian Carl Glaubrecht claimed to be able to harmonize science and religion on the basis of facts. His "new empirical philosophy of nature" converged with scripture (Glaubrecht 1878/1881, vol. 2, 6f.), because of the consilience of two kinds of facts: empirical facts and "facts of faith" related either to external sensory experience and thus to knowledge, or to the internal experience of transcendental things and thus to the "certainty of belief." The certainty of belief was much stronger than that of experience, since it affected the whole life (Glaubrecht 1878/1881, 9f., 301). Though the bible was not a textbook about nature and its content had been accommodated to human understanding, in cases of seeming incompatibility it was the *interpretation of scientific facts* rather than the interpretation of scripture, that had to be rejected (Glaubrecht 1878/18816f., vol. 1, IVf.; Hübner 1966, 38f.).

A highly instructive example for boundary disputes conducted on the same epistemological and ontological level is that of Otto Zöckler. Zöckler was most influential, for he edited the encyclopedic "Handbuch des theologischen Wissensganzen in seiner historischen Entwicklung und organischen Gliederung" (Zöckler 1883), apologetic journals like "Der Beweis des Glaubens" (1865–1906), the "Evangelische Kirchenzeitung" (1882–1892), and wrote literally hundreds of articles for the "Realencyklopädie für protestantische Theologie und Kirche," including the ones on "Jesus Christ," "Man," and "Creation" (see Gregory 1992, 112–159; Rohls 2007). Compared to most other contemporary Protestant accounts of the relation between science and religion, Zöckler's "biblical physics" was more radical and conservative: science and religion were concerned with the same objects, but scripture had absolute authority, the certainty of faith comprising the uncertainty of scientific knowledge. Science could illustrate religious truths, but there was little scope for it to contribute to real knowledge, which was solely about *essence*, about the – teleological or eschatological – *meaning* of things.

Zöckler spotted materialists as the primary foes of a revelatory natural theology. Albeit scientifically materialism was "totally dead," natural theology had to integrate sensual objects and evidences of experimental science, fighting carnal realism with pneumatic realism of scripture (Zöckler 1860, 281–283). The "Theologia Naturalis" aimed at the "verification of the fundamental consilience of the book of nature and of revelation" (ibid., III), trying to conceive God from nature without following the principles of a *theologia rationalis*. Instead of being based on a perception of God derived solely from "explanation and observation of an objective nature," it rather had to be based on revelation, on the principle of a "hopeful expectation" of the coming of the realm of Christ (ibid., 2). Since the unaided mind was not capable of true knowledge, natural revelation was but a "preliminary and shadowy step," while natural theology extended and elucidated the biblical and ecclesiastical teaching of faith. It followed the doctrine of *credo ut intelligam* in order to "illustrate the book of revelation by means of the book of nature, and to interpret the latter with help of the former" (ibid., 6). Zöckler's negative account of unaided human perception and rationality corresponded with a contempt for the world, necessitated by the principle of *hopeful expectation*, that called for a positive-anagogical comparison between this world and the next (see ibid., 195–199). The method of natural theology was to progress from an analogical–symbolical to an anagogical–typical consideration of nature. Promoting a factual symbolism in the tradition of the Augustinian *allegoria in factis*, Zöckler's method relied on the immediate tie between nature and scripture in biblical imagery, symbolism and metaphorical language as the "sometimes objective, sometimes absolute norms and tests for the interpretation of nature" (Zöckler 1860, 203).[4] Zöckler thus demanded scientific explanations of nature to follow the principles of biblical hermeneutics, in order to detect unperceivable patterns of natural phenomena that corresponded to types of symbolical, allegorical and parabolical representations used in scripture (Zöckler 1860, 204–206). This "positive criticism of biblical symbolism" resulted in the insight that "[a]ll natural beings in their innermost divinely determined essence match with the meaning implied by the symbolism of Holy Scripture," the essence of all natural creatures being their eschatological and teleological character, revealed by "biblical physics" (ibid., 220). Based on observation and experiment, the sciences only helped to extend biblical symbolics to all natural things – biblical physics provided the method to understand the true essence of natural phenomena, natural hermeneutics followed the principles of biblical hermeneutics (see ibid., 239–248).

In his "Geschichte der Beziehung zwischen Theologie und Naturwissenschaft" (Zöckler 1877, 1879), intended to assist in the development of biblical hermeneutics and comparative history of religion (vols. 1, 7), Zöckler discussed the relationship between biblical exegesis and the rational progress of knowledge. He not only acknowledged scientific progress, but also hinted at the dependence of exegesis on the peculiar knowledge and world views of the time: a historic perspective on the interpreters of scripture had to be taken up, since their Christian faith did not put them outside the specific world views of their time. Still the consequences that Zöckler drew from the necessity of applying historical–critical methods to scripture differed from those of Strauss, since Zöckler conceived historical developments as a part of divine eschatology. According to Zöckler scripture provided eternal truths about the essence of natural things, but these truths appeared in a specific historical form depending on the stage of history. Therefore, biblical hermeneutics had to analyze the relation between the (active) *accommodation* of the word of God and the (passive) *adaptation* of the prevalent patterns of thought: the history of exegesis showed that the biblical account of the world might not only have been consciously and deliberately altered in order to enhance its comprehensibility – it also necessarily could not provide empirical knowledge of man and nature beyond the scientific development of the time. With this, Zöckler seemed to present a Christianized Hegelian theory of development, wherein progressive stages of exegetic truth corresponded directly to the enfolding of salvation history. Science, defined as the

[4] see St. Augustine, De Trinitate 15,9, CChr. SL 50A, 482.

experimental and descriptive investigation into nature, was also part of this history: the scientific-utilitarian revolution of the late eighteenth century corresponded to a stage in salvation history, because the biblical promise of man's mastery over nature had now finally been achieved (Zöckler 1877, 1879, vol. 1, 72f.).

Still Darwin's theory of natural selection was but a transitory phenomenon following the naturalistic Zeitgeist, which would once be substituted by a higher stage of knowledge (see ibid., vol. 2, 697, 779). Zöckler harshly judged the Darwinist "Hypothesengebäude" to be a "chain of pseudo-arguments," "scientifically untenable," and "pathological," as it lacked the support of *observed evidence* (Zöckler 1903, 619f.). Following a neo-Baconian epistemology that based truth solely on observation and experiment, Zöckler made use of the non-experiential character of the theory of descent to point out its "insurmountable flaws" (ibid., 620; Zöckler 1877, 1879, 736f.; Mayr 1982, 71–73). Since empirical observation had never shown anything other than constant species, any alleged similarities between man and animal had to be dismissed as mere "products of imagination," invented to fit in with the "scientific novel" of Darwinism, instead of relying on "sober observation" (Zöckler 1903, 621). In the end, Darwinism, according to Zöckler's regained self-confidence, turned out to show all the features of literary imagination: it was constructed as a tantalizing narrative based on imagination, *fiction* as opposed to the *facts* that supported the biblical account of nature.

Even though Zöckler regarded Darwin's theory as an indicator of human perfection, because it strengthened the idea of historical progress and undermined polygenetic theories that were opposed to the monogenetic descent from Adam and Eve (Zöckler 1877, 1879, vol. 2, 779), it was wrong in respect to the (common) descent of the organic world and in questioning the exceptional position of man. Racial differences were but the product of moral brutalization and thus the effect of a degeneration of man, originally the *imago dei* (Zöckler 1860, 584–614). When Zöckler proposed his thoroughly anti-Darwinian theory of speciation by moral decay in 1903, neo-Lamarckian, neovitalistic and other non-Darwinian theories had gained more and more ground, stressing the finality of natural processes (see Bowler 1985, 1992). Thus, it had become much easier to challenge the alleged truth of the theory of descent by referring to opposing interpretations of empirical data (see Zöckler 1903, 617). Zöckler asserted a direct causal effect of original sin onto the development of mankind, reconciling biblical and scientific chronology: the "principle of sin or the spirit of Cain" had an accelerating effect on the diversion of the originally united mankind into different races, which had not been taken into account by naturalistic anthropologists who believed in "myriads of years" of development (Zöckler 1903, 624, referring to Gen. 11, 1–3 and 1. Jo. 3, 12–14). The idea that the 'principle of sin' had the status of an additional factor at work in evolution as a process that could directly be intervened and directed by God could be linked to the recent outcomes of experimental embryology.

The assumption of a concerted influence of material and non-material factors was familiar to Zöckler, as it had been reintroduced by neovitalistic theories in order to explain peculiarities in embryogenetic development: in the 1890s experiments had shown that the prevailing mechanistic theory of development failed to explain the teleomorphic character of organismic differentiation processes.

Neovitalistic authors like Driesch (1894) thus launched a non-material 'additional factor' that supposedly led organismic developments towards their intended *teloi*, claiming to adhere solely to the scientific authority of empirical experiments. With Driesch and others, the explanatory gap in mechanistic approaches could now be filled by an immanent principle of organic formation or even a transcendent principle like divine interference. Encouraged by these new findings about ontogenetic development, conservative Protestants like Zöckler ultimately dismissed Darwinism, holding that the biblical references to nature and man were correct, even if they had to be interpreted with care, as revealed meaning differed from literal meaning. Science might be able to deliver empirical knowledge about nature, but this knowledge had to be interpreted and warranted in order to be understood: it only formed the first step in the process of gaining knowledge. As human reason since the Fall was flawed, a correct interpretation of empirical facts depended on the assistance of revelation. Without the help of the redeemer, knowledge could not be gained, and no religion of humanity could ever prevail over the "blood-dripping specter of nihilism" (Zöckler 1903, 629). In Zöckler's biblical physics, religious meaning had gained hegemony over all kinds of explanations of natural phenomena.

## 3.4 Epistemological Complementarity and Inner Experience

Many liberal theologians subscribed to the idea of a peaceful coexistence of the realms of explaining and understanding on basis of romantic concepts of complementarity. The idea of an epistemological unity of knowledge had been one of the main principles of early nineteenth century *Naturphilosophie*. According to Johann Wolfgang von Goethe, knowledge about nature was built on the mutual relation of analysis and synthesis, romantic naturalists pursued Alexander von Humboldt's idea of a complementary supplementation of science and aesthetics (Kleeberg 2007b). Many theologians referred to natural philosopher Gotthilf Heinrich von Schubert's (1780–1860) dualistic epistemology of nature, a romanticized version of natural theology inspired by Friedrich Wilhelm Schelling (1775–1854), who distinguished between two forms of perception of nature – the sober scientific 'day-view' on individual natural phenomena and the 'night-view' of nature, the contemplation of nature hinting to the divine *oeconomia naturae* (von Schubert 1835; von Schubert 1845, 540–546; see Gregory 1995; Kleeberg 2005, 43–46). The idea that the transcendental dimensions of the world could be grasped directly marked not only the Liberal Protestantism of Friedrich Daniel Schleiermacher (1768–1834) and his romantic "theology of consciousness," but the concept of *Ahnung* (intuitiveness) as the epistemological mode of "unmediated knowledge" as well, which Jakob Friedrich Fries (1773–1843) had introduced (see Pfleiderer 1992, 108–114). For Fries, to believe meant to deny all limitations of knowledge, while 'Ahnung' formed a necessary prerequisite for real scientific knowledge, since it was "the feeling that the eternal was reflected, albeit in an imperfect and restricted manner, in the finite" (Gregory 1992, 41). Hence, *Ahnung* showed that causal explanations were comprised within a realm of *absolute* causes, which the beauty and harmony of nature reflected. The individual and subjective dimension of reveling in nature was understood to bring about a kind of unmediated access to the truth of nature as a whole – a truth *felt*, able and necessary to frame the detailed knowledge of scientific or philosophical analysis of the phenomena. In this sense, scientific reasoning and aesthetic contemplation formed two sides of an epistemological coin – an idea that allowed to integrate knowledge and belief into mediation theology *and* into science (see Kleeberg 2005; Kleeberg 2007b, 199–201).

Thus many liberal German theologians shared with biologists a common tradition of Idealism and Romanticism. If biologists therefore did not openly challenge Christian religion as such – as Haeckel and his followers did – their positions could often easily be reconciled with religion, especially if they came from morphology, a discipline with a disposition towards teleological explanations. One of the most elaborated attempts at 'mediation theology' was undertaken by Rudolf Schmid (see Gregory 1992, 53, 160–198). As Rohls (2007, 125f.) has pointed out, Schmid declined materialistic interpretations of nature, but nevertheless was convinced not to be in conflict with Darwin when integrating causal development into a teleological theism. In his "Die Darwin'schen Theorien und ihre Stellung zur Philosophie, Religion und Moral" (Schmid 1876), Schmid claimed that the absolute peace between the freedom of scientific investigation and the "unwithered maintenance of all our religious properties" was due to "one function of the mind directly depending on the other" (VIf.). Science and religion formed an epistemological whole of complementary knowledge, based on mutual supplementation of different perspectives – what science called 'causality,' religion termed 'divine acts.' Religion would have to expel accommodated scientific ideas, if these were proven wrong, just as science would have to, concerning the religious insights it had picked up (Schmid 1876, 236). Schmid thought of religious and scientific truths as following the same procedures of justification. Provided scientific *truths* were not mixed up with their natural philosophical *interpretations* (ibid., 4), science and religion pointed to the same fundamental truth – God. The advancement of scientific knowledge about the development of the universe simply amplified our knowledge about divine actions (ibid., 7). Similarly the Swiss reform theologian Heinrich Lang in 1873 argued that religion and natural sciences should not mutually restrict their explanations, as there was a unity of mind and matter in God, alluding to the complementarity of subjective religious feelings and objective scientific reasoning discussed above.

When Schmid in "Das naturwissenschaftliche Glaubensbekenntnis eines Theologen" (Schmid 1906, 79f.) proposed a teleological interpretation of nature, he, like Zöckler and many others, was encouraged by the return of Lamarckist and vitalist scientific approaches since the 1880s. His attempt to combine theology with vitalism *and* the theory of descent was not too peculiar, as indeed many followers of Darwin now held teleological positions. In Britain, where natural theology following Paley's popular formulation of the argument from design was strongest, even the majority of the scientists themselves "retained the view that the material universe must have a moral purpose" (Bowler 1992, 178; see Robson 1990; Brooke et al. 2001). Thus it was convenient to refer to the second outstanding father of the theory of natural selection, Alfred Russel Wallace, who at the end of the 1880s had reintroduced a theistic notion to his interpretation of nature in order to explain the higher faculties of man and to fight off the idea of blind necessity ruling nature (Wallace 1889, 477f.). As long as *general providence* rather than *direct interference* was involved, Wallace and other Darwinists had no problems with integrating religion and evolutionary theory (see Wallace 1889, 473; Smith 1972). Just like them, Schmid based his concept of an interdependency of science and religion, reason and faith, on the idea of an underlying plan of nature. Yet he regarded religion as superior to science, claiming that in the end general providence would lead to mind prevailing over matter. He conceived the agency of this plan as an immaterial force affecting nature from the outside. Ultimately, this force was part of the divine plan (Schmid 1876, 256f.; see Rohls 2007, 126). If teleology as the crucial principle of any kind of religion could be saved or even integrated with Darwinism, religion and science could peacefully coexist. If biblical hermeneutics maintained that the essence of scriptural references to nature lay in accounts of nature's teleological structure, while literal interpretations were dismissed, science could provide explanations of processes, while religion answered questions about origins, ends, and thus ultimately: meaning.

An interesting figure beyond academic theology is Emil Pfennigsdorf (1868–1952), professor for practical theology in Bonn. The Protestant pedagogue recommended his popular bestseller "Christus im modernen Geistesleben" to candidates for confirmation as a means to fight the growing number of anti-Christian intellectual currents in the "severe struggle for Weltanschauung" (Pfennigsdorf [1899] 1907, v–viii). Pfennigsdorf had a thoroughly positive view on science: The nineteenth century was the century of new scientific insights and enormous technical progress, the century in which man would fulfill God's command to be the master over nature to a new extent, the steamboat accelerating proselytization. Accordingly, the deep breach between the sciences and Christianity was not the result of science as such, but of new false interpretations of the autonomy of man and nature, which led to immorality and arbitrary action. These interpretations were based on the materialist conviction that the universe was ruled by blind chance and necessity, the latest example for the new scientific religion being Haeckel's Monism (ibid., 24–29).

This critique is illuminating as it gives a clue about the epistemological values that Pfennigsdorf held to be relevant in science *and* religion: thoroughness, accuracy, necessity of proofs and rational justification. But unlike science, only religion was able to bridge the gap between creative and sentient man and cold and insentient nature, because it was – in this respect similar to art – based on personal experience (ibid., 2). In questions of epistemology, Pfennigsdorf followed the systematic theologian Martin Rade, who edited the influential journals "Christliche Welt" and "Zeitschrift für Theologie und Kirche." Rade regarded the differences between scientific and religious interpretations of man and nature as differences *of perspective*, based on *diverging modes of experience* (Rade 1898, 32–34). He saw no conflicts between science and religion arising, as long as both stayed within their ontological, methodological, and explanatory boundaries. These boundaries were constituted along four distinctions: (1) the visible versus the invisible world, (2) observation, experiment, and calculation versus authority, (3) the individual part versus the whole, and (4) causality versus teleology (Rade 1898, 21–26, see 36f.). Similarly, Pfennigsdorf distinguished between the "sensible, perceivable, impersonal" objects of the sciences and the "invisible spiritual world [...] behind, above and within the visible world" (Pfennigsdorf [1899] 1907, 32). As only the materialistic interpretation of Darwin's theory would create an antagonism, so would only a "mechanical interpretation of Holy Scripture" (ibid., 28), that is, a literal exegesis as opposed to the spiritual and moral interpretation that served to elevate humanity.

On this basis, Pfennigsdorf called it a delusion to think *belief* would contradict *science* (ibid., 28, 32), alluding to Eberhard Dennert, editor of the journal "Glauben und Wissen," and founder of the "Keplerbund zur Förderung der Naturerkenntnis" (Pfennigsdorf [1899] 1907), a reaction to the founding of the "Monistenbund" (Pfleiderer 1906). In "Bibel und Naturwissenschaft" (Dennert 1904, 11–20), also understood as a guide for the youth, Dennert distinguished between three stages in the process of gaining knowledge: sensory perception (which might be deluded), reasoning (which is never absolute, but always human), and comparison with (historical) experience. Since the history of human knowledge was related to a history of errors, every stage of the epistemological process contained uncertainties, rendering absolute knowledge impossible. There was no difference between knowledge and belief – knowledge being more or less warranted belief, and vice versa (Dennert 1904, 20f.). Yet the certainty of religious belief outrivaled that of scientific knowledge, since it formed an integral part of human life and morality, supported by inner experience and revelation (ibid., 22–28). Thus, its aim was different from that of scientific knowledge, which found its boundaries in the external world (ibid., 53).

Like Dennert, Pfennigsdorf distinguished between the aims of science and religion, both following different epistemologies. The Christian searched for the aim of things, asking 'to which end?,' ultimately looking for the meaning of life and death, a question that science could never provide an answer to. Science on the other hand investigated the causes of phenomena, explaining them by (experimentally) recurring to other known phenomena. Yet science and religion were both based on facts, and even the epistemological status of material scientific and mental religious facts was the same (see Pfennigsdorf [1899] 1907, 33–35). This line of reasoning is revealing, as it displays an amalgam of Platonic-Augustinian thoughts and modern neuro-physiological insights: since the early nineteenth century, physiological concepts on the subjectivity of perception had evoked epistemological uncertainties that scientists tried to dissolve by either assuming an innate natural knowledge like Haeckel and others, or by enforcing a "mechanical objectivity" that tried to push back the flawed human element in research (see Daston and Galison 1992; Kleeberg 2004). When prominent physiologist Emil Du Bois-Reymond put forward his formula of scientific agnosticism *ignoramus et ignorabimus* (Du Bois-Reymond [1872] 1882), it perfectly met the needs of apologetics. Theologians alluded to his statements about the insurmountable limits of scientific knowledge, which Du Bois-Reymond had conceived to be the consequence of the narrowness of the senses, the limits of reason and the inexhaustibility of the world. Reason could never grasp the *riddles of the universe*, since every answer only led to more questions. It was precisely this "inexhaustibility of the world" that was "a sign of its divine origin" since it corresponded to the principle of abundance of natural objects. If scientists tried to solve questions of meaning, they left the boundaries of scientific explanation and entered the realms of belief, presenting convictions that were mere superstition as opposed to religious knowledge based on the "experienced realities" of faith (Pfennigsdorf [1899] 1907, 39, see 35–42).

Pfennigsdorf linked this position to the ideas of yet another romantic theorist of epistemological complementarity, Gustav Theodor Fechner, who from his research on the physiology of aesthetic perception had drawn dualistic consequences that followed in the tradition of the neo-Platonic differentiation between *mundus intelligibilis* and *mundus sensibilis* (see Fechner [1879] 1904). Citing Fechner, Pfennigsdorf regarded the appearance of objects in the perceptionally restricted human mind as a "weak reflection of the rich variety of the external world" (Pfennigsdorf [1899] 1907, 36f.). This epistemological uncertainty had important consequences for the status of belief in contrast to knowledge: even in everyday life all knowledge rested on belief. Due to the physiology of the senses, human perception could not obtain true knowledge from empirical evaluation – since there was no ultimate certainty about the world, all knowledge was based on trust (see ibid., 86). While Christian belief could unite individual facts to a harmonic whole, science only explained mechanical relations. Following the philosopher Rudolf Hermann Lotze, Pfennigsdorf now described natural laws as scientific *constructions* to explain the uniformity of natural phenomena and processes: Natural laws were nothing but "*Denkformeln*, used to bring to mind the regularity of natural processes" (ibid., 42; see Lotze 1854, vol. 3, 481). With Lotze, Pfennigsdorf followed the Kantian-Friesian epistemological tradition of the *coherence theory of truth*. According to this theory, as Gregory (1992, 20–22) has pointed out, the relevant criterion to judge the truth of a proposition was the inner coherence of a system of beliefs and not the relation to the external world. Thus there could be a truth about nature independent from science, a different and even more coherent and certain truth revealed by God. Scientific constructions could even be regarded as unconfirmed speculation, unless they were understood as "tools in the hand of a higher being" (Pfennigsdorf [1899] 1907, 42).

Following from that, as science only knew the intermediate but never the last and ultimate causes, it was not authorized to pose any statements about divine interference in natural processes – only the "faithful human" who had had a religious experience discerned the glory of God in nature (Pfennigsdorf [1899] 1907, 44): judged on scientific grounds, Darwin's theory was justified, since he never attempted to explain anything beyond the realm of the empirical things. German materialists like Ludwig Büchner or "development-fanatic" Haeckel used Darwin's theory to revitalize their position, but as science only dealt with the visible world, Haeckel could not give any scientific answers to questions about the existence of God, his atheism merely being belief. While Monism opposed the facts of nature and mental life, the biblical account of the qualitative difference between animal and man was based on facts. Not only man's mental side – language, science, art, and religion – but also his material side was rightly considered in the allusion to man having been formed from clay (see ibid., 54–63, 68–70). Pfennigsdorf clearly dismissed any literal interpretation of the bible, but upheld its fundamental concept of life to be empirically correct: "Christian Belief completes the theory of development" (ibid., 72). Whoever believed in the literal meaning of the bible without believing Christ was a "Pharisaic literalist" – the bible was written and compiled by human beings for human beings, containing "Christian truth, coated with the veil of human transmission" (ibid., 299f.). The knowledge about nature that scripture offered was not on direct display, neither was the historical truth of its parts: the ultimate meaning of the bible lay in its aim to arouse faith. And in that, Pfennigsdorf states alluding to Luther, it depended on belief in Christ: "We believe in the bible, because we believe in Christ" (ibid., 291).

## 3.5 Modes of Experience, Modes of Science

Theologians concerned with questions of morality and pedagogy for the most part seem to have regarded the realms of science and religion as independent, recommending an (explanatory) dualism that mirrored the different objects of theological and scientific investigations. Related to different sets of questions, explaining and understanding did not come into conflict, since both methodologies were restricted to specific areas. Albrecht Ritschl, exponent of the Union of Prussian Regional Churches and highly influential in Prussian church politics, claimed that biblical hermeneutics ultimately had to be based on the belief in Christ and on moral practice. Gregory has shown that with his sharp differentiation between religion, metaphysics and science, Ritschl and his Göttingen school around 1870 signaled the death of theological positions mediating between science and religion (see Gregory 1992, 54–63). According to his "Die christliche Lehre von der Rechtfertigung und Versöhnung," the natural world and the whole biblical account of creation had to be interpreted as a "relative necessity" (Ritschl [1870, 1874] 1895, vol. III, 266), a means to the end of improving human morality. For Ritschl and his school of thought, it was not sufficient to regard the bible as revealed, since only the interpreted books of the bible were able to lay the grounds for a dogmatic or positive knowledge of Christianity. Yet biblical hermeneutics were not to be constrained by ecclesiastical laws or specific dogmatics, especially if these were subject to historical change: the sole and indispensable prerequisite was moral practice, as other criteria for the right interpretation differed from denomination to denomination (Ritschl [1870, 1874] 1895, vol. II, 5). The biblical account of creation accordingly was to be understood as a *moral narrative*, residing on a different explanatory level than the one relevant in science: though the divine creation of nature was an apodictic truth, this kind of true knowledge could not be warranted, proved or challenged by any means or methods familiar from the sciences, but only in respect to the morality as the divine *telos* of the bible. Biblical hermeneutics hence ultimately had to follow along the lines of tropological (and allegorical) exegesis.

With this approach, Ritschl opened up the possibility of preserving a biblical truth that could not be confronted by any of the new insights of the natural sciences in general, or the theory of descent in particular. Following the *coherence theory of truth*, he rejected any kind of correspondence not only between nature and scripture, but between scientific truth and reality (see Gregory 1992, 59–61). But what might seem to have been a successful apologetic strategy concerning the interpretation of scripture in the end only led to shifting the struggle between science and religion to a different field of discussion: Ritschl's biblical hermeneutics again ultimately relied on the idea of divine teleology. Scientific knowledge could not provide answers to ultimate questions, unless transgressing the "narrower area of nature" (see Ritschl [1870, 1874] 1895, vol. III, 543). Yet any interpretation of scripture would have to be based on the ultimate meaningfulness of nature, which had to be defended against any kind of argument from contingency, necessity, or chance. Biblical hermeneutics was to correspond to a kind of natural hermeneutics that detected the numinous element in the empirical world.

Ritschl's disciple Max Reischle (1858–1905), professor for practical theology in Giessen who coined the term 'religious pedagogy,' proposed a *theologia viatorum* that included the idea of a subordination of nature to morality (Reischle 1898, 42; on pedagogy Reischle 1889). Concerning method, Reischle pointed to fundamental differences between scientific and theological endeavors, following Dilthey's differentiation between quantitative explanations and the mutual understanding of human beings (Reischle 1898, 25). But nevertheless, Darwinism had to subordinate its *etiologic empirical* explanations to a *universal teleological* framework. Comprised within religion, science had to accept that all material substances, forces, and developments had their ultimate basis in the *incognizable realm* of divine *providentia generalis*, while religion had to refrain from transgressing its boundaries towards the *empirical realm*. In the end, the indispensable precondition for a peaceful coexistence between science and religion was again that development be interpreted within a teleological framework (Reischle 1898, 13–17; see Rohls 2007, 127). The current evolutionist interpretation of natural development as a progress towards more and more perfection, which Reischle explicitly described as an amalgam of idealistic and naturalistic evolutionism, indeed fitted well with this idea (see Reischle 1898, 9, 41; Reischle 1902).

The teleological interpretation of evolutionism becomes even more evident from Ritschl's disciple Rudolf Otto (1869–1937), professor for systematic theology in Göttingen. In his "Naturalistische und religiöse Weltansicht" (Otto [1904] 1929) Otto claimed that there was a fundamental difference between science and religion, the first referring to immanent causes in nature, the second referring to divine purposes, realized in nature. Like Pfennigsdorf, Otto maintained that science and religion followed different epistemologies, the faithful knowing with a higher degree of certainty as a result of a specific state of mind. The relevant truth criteria – inner experience, conscience, and mind – made clear that to know religious truth meant to have a specific attitude towards life *merus enthusiasmus* (Otto 1905, 9f.). This perspective differed from a perspective based on chance and necessity. Hence when Otto dismissed Darwinism, it again was not because of the idea of development, but due to Darwin's description of development as contingent and undirected: Darwin's theory "is only downright anti-theological in that it is anti-teleological"

(Otto [1904] 1929, 107; cit. from Rohls 2007, 128). Accordingly, Otto discriminated between Darwin's theory of natural selection on the one hand, and the theories of Goethe, Schelling and Hegel on the other, which all had used the concept of development to establish the unity of nature. While Monistic evolutionism had referred to Goethe as well in order to claim the unity of nature, Otto explicitly conceived development as a teleological process, according to which every higher developmental stage was a total enfolding of a lower stage, and man had to be considered a "total realization of what had already been implemented at the lowest level as something potential" (Otto [1904] 1929, 98; cit. from Rohls 2007, 128). Development was not a descent from the lower level in a passive process of adaptation by natural selection and thus effect of chance and necessity, but a "creative reorganization," a purposeful process of general providence (Otto 1905, 28f., 59). The transcendental teleological dimensions of the world could be grasped directly, Otto stated following Fries, by means of *Ahnung* as the epistemological mode of "unmediated knowledge."[5] *Ahnung* showed that causal explanations were comprised within the realm of the *absolute*. Christian religion helped to understand the 'absolute' as the divine: God had not created a finished world, but a world coming into being, he had set the world as "will to mind [Geist]" (Otto [1904] 1929, 282; Rohls 2007, 129).

Like Otto and Reischle, Karl Beth, professor of systematic theology in Vienna and one of the founding fathers of psychology of religion, proposed an idealistic interpretation of evolution. Affirmed by contemporary Neo-Lamarckism and Neo-Vitalism, he went along with Darwinism as far as development and descent were concerned, but criticized its core idea of natural selection for "violently substituting" contrivance with chance.[6] Instead, Beth subscribed to the Neo-Vitalistic idea of evolution as a teleological epigenetic process that allowed to trace back creation and development to "direct divine action" (Beth 1907b; Beth 1909, 126; see Altner 1965, 13–24). He vindicated miracles as a specific kind of direct interference, opposing to a history of religion that "regards all religious data as subjective" and thus miracles as the "outcome of imagination to which there is no reality" (Beth [1905] 1907, 60f.). Miracles could not be discussed within the framework of fact or fiction – they depended on religious experience. Like Otto, Beth pointed to the necessity of entering the right state of mind as a precondition for accepting religious truths. Thus concerning biblical hermeneutics, Beth, who had studied with Dilthey, related the correct interpretation of scripture and the miraculous events it narrated to the articulation of religious experiences that opened them for understanding (see Beth [1905] 1907, 16–21, 36). He also seems to have followed Dilthey in his attempt to understand individual actions and experiences by contextualizing them against the contemporary state of awareness of an epoch (see Beth 1907a, 64–84, referring to Dilthey 1891/92).

To Beth, theology could not simply be grouped amongst the humanities, restricting itself to the interpretation of the bible and leaving explanations of nature to the sciences. His answer to the question of whether Christianity was endangered by the

[5] See Otto (1909, 40f., 83); Otto (1905, 57–60); Pfleiderer (1992, 108–114); Gregory (1992, 38–42).

[6] See Beth (1909, 95, 33). Beth (1909, 106–108) argues against Haeckel, following the Neo-Vitalism of Johannes Reinke and Carl Nägeli.

modern sciences was negative, because since science explored matter and forces of nature in all its perceivable nuances, there was an equally important "*meaningful rest* of the real world:" "This is the domain of the Geist, of intellectual activity and striving, of the sensually not-perceivable driving forces of coming into being and passing away; these are the factors that – non-analyzable – bring about and spur on life and history, and who knows, how far beyond human history they operate! Hence science has its boundaries there, where the world of external phenomena has it." The "main problem of science" was "to identify the boundaries between the realms of nature and Geist" (Beth 1907a, 304). Following Schleiermacher and Ritschl, Beth defined religious knowing as a process within human consciousness, in which "an immediate certainty imposes itself" (ibid., 248). Thus scientific and religious knowledge differed in their epistemological mode – one empirical, one based on faith, "they do not compete with each other:" "Religious knowledge as such has no direct interest in empirical investigations in nature. It calmly watches this different examination of nature. [...] Both can walk side by side as strangers" (ibid., 305).

To this point, Beth's ideas resemble those of mediation theology mentioned above, but nevertheless he harshly criticized the "new-romantic standpoint" of a peaceful coexistence between science and religion, citing his former professor for systematic theology from Berlin Otto Pfleiderer: "Religion and science, they say, should separately and peacefully coexist [...]. Science should confine itself to causal knowledge of the relation of finite things and processes, while religion has nothing to do with knowledge, neither of god nor of the world, but with the experiences of the mind, of our inner life" (Pfleiderer 1906, 437; Beth 1907a, 259). But *religious* was not *theological* knowledge – the "scientific [wissenschaftliche] knowledge of religious statements" (Beth 1907a, 257; see Lipsius 1879, §72, 68). Theology was more than just religious experience, and a meditative approach would cut off theology from the other sciences. At the same time, theology was more than a psychological or historical discipline amongst the humanities. Beth disapproved of Reischle and others from the school of Ritschl, who incorporated psychological methods into theology as a *Wissenschaft* that "conducts historical research and leaves aside questions about the objective reality of the objects of faith" (Beth 1907a, 291; see Beth 1904, 6f.; Ritschl [1870, 1874] 1895, vol. III, 188; Reischle 1889).

Similarly, systematic theologian Ernst Troeltsch's modern historical analysis of Christianity was no possible resort. Troeltsch, whose broad interest in cultural, historical, social and philosophical aspects of religion had been influenced by Dilthey, Max Weber and the Neo-Kantianism of Heinrich Rickert and Wilhelm Windelband, in 1901 proposed the idea of a parallel evolution of human self-consciousness and the idea of religion: He regarded religion as the product of the ideal powers of human civilization, answering 'that,' 'how,' 'under which conditions,' but not 'because of which inner necessity' religion evolved. For Beth (1904, 2–5, 28–30), Troeltsch had thus overstretched his inductive approach though he was right – this was Strauss' legacy – that the essence of Christianity had to be identified by means of historical research. But even though inductive historical analyses were generally regarded as a precondition of biblical hermeneutics

(Heinrici 1899, 726), Beth insisted that neither historical, nor psychological approaches provided a sufficient basis for theology: Theology was based on "autonomous theological knowing" (Beth 1907a, 292) and thus was more than just a discipline from the humanities. It didn't simply use a specific method to understand the meaning of nature that wouldn't collide with scientific explanations, since both reflected the basic principles of nature like causality and teleology (ibid., 310f.). In this respect, theological knowledge – unlike religious knowledge – argued "on the same lines" and with "the same theoretical function" as scientific knowledge in that it reflected on the reality of its scientific objects (ibid., 306): It was a *positive* science focusing on the irremovable factor "revelation" as its scientific object, though unlike the positive sciences it had to prove the validity of its object in practice, which made it one of humanities like law and philosophy (ibid., 11).

In his attempt to establish a "modern positive theology," Beth followed the influential systematic theologian Reinhold Seeberg, conservative adversary of another of Beth's theological professors from Berlin, church historian Adolf von Harnack: Based on historical understanding of present theology in its different manifestations, old religious truths should be taught on the basis of modern questions, problems, and methods (Beth 1907a, 137). But while the modern humanities in reaction to Hegel's intellectualism were caught up in positivism, relied on observation, analysis and description, and lost their ambition to answer ultimate questions, theology had to preserve the practice of "seriously calm thinking" without falling into the extremes of intellectualism or empiricism.[7] Following Pfleiderer and Seeberg, Beth demanded that "theology entered into close contact to the sciences [Wissenschaften] in their entirety, not alone to history or philology, but primarily to the 'real' sciences of research and processing of research and to philosophy as well" (ibid., 298f.). The religious interest in the question about the origin of man lay in his eternal purpose and him being God's own likeness, while the sciences studied it empirically. It added extra dimensions to scientific objects, discussed them from a different perspective, just like linguistics and psychology were able to provide answers to the question about human origins differing from those given by the sciences: "if you admit the difference between the sciences and the humanities, you will have to admit them to the solutions of this problem" (Beth 1909, 105).[8]

The next generation of Protestant theologians more or less agreed that science and religion related to separate realms, asking either 'how?' or 'why?'-questions.

---

[7] Beth (1907a, 40f.), referring to Eucken (1904, 378). See Beth (1913, 244–246). Similarly, Seeberg's pupil Girgensohn (1900, 464f.), professor for systematic theology in Dorpat (Estonia), argued that the basis of Christianity still was religious experience, even though the "spirit of exact science has extended its scope to the humanities as well; historical-critical research methods rule whole theology, whether positive or liberal."

[8] Beth follows the Wasmann (1907, 34), who had been wrongly criticized by Wobbermin (1908, 152), who in 1915 followed on Troeltsch's chair for dogmatics in Heidelberg, for allegedly leaving the question about the origin of man to be decided only by theology: Wasmann's position would – on the contrary – underpin Wobbermin's position that follows Dilthey's differentiation between sciences and humanities. The idea that theology simply analyzed value-statements can be found in Häcker (1907, 23f.); Beth (1909, 104).

Their main protagonists like Karl Barth or Rudolf Bultmann subscribed to Rade's and Otto's ideas about diverging modes of scientific and religious experience, about two independent epistemological realms that could peacefully coexist. This is, what Gregory has termed the "rise of de-natured theology" (see Gregory 1992, 5–10). Still for the earlier theologians discussed here, the boundaries between science and religion were not at all settled. The notion of incompatibility between the two realms, both referring to the same objects and working on the same epistemological level, can be found with those who argued from the absolute authority of scripture. According to orthodox Protestants like Zöckler, science merely helped to illustrate revealed meaning, the certainty of faith comprising and overruling scientific truth. To them, theology was not merely a *Geisteswissenschaft* – it was the fundament of all knowledge, of the sciences and the humanities.

The idea of a peaceful coexistence between the realms of explaining and understanding formed the basis of mediative positions that followed romantic concepts of complementarity. Liberal theologians concerned with questions of morality and pedagogy like Ritschl and his school seem mainly to have regarded the realms of science and religion as independent, and recommended an (explanatory) dualism that mirrored the different objects of theological and scientific investigations. Related to different sets of questions, explaining and understanding were restricted to specific areas. Still most other Protestant theologians proposed a dualistic monism, highlighting the authority of the whole that *comprised* causal explanations of how nature worked as well as attempts to understand its universal meaning. Here, scientific *explanations* did not per se pose a threat to religion – only if scientists started to *interpret* nature's ultimate meaning, only if they practiced an atheistic form of *natural hermeneutics*, they had to be attacked. It was not the interpretation of the bible that these theologians cared about most: they had narrowed down their understanding of the range of scriptural facts in the course of the nineteenth century from nature to the organic world, to man and finally human action and morality. But with dualism and teleology questioned, the very fundaments of faith were at stake. The idea of a *telos* of nature existing in a spiritual or mental realm opened up the possibility of a higher force guiding nature, and therefore reassured the crucial doctrine of man as the *imago dei* and of a meaningful universe, the possibility of a religious answer to the ultimate questions. The scientific discourse of the late nineteenth century even allowed for interpretations of nature in terms of the teleological structure of human action, some of these were even considered state of the art.

In the course of their attempts to defend these basic principles, some of the authors put forward methodological objections against the possibility of merely quantitative accounts of nature and human beings, used arguments hinting at the constructivist character of scientific accounts of nature, or at the hermeneutical difference between explaining and understanding, especially in regard to the difference between human action and natural events. As the gap between science and humanities widened, *biblical hermeneutics* as a theory in the academic discipline of theology, being a part of the humanities, could be related to the more general hermeneutic approaches in philosophy that stressed the sole competence of humanities for questions of *meaning of texts*. If it was willing to abandon the idea that scripture gave a correct account of the objects of the sciences, biblical hermeneutics was able to retreat to the standpoint of the humanities in general, interpreting expressions of personal experience. By the end of the nineteenth century, more or less all Protestant theologians agreed on the dismissal of literal interpretations of the bible, even though most of them did not follow Troeltsch and Harnack on their way towards a modern sociology and history of religion. Still most of the theologians discussed here pointed to the specificity of *religious* experience and its (biblical) expression, not easily ranking theology amongst other humanities. If the art of hermeneutics was triggered by the desire to keep scriptural contents alive, as Heinrici (1899, 718) put it, the precondition for a correct interpretation was to be sought in analogous experience, similar to congeniality: "The unmusical ear doesn't hear music [...]. We only understand what we love. [...] He who has not experienced the religious-moral power of faith, judges faith as vestigial knowledge or narrow-minded enthusiasm."

## References

Altner G (1965) Schöpfungsglaube und entwicklungsgedanke in der protestantischen theologie zwischen ernst haeckel und teilhard de chardin. EVZ, Zürich
Beth K (1904) Das Wesen des christentums und die moderne historische denkweise. A. Deichert, Leipzig
Beth K ([1905] 1907) The Miracles of Jesus. Originally published as Die Wunder Jesu. Eaton & Mains, New York
Beth K (1907a) Die moderne und die prinzipien der theologie. Trowitzsch & Sohn, Berlin
Beth K (1907b) Empirische Teleologie. Die jüngste Wendung in der Biologie, published as Neue kirchliche zeitschrift, vols 1 and 2
Beth K (1909) Der entwicklungsgedanke und das christentum. Runge, Berlin
Beth K (1913) Die entwicklung des christentums zur universalreligion. Quelle & Meyer, Leipzig
Bowler PJ (1985) The Eclipse of Darwinism. Anti-Darwinian Evolution Theories in the Decades Around 1900. The John Hopkins University Press, Baltimore, MD
Bowler PJ (1992) The Non-Darwinian Revolution: Reinterpreting a Historical Myth, 2nd edn. John Hopkins University Press, Baltimore, MD
Brooke JH, Osler M, van der Meer JM (eds) (2001) Science in theistic contexts: Cognitive dimensions. Osiris, 16. Chicago University Press, Chicago, IL
Darwin C ([1859] 1860) Über die entstehung der arten im thier- und pflanzenreich durch natürliche züchtung oder die erhaltung der vervollkommneten rassen im kampfe um's daseyn. Translated by Georg Heinrich Bronn. Schweizerbart, Stuttgart
Daston L, Galison P (1992) The image of objectivity. Representations 40:81–128
Dennert E (1904) Bibel und naturwissenschaft. Gedanken und bekenntnisse eines naturforschers, 2nd edn. Kielmann, Stuttgart
Dilthey W (1891/92) Auffassung und Analyse des Menschen im 15. und 16. Jahrhundert. Archiv für Geschichte der Philosophie IV(4):604–651; V(3):337–400
Dilthey W ([1900] 1974) Die Entstehung der Hermeneutik. In: Die geistige welt: einleitung in die philosophie des lebens: 1. Hälfte, abhandlungen zur grundlegung der geisteswissenschaften. gesammelte schriften, vol 5. 6th ed. Vandenhoeck & Ruprecht, Göttingen, pp 317–338
Dilthey W ([1910] 1981) Der aufbau der geschichtlichen welt in den geisteswissenschaften. Suhrkamp, Frankfurt am Main
Driesch H (1894) Analytische theorie der organischen entwicklung. Engelmann, Leipzig
Du Bois-Reymond EH ([1872] 1882) Über die grenzen der naturerkenntnis. Veit, Leipzig

Eucken R (1904) Die geistigen Strömungen der Gegenwart. 3rd revised edn. Originally published as Die grundbegriffe der gegenwart. Veit & Co., Leipzig
Fechner GTh ([1879] 1904) Die tagesansicht gegenüber der nachtansicht. 2nd edn. Breitkopf und Härtel, Leipzig
Froböß G (1903) Lutheraner, separierte. In: Hauck A (ed.) Realencyklopädie für protestantische theologie und kirche, vol 12, 3rd edn. Hinrichs, Leipzig, pp 1–19
Girgensohn K (1900) Noch ein Wort zur Forderung einer modernen positiven Theologie. Mitteilungen und Nachrichten für die evangelische Kirche Russlands 21:451–466
Glaubrecht C (1878/1881) Bibel und naturwissenschaft in vollständiger harmonie, nachgewiesen auf grund einer neuen empirischen naturphilosophie. 2 vols. H. Schultze, Leipzig
Gregory F (1992) Nature Lost? Natural Science and the German Theological Traditions of the Nineteenth Century. Harvard University Press, Cambridge
Gregory F (1995) Gotthilf Heinrich Schubert and the dark side of natural science. Internationale Zeitschrift für Geschichte und Ethik der Naturwissenschaften, Technik und Medizin 3:255–269
Haeckel E ([1866] 1988) Die generelle morphologie der organismen. Allgemeine grundzüge der organischen formen-wissenschaft, mechanisch begründet durch die von Charles Darwins reformierte deszendenz-theorie. 2 vols. de Gruyter, Berlin
Häcker W (1907) Naturwissenschaft und theologie. Mohr, zwei referate von valentin und walter häcker. Tübingen
Heinrici G (1899) Hermeneutik. In: Hauck A (ed) Realencyklopädie für protestantische theologie und kirche, vol 7. 3rd edn. Hinrichs, Leipzig, pp 718–750
Howard TA (2006) Protestant Theology and the Making of the Modern German University. Oxford University Press, Oxford
Hübner J (1966) Theologische und biologische entwicklungslehre. Ein beitrag zum gespräch zwischen theologie und naturwissenschaft. C.H. Beck, München
Kleeberg B (2004) Vor der Sprache. Naturalistische Konzepte objektiver Wahrnehmung. In: Crivellari F et al (eds) Die medien der geschichte. Historizität und medialität in interdisziplinärer perspektive. UVK, Konstanz, pp 85–108
Kleeberg B (2005) Theophysis. Ernst Haeckels monistische philosophie des naturganzen. Köln, Böhlau
Kleeberg B (2007a) God-nature progressing. Natural theology in German Monism. Sci Context 20(3):537–569
Kleeberg B (2007b) Ideal (geometrical) types and epistemologies of morphology. In: Fiorentini E (ed) Observing Nature – Representing Experience. The Osmotic Dynamics of Romanticism 1800–1850. Reimer, Berlin, pp 187–203
Kübel RB (1883) Apologetik. In: Zoeckler O (ed) Handbuch der theologischen wissenschaften in encyclopaedischer darstellung, vol 3, 2nd edn. Beck, Nördlingen
Kuhn TK (1907) Strauß, David Friedrich. In: Hauck A (ed) Realencyklopädie für protestantische theologie und kirche, vol 19. Hinrichs, Leipzig, pp 76–92
Lang H (1873) Die Religion im Zeitalter Darwins. Reform 2:1–8
Lipsius RA (1879) Lehrbuch der evangelisch-protestantischen dogmatik, 2nd edn. Schwetschke, Braunschweig
Lotze H (1854) Mikrokosmus. Ideen zur naturgeschichte und geschichte der menschheit, 3 vols. Hirzel, Leipzig
Luthard CE ([1864] 1897) Apologetische vortraege über die grundwahrheiten des christentums. 12th to 14th ed. Dörfling und Franke, Leipzig
Luthard CE ([1880] 1908) Die modernen Weltanschauungen und ihre praktischen konsequenzen. Über fragen der gegenwart aus kirche, schule, staat und gesellschaft; im winter 1880 zu leipzig gehalten. 4th edn. Dörffling und Franke, Leipzig
Luthardt CE ([1880] 1967) Wirkungen des Kulturkampfes. In: Höfele KH (ed) Geist und gesellschaft der bismarckzeit [1870–1890] (= Quellensammlung zur Kulturgeschichte, vol. 18). Musterschmidt, Göttingen et al, pp 381–383

Mayr E (1982) The Growth of Biological Thought. Diversity, Evolution and Inheritance, 2nd edn. Belknap Press of Harvard University Press, Cambridge, MA
Otto R ([1904] 1929) Naturalistische und religiöse weltsicht. 3rd edn. Heinrich Laupp jr, Tübingen
Otto R (1905) Naturalistische und religiöse Weltsicht. Ph.D. thesis, University of Tübingen
Otto R (1909) Kantisch–Fries'sche religionsphilosophie und ihre anwendung auf die Theologie. Zur einleitung in die glaubenslehre für studenten der theologie. Mohr, Tübingen
Pfennigsdorf E ([1899] 1907) Christus im modernen geistesleben. Christliche einführung in die geisteswelt der gegenwart. Der gebildeten evangelischen jugend und ihren freunden dargeboten. 10th edn. Bahn, Schwerin
Pfleiderer G (1992) Theologie als wirklichkeitswissenschaft. Studien zum religionsbegriff bei Georg Wobbermin, Rudolf Otto, Heinrich Scholz und Max Scheler. Mohr, Tübingen
Pfleiderer O (1906) Religion und religionen. J.F. Lehmann, München
Rade M (1898) Die Religion im modernen geistesleben. Mit einem anhang über das märchen von den drei ringen in lessings nathan. Freiburg im Breisgau et al, Mohr-Siebeck
Reischle M (1889) Die Frage nach dem wesen der religion. Grundlegung zu einer methodologie der religionsphilosophie. Mohr, Freiburg im Breisgau
Reischle M (1898) Christentum und entwicklungsgedanke. Hefte zur christlichen Welt, vol 31. Mohr, Leipzig
Reischle M (1902) Wissenschaftliche Entwicklungserforschung und evolutionistische Weltanschauung in ihrem Verhältnis zum Christentum. Zeitschrift für Theologie und Kirche 12:1–43
Ritschl A ([1870, 1874] 1895) Die christliche lehre von der rechtfertigung und Versoehnung. 3 vols., 4th edn. Adolph Marcus, Bonn
Robson JM (1990) The fiat and the finger of god: the Bridgewater treatises. In: Helmstadter RJ, Lightman B (eds) Victorian Faith in Crisis: Essays on Continuity and Change in Nineteenth-Century Religious Belief. MacMillan, London
Rohls J (1997) Protestantische theologie der neuzeit, vol 1. Mohr-Siebeck, Tübingen
Rohls J (2007) Darwin und die Theologie. Zwischen Kritik und Adaption, in Kurt Bayertz et al (eds) Weltanschauung, philosophie und naturwissenschaft im 19. Jahrhundert Meiner, Hamburg, pp 107–113
Schmid R (1876) Die darwin'schen theorien und ihre stellung zur philosophie, religion und moral. Paul Moser, Stuttgart
Schmid R (1906) Das naturwissenschaftliche glaubensbekenntnis eines theologen. Ein wort zur verständigung zwischen naturforschung und christentum, 2nd edn. Kielmann, Stuttgart
von Schubert GH (1835) Ansichten von der nachtseite der naturwissenschaft. 3rd edn. Hausmann, Stuttgart
von Schubert G (1845) Spiegel der natur: ein lesebuch zur belehrung und unterhaltung. Palm & Enke, Erlangen
Smith R (1972) Alfred Russel Wallace: philosophy of nature and man. Br J Hist Sci 6:177–199
Strauss DF ([1835/1836] 1895) Das leben jesu für das deutsche volk bearbeitet, Teil I–II, 8th edn. Strauss, Bonn
Strauss DF (1840/41) Die christliche glaubenslehre in ihrer geschichtlichen entwicklung und im kampfe mit der modernen wissenschaft. 2 vols. Osiander, Tübingen
Strauss DF ([1872] 1873) Der alte und der neue glaube. Ein bekenntnis, 6th edn. Strauss, Bonn
Troeltsch E (1901) Die absolutheit des christentums und die religionsgeschichte. vortrag, gehalten auf der versammlung der freunde der christlichen welt zu mühlacker am 3. Oktober 1901, Erw. u. mit einem vorwort versehen von ernst troeltsch. Mohr, Tübingen and Leipzig
Wallace AR (1889) Darwinism. An Exposition of the Theory of Natural Selection with Some of its Applications, 2nd edn. MacMillan, London
Wasmann E (1907) Der kampf um das entwicklungs-problem in Berlin. Ausführlicher bericht über die im Febr. 1907 gehaltenen vorträge u. über d. diskussionsabend von erich wasmann. Herder, Freiburg im Breisgau
Wobbermin G (1908) Der Kampf um die Entwicklungslehre. Theologische Rundschau XI(5): 143–154

Zöckler O (1860) Theologia naturalis. Entwurf einer systematischen naturtheologie vom offenbarungsgläubigen standpunkte aus. Heyder & Zimmer, Frankfurt am Main

Zöckler O (1877, 1879) Geschichte der beziehungen zwischen theologie und naturwissenschaft mit besonderer rücksicht auf schöpfungsgeschichte. 2 vols. Bertelsmann, Gütersloh

Zöckler O (ed) (1883) Handbuch des theologischen wissensganzen in seiner historischen entwicklung und organischen gliederung, 3 vols. C.H. Beck, Nördlingen/München

Zöckler O (1903) Mensch. In: Hauck A (ed) Realencyklopädie für protestantische theologie und kirche, vol 12, 3rd edn. Hinrichs, Leipzig, pp 616–629

# Chapter 4
# How Wilhelm Dilthey Influenced Popular Science Writing: Kurd Laßwitz's "Homchen. Ein Tiermärchen aus der oberen Kreide"

**Safia Azzouni**

In the second half of the nineteenth century, popularization of science appeared to be a necessity of the time. Scientists and politicians discussed the pros and cons of making scientific knowledge accessible to the public. The question of how and by whom popularization should be done was a common topic in newspapers and magazines of the time. Even though museums as well as zoological and botanical gardens played an important role in disseminating knowledge, it can be said that the popularization of science basically was (and probably still is) a "language-based event".[1]

A large variety of writings successfully made scientific findings an everyday companion of the middle-class family around 1900 (Sarasin 1995). It was written not only by scientists but by professional popularizers. They did not limit themselves to mere simplifications of scientific facts or terminology. These authors – journalists, poets or former scientists – often had their background in two fields: science and literature. They were go-betweens, who made popular science literature a genre of its own.

In order to grasp the different ways of writing popular science, one has to take into account not only the development of science itself but also the contemporary views on education, poetics and philosophy. In this paper, I want to examine the relationship between popular science literature and the concept of the human sciences in Germany around 1900. Concerning the latter, I focus on Wilhelm Dilthey's methodological distinction between science and the humanities (hereafter "Geisteswissenschaften"). Then as an example for popular science literature I take "Homchen: Ein Tiermärchen aus der oberen Kreide" (Homchen: A Fairy Tale with Animals from the Upper Cretaceous), a work by one of Dilthey's former philosophy students, the popularizer Kurd Laßwitz, still known nowadays as "the German Jules Verne",[2] that is to say, one of the first authors of science fiction in Germany.

S. Azzouni (✉)
Max Planck Institute for the History of Science, Boltzmannstr. 22, Germany 14195, Berlin
e-mail: sazzouni@mpiwg-berlin.mpg.de

[1] "wesenhaft ein sprachliches Ereignis" (Daum [1998] 2002, 242). Unless otherwise stated, the English translations are mine.

[2] "der deutsche Jules Verne" (Willmann 2002, 97).

U. Feest (ed.), *Historical Perspectives on Erklären and Verstehen*, Archimedes, Vol. 21,
DOI 10.1007/978-90-481-3540-0_4, © Springer Science+Business Media B.V. 2010

## 4.1 Popular Science

Now the first question to be answered is: What is meant by "popular science" or "the popularization of science"? In this paper the two terms will be used in a quite synonymous way. A rather general and traditional definition would be that popularizing science means making scientific expert knowledge accessible to a non-scientific lay public. This definition clearly distinguishes between the sphere of the scientific expert who produces knowledge and the sphere of the non-scientific layperson who only receives knowledge: that is to say the active part of the knowledge producer is opposed to the passive part of the knowledge acquirer (see Whitley 1985). In this case popularization is seen as the transfer of knowledge about single scientific facts or theories from the sphere of its production to the sphere of reception, for example by giving simplified translations of Greek or Latin scientific terms into everyday language or, more generally speaking, by explaining the scientific theories in question. This definition only holds to a certain extent. It mainly applies to school teaching and to educational tools like textbooks. Around 1900 the debates about popular science were often linked to the discussion of teaching and curricula in secondary schools (see, e.g., Bölsche 1913, 296). Nevertheless popularizing science was and is different from teaching school children. It takes place outside academic institutions and it generally addresses a grown-up public.

In the 1980s Terry Shinn and Richard Whitley coined the term "expository science" mainly in regard to the popularization of science in museum exhibitions, zoological gardens and in public lecture halls. For Shinn and Whitley the most important aspect of popular science is that it takes place in the public sphere. Thus they focus on the space between the two spheres of science and non-science that the traditional definition of popular science so carefully keeps distinct. The public sphere is the place where scientists and non-scientists, experts and laypeople meet. Here they productively interact by doing expository or popular science (see Shinn and Whitley 1985; Felt 1996). In this interactive model or definition popular science is seen as a "low science" in contrast to the "high science" done only by scientists.

I prefer keeping the terms "popular science" and "popularization" because those are in fact the terms used around 1900. I nevertheless want to take up what I think is the most important element in the interactive model: Popularizing science is a special way of producing knowledge. But what kind of knowledge is produced by the popularization of science?

In a famous talk in 1886 Werner von Siemens characterised the nineteenth century as "the scientific age" (Siemens [1886] 1891). Science and technology had become more and more important in almost every part of social life. Unfortunately the increasing specialization of science and the diversity of its disciplines made it difficult for most people – even scientists – to keep track of the developments in current scientific knowledge. Science consequently was a common topic in newspapers, magazines and books.

One of the most successful German popularizers around 1900, the former poet Wilhelm Bölsche, once stated his view on how to popularize science:

> Popularizing simply means to me: to recast things. The must be recast in an artistic form due to aesthetic laws of effect. [...] The chemist just writes $H_2O$ on the blackboard; for the layperson the water has to rush.[3]

Here we can see that popularizing science does not mean teaching scientific terminology. It obviously does not even mean explaining single scientific facts. In contrast popular science should assemble the scattered pieces of knowledge and try to reunite them in order to give a picture of the whole thing in question. That is to say, popularizing science means contextualizing or even recontextualizing the pieces of scientific knowledge. The popularizer has to make the readers understand $H_2O$ as a real thing belonging to the world they are living in. Thus popular science is meant to produce knowledge about nature, reality, life or briefly: the world. It is now interesting to see that the development of popular science writings in Germany paralleled the rise of the Geisteswissenschaften.

## 4.2 Dilthey's Concept of Geisteswissenschaften and Poetry: Understanding and the Lived Experience

When Dilthey published the first part of his "Einleitung in die Geisteswissenschaften" in 1883, he intended to introduce Geisteswissenschaften as a necessary complement to science, no less scientific but offering a different approach to knowledge.

Dilthey seems to draw a clear line between science and Geisteswissenschaften: Science as an exact empirical discipline looks for the laws of nature and tries to find the terms with which they can be grasped. In order to reveal the causes of natural phenomena, science employs the method of explanation. Explanation relies on preliminary hypotheses that have to be verified, by taking into account the countable and measurable details of the material or sensory world.

In contrast, Geisteswissenschaften are occupied with the output of the human spirit. History and psychology are their basic disciplines; understanding is their fundamental method. Understanding means interpreting reality. This can only be done because reality is not just an aggregation of objective physical data from the outer world. Reality is given by human experience. In this context reality is another word for life as a whole. Its smallest unit is the lived experience ("Erlebnis"). The human mind does not collect data but it subjectively connects things, events, situations, feelings and metaphysical thoughts. The context of life ("Lebenszusammenhang") cannot be explained in terms of material causes and effects. We only get to know it by understanding the lived experience of the individual human being:

---

[3] "Für mich heißt Popularisiren einfach: die Dinge ganz umgießen. Sie müssen in eine Kunstform umgegossen werden, nach ästhetischen Wirkungsgesetzen. [...] Der Chemiker schreibt $H_2O$ an die Tafel; für den Laien muß das Wasser rauschen" (Bölsche 1889, 91).

> For knowledge may not posit a reality that is independent of lived experience. It may only trace back what is given in experience and in lived experience to a system of conditions by which experience becomes comprehensible. (Dilthey [1883] 1989, 202)[4]

Dilthey did not of course suggest replacing science by Geisteswissenschaften. His concept of Geisteswissenschaften was instead an objection to positivist thinking, e.g., to the way John Stuart Mill modeled moral sciences upon natural science (see Mill [1843] 1996, 833–952; Anderson 2003). Dilthey argued that the explanatory mechanisms of science like hypotheses, laws and technical terms are nothing but abstractions from lived experience.

> The conditions sought by the mechanistic explanation of nature explain only part of the contents of external reality. This intelligible world of atoms, ether, vibrations, is only a calculated and highly artificial abstraction from what is given in experience and lived experience. [...] As science advances, the differentiation of conceptual knowledge, which is a process of abstraction, can disregard more and more of the elements of this living reality; nevertheless, the indissoluble core remains. (Dilthey [1883] 1989, 203, 234)[5]

Here is the point where explanation as the method of science and understanding as the method of Geisteswissenschaften meet: Explanation depends on understanding.[6] Geisteswissenschaften examine the manifestations of lived experience: religion, art, music and literature. Dilthey outlines the creative force of the lived experience in his psychological and historical essay "Die Einbildungskraft des Dichters: Bausteine für eine Poetik", published in 1887.

Dilthey assumes that "[t]he poet's creative work always depends on the intensity of lived experience" (Dilthey [1887] 1985, 5).[7] The poet has the characteristics of the genius. These characteristics are related to other psychic phenomena like dreams, hallucinations, even lunacy. The poet is a person especially gifted with the ability to experience life deeply and to transform this experience by means of his imagination. Dilthey distinguishes poetic from scientific imagination. Both of them transform the lived experience into something that goes beyond all experience. Scientific imagination's construction of hypotheses is described as a voluntary act of logic (Dilthey [1887] 1985, 75; Dilthey [1887] 1994, 145), whereas poetic imagination works irrationally and relies on feeling. The poet therefore provides us

[4]"Denn das Erkennen vermag nicht an die Stelle von Erlebnis eine von ihm unabhängige Realität zu setzen. Es vermag nur, das in Erleben und Erfahren Gegebene auf einen Zusammenhang von Bedingungen zurückzuführen, in welchem es begreiflich wird" (Dilthey [1883] 1990, 369).

[5]"Die Bedingungen, welche die mechanische Naturerklärung sucht, erklären nur einen Teilinhalt der äußeren Wirklichkeit. Diese intelligible Welt der Atome, des Äthers, der Vibrationen ist nur eine absichtliche und höchst kunstvolle Abstraktion aus dem in Erlebnis und Erfahrung Gegebenen. [...] Der Differenzierungsprozeß der Erkenntnis in der fortschreitenden Wissenschaft kann daher als Vorgang der Abstraktion von immer mehr Elementen dieses Lebendigen absehen: jedoch der unlösliche Kern bleibt" (Dilthey [1883] 1990, 401).

[6]"Und das Erklären hat wieder die Vollendung des Verstehens zu seiner Voraussetzung" (Dilthey [1900] 1990, 334).

[7]"Das Schaffen des Dichters beruht überall auf der Energie des Erlebens" (Dilthey [1887] 1994, 130).

with something closer to the heart of the lived experience. The poet's transformation of the lived experience is not limited to a mimetic depiction of a real event. It is a symbolic representation of types of people, actions or emotions, derived from similar experiences:

> It is the mark of the great writer that his constructive imagination produces – out of elements of experience and based on analogies with experience – a type of person or plot which surpasses experience and yet through which we nevertheless come to understand it better. (Dilthey [1887] 1985, 68)[8]

Even though the poet is historically situated in and determined by his time, he is able to represent life and the process of life as a whole in his work: "Thus there arises, so to speak, a rounding-off of life" (Dilthey [1887] 1985, 156).[9] His poem, play or narrative is an objectivation of the single and subjective experience.

An important point in Dilthey's argumentation is the necessary relationship between the poet and his audience. It is a simple fact that living, in the sense of experiencing life, is a built-in feature of human nature. Poetry, i.e., the poet's transformed representation of the lived experience, can therefore reach out and evoke the corresponding feelings in the reader. By the act of reading, the reader relives the experience and consequently understands it. Poetry enables the reader to understand life as a whole. That is to say, it satisfies one of the fundamental emotional and metaphysical needs of the people. The process of understanding can only be completed if the reader participates in it by the activity of his own imagination.

## 4.3 Opaque Poetics: Kurd Laßwitz and Wilhelm Dilthey

After the Franco-German war of 1871, mathematician and physicist Kurd Laßwitz studied philosophy with Wilhelm Dilthey at the University of Breslau. In 1873 Laßwitz completed his PhD in philosophy. He tried to pursue his academic career and published several articles on physics in scientific journals, but he never succeeded in obtaining a teaching position at a university. Instead he became a teacher for mathematics, physics and geography at a German secondary school, the Ernestinum in Gotha.

Already as a student, Laßwitz had started writing poems, short utopian stories and popular science articles for magazines. In his dissertation "Über Tropfen, welche an festen Körpern hängen und der Schwerkraft unterworfen sind" (Laßwitz 1873), Laßwitz made two important statements concerning the relationship of science and poetry and the importance of popular science: "The Weltanschauung given by science

[8]"Das ist das Merkmal des großen Dichters, daß seine konstruktive Phantasie aus Erfahrungselementen, getragen von den Analogien der Erfahrung, einen Typus von Person oder Handlung hervorbringt, der über die Erfahrung hinausgeht und durch den wir diese doch besser begreifen" (Dilthey [1887] 1994, 139).

[9]"[S]o entsteht gleichsam die Rundung des Lebens" (Dilthey [1887] 1994, 224).

contains a lot of poetic elements. [...] Science could and should be popularized."[10] In 1882 he won an award for the best popularization of Kantian philosophy for his monograph "Die Lehre Kants von der Idealität des Raumes und der Zeit, im Zusammenhange mit seiner Kritik des Erkennens allgemeinverständlich dargestellt". The jurors were Ernst Laas, Max Heinze and Wilhelm Wundt.

Laßwitz and Dilthey had probably stayed in contact since Laßwitz's student years, but the letters preserved only cover the period between 1883 and 1909. In a postcard dating from February 1883, Dilthey asks Laßwitz to proof-read "a printed sheet about the representations of nature in the Middle Ages",[11] which likely became a part of Dilthey's "Einleitung in die Geisteswissenschaften". In later years, Dilthey encouraged Laßwitz to finish his main scientific work on the history of atomism, the "Geschichte der Atomistik vom Mittelalter bis Newton" that finally came out in 1890.[12] Furthermore, he seems to have tried in vain to use his influence to get Laßwitz a university chair.[13]

On the occasion of the edition of Kant's works by the German Akademie der Wissenschaften, Dilthey, who was responsible for the enterprise, chose Laßwitz to edit some of Kant's pre-critical writings in 1897. The two volumes in which Laßwitz participated were published in 1902 and 1905 (see Kant 1902/1905). Laßwitz hesitated to undertake the editorial work and carefully negotiated the contract with Dilthey. Laßwitz even tried at one point to rid himself of this responsibility. In November 1901 he wrote in a letter to Dilthey that his work as a teacher and his literary activity took up all of his time and that he could not give up either on account of following his own vocation:

> You think that the work of making a critical edition of the Kantian text objectively is of bigger value to science than one's own systematic production. This I do not deny. But others can do this better than I; everyone just has to ask himself for which task he feels qualified.[14]

[10] "Die durch die Naturwissenschaft gegebene Weltanschauung enthält in reichem Maße poetische Elemente. [...] Die Naturwissenschaft kann und soll popularisirt werden"(cited in Schweikert 1979, 983).

[11] "einen Bogen, der über die Naturvorstellungen des Mittelalters handelt" (Forschungsbibliothek Gotha (FB-Gotha), Laßwitz papers, Postcard from Dilthey to Laßwitz, February 15, 1883, Chart. B 1963, Bl. 54).

[12] See FB-Gotha, Laßwitz papers, Letter from Dilthey to Laßwitz, December 1888, Chart. B 1962a, Bl. 156.

[13] Between 1890 and 1894 Dilthey at least nominated Laßwitz for chairs in philosophy at the universities of Breslau, Bonn and Würzburg FB-Gotha, Laßwitz papers, Letter from Dilthey to Laßwitz, April 2, 1890, Chart B 1962a, Bl. 158f.; Postcard from Dilthey to Laßwitz, without date, Chart B 1962a, Bl. 160; Letter from Dilthey to Laßwitz, May 2, 1894, Chart B 1962a, Bl. 161f.).

[14] "Sie werden meinen, die Arbeit, den Kantischen Text kritisch herzustellen, sei objektiv für die Wissenschaft ein größerer Wert als eigne systematische Produktion. Das bestreite ich nicht. Aber das können andre besser als ich; es muß sich eben jeder fragen, wozu er sich berufen fühlt" FB-Gotha, Laßwitz papers, Letter from Laßwitz to Dilthey [draft], November 28[?], 1901, Chart. B 1962a, Bl. 154).

Dilthey was interested in Laßwitz as a philosopher coming from a scientific background. In general, he did not show any interest for either Laßwitz's literary work or his popular science writings. There is but one exception: Dilthey was very pleased with the way Laßwitz popularized the works of his former teacher: "I am especially indebted to you because as my former scholar and my friend you sympathetically and kindly pave the way for works of mine that are difficult to access."[15]

Laßwitz treated Dilthey`s recent publications in several articles for newspapers and magazines. In 1887, Kurd Laßwitz reviewed Dilthey's essay "Die Einbildungskraft des Dichters: Bausteine für eine Poetik". In this review he approved of Dilthey's poetics because he understood it as a possible link between science and Geisteswissenschaften. In his reading of the essay, poetics is part of Geisteswissenschaften but nevertheless has an explanatory potential that has to be developed further in order to make poetics become science:

> Poetics, i.e., the theory of poetry, seems to have the best chance among the different fields of aesthetics to attain the position of a causal explanatory science. [...] Thus poetics is the most promising of all Geisteswissenschaften regarding causal methodology.[16]

Here Laßwitz gets right to a point that stays rather opaque in Dilthey's own poetics. Dilthey gives a comprehensible description of what he thinks poetry and the creative work of the poet to be, but the status and the method of poetics is quite unclear. The methodological dichotomy of explanation and understanding established in the "Einleitung in die Geisteswissenschaften" appears to be indistinct in the essay on poetics.

Dilthey discards the normative function of poetics. Poetics should not postulate or deal with an ideal concept of poetry and beauty. He wants poetics to explore and trace the "life-forms of poetry"[17] (Dilthey [1887] 1985, 59). Dilthey's concept of poetics as a discipline obviously corresponds to literary studies, and for this reason poetics must belong to Geisteswissenschaften even if it is never referred to by this term in the text. Poetics should help in understanding poetry as a manifestation of the lived experience. So one could guess that it has a hermeneutic function, hermeneutics being for Dilthey the "artistic teaching of understanding written statements".[18] On the contrary Dilthey wants poetics to explore and indeed explain the creative work of the poet. According to Dilthey, it is the causal nexus of human nature leading to writing poetry that has to be revealed by poetics.

---

[15] "Besonderen Dank schulde ich Ihnen daß Sie als alter Schüler und Freund, aus Ihrem Verständniße heraus meinen schwer zugänglichen Arbeiten freundlich die Wege bahnen" FB-Gotha, Laßwitz papers, Letter from Dilthey to Laßwitz, December 1888, Chart. B 1962a, Bl. 156f.).

[16] "Von den verschiedenen Gebieten der Aesthetik scheint die Poetik, die Theorie der Dichtkunst, die beste Aussicht zu haben, auf jenen Standpunkt einer ursächlich erklärenden Wissenschaft geführt zu werden. [...] So verspricht die Poetik von allen Geisteswissenschaften der kausalen Behandlung den nächsten Erfolg" (Laßwitz 1887b, 537).

[17] "Lebensformen der Poesie" (Dilthey [1887] 1994, 130).

[18] "Kunstlehre des Verstehens schriftlich fixierter Lebensäußerungen" (Dilthey [1900] 1990, 332f.).

The individual poet has to be treated as an object of psychology and history. In this context, autobiography plays an important role as a document of individual experiences. The first step would be to look at the psychological and psychophysical causes and effects that determine the creative process. Dilthey extensively discusses Gustav Theodor Fechner's "Vorschule der Ästhetik" and he seems to at least partly agree with it. So poetics would be founded on psychology but for further examination the method of research should change. Thus in a second step, the knowledge gained so far about the poet's work should be reconsidered under the aspect of empirical literary and social history. Dilthey also proposes to draw upon ethnological research in order to find the original roots of epic, drama and lyric in human nature:

> Psychology has dominated our previous discussions. Now that we have obtained a foundation for poetics, our method changes and a literary–historical empirical approach will guide us. In accord with the spirit of modern scholarship, it must encompass the entire field of literature and seek elementary structures, especially in the artistic works of primitive peoples. (Dilthey [1887] 1985, 127f)[19]

Poetics should not just describe but explain. It is meant to combine and assimilate psychological and historical knowledge and methodology in order to resemble contemporary modern science:

> As much as contemporary poetics owes to the two older methods [...], poetics must still take a decisive step in order to become a modern science. Poetics must recognize the productive factors, study their effects under varying conditions, and solve its practical problems by means of this *causal knowledge. Knowledge of technique is based on a causal approach, which not only describes the composition of poetic products and forms, but really explains them.* (Dilthey [1887] 1985, 132)[20]

This is indeed a contradictory and puzzling suggestion, made by Dilthey only 5 years after the publication of his "Einleitung". Just how does this fit into the concept of Geisteswissenschaften that is defined by the method of understanding and opposed to the scientific method of explanation?

The status of psychology and history previously established as basic Geisteswissenschaften seems to be turned upside-down with regard to poetics. Their methods are employed for explaining, and this explanation should finally lead to understanding. Here science and Geisteswissenschaften do not seem to interact

---

[19] "In den bisherigen Entwicklungen herrschte die Psychologie vor. Nachdem nun eine Grundlegung der Poetik gewonnen ist, ändert sich die Methode. Die literarhistorische Empirie hat jetzt die Führung. Sie muß dem Geiste der modernen Forschung entsprechend, das ganze Gebiet der Dichtung umfassen und gerade bei den Naturvölkern die elementaren Gebilde aufsuchen" (Dilthey [1887] 1994, 197).

[20] "So viel die Poetik den beiden älteren Methoden verdankt [...]: sie muß den entscheidenden Schritt tun, eine moderne Wissenschaft zu werden; sie muß die hervorbringenden Faktoren erkennen, ihr Wirken unter wechselnden Bedingungen studieren und vermittels dieser Kausalerkenntnis ihre praktischen Aufgaben lösen. Die Erkenntnis der Technik [des Dichters, S. Az.] gründet sich auf eine Kausalbetrachtung, welche die Zusammensetzung der poetischen Gebilde und Formen nicht nur beschreibt, sondern wirklich erklärt" (Dilthey [1887] 1994, 201).

as complements but rather overlap to a certain extent.[21] Apparently Kurd Laßwitz chose not to be puzzled but to welcome Dilthey's poetics as an attempt to bridge the gap between science and Geisteswissenschaften. In 1887, i.e., at the time when Dilthey's essay was published, Laßwitz reflected upon the differences between the scientific and the poetic way of looking at nature (Laßwitz 1887a). In fact he worried about his own situation as a writer equally attracted by science and poetry. Laßwitz felt the need to distinguish between the two in order to find an intermediary solution. You can see from the way in which he outlines this problem that he is concerned about his role as a popularizer and about the question of how to write popular science:

> The fact that in some people the features of a scientist and a poet are combined does not solve the problem, but rather the problem is exactly how this combination of the opposite interests of mankind has its ground in human nature. Unfortunately a confusion of the two perspectives can be found in some popular representations of nature.[22]

In this context, the Diltheyan way of seeing reality as basically given (and as a consequence understood) by the lived experience was very attractive to Laßwitz, who refers several times to Dilthey's "Einleitung in die Geisteswissenschaften". In his argumentation Laßwitz interestingly replaces "reality" by "nature" and therefore concludes that nature is related to the productivity of the human mind:

> Nature is experienced by man and it is thought by man. [...] With the eyes of the scientist we see the objective organisation in the world; with the eyes of the artist we feel the subjective organisation of our free self. It is no contradiction that we own both features but instead it is the condition of our existence.[23]

Laßwitz's reflections here very much resemble his descriptions of Gustav Theodor Fechner's concept of psycho-physical parallelism. Around 1890 Laßwitz was also working on a monograph about Fechner's life and works.[24] Parts of the book are similar to an English article by Laßwitz entitled "Nature and the individual mind" that was published in the American journal "The Monist" in 1896 (Laßwitz 1896b).

---

[21] In his study on Dilthey, Rudolf Makkreel describes the essay on poetics as a continuation and elaboration of the "Einleitung". In his view, there is no contradiction, because poetics as well as Geisteswissenschaften were supposed to offer a different kind of explanation than science. He outlines the relation between Dilthey's poetics and his work on descriptive and explanatory psychology (see Makkreel 1991). On Dilthey's critique of explanatory psychology and the importance of his critique for his concept of understanding, see Feest (2007).

[22] "Daß in manchen Menschen Forscher und Dichter vereint sind, ist keine Lösung des Problems, sondern es ist eben das Problem selbst, wie dieses Zusammen der beiden entgegengesetzten Interessen der Menschheit in der Menschennatur sich begründe. Eine Confusion beider Standpunkte findet sich in manchen populären Naturdarstellungen in bedauerlicher Weise" (Laßwitz 1887a, 278f.).

[23] "Natur wird vom Menschen erlebt und sie wird vom Menschen gedacht. [...] Sehen wir mit dem Auge des Forschers, so sehen wir die objective Ordnung in der Welt; sehen wir mit dem Auge des Künstlers, so fühlen wir die subjektive Ordnung unseres freien Ich. Daß wir beide Fähigkeiten besitzen, ist kein Widerspruch, sondern die Bedingung unseres Daseins [...]" (Laßwitz 1887a, 288).

[24] The book was published in 1896 but in the preface Laßwitz states that he had already finished writing it in 1893 (see Laßwitz 1896a, VI).

In this article Laßwitz outlines how the physical and the psychical aspect of nature depend on the experience of the individual mind:

> Experience, therefore, is the form in which the determined contents of time and space confront us. We know it only in our individual ego and nowhere else in the world; [...] If we wish to emphasise simple the determinateness by law of a spatial system, as is done in natural science, we call the event in question a physical one; the same event is called psychical, where, as in psychology, it is presented as a component part of the experience of an individual mind. (Laßwitz 1896b, 406, 407)

Experience necessarily precedes the dualistic separation of nature, reality or the world into objective/physical and subjective/psychical elements:

> But whence experience originates, cannot be explained, because it is the original fact to which all explanations revert; it is the given occurrence, the phenomenon, which must be assumed. (Laßwitz 1896b, 406)

From the monograph on Fechner we can see that Laßwitz uses "experience" in the sense of "Erlebnis" (Laßwitz 1896a, 150f.): that is to say in the Diltheyan sense of the lived experience. Laßwitz, who was looking for a combination of Fechnerian and Kantian thought,[25] possibly found exactly this in the notion of the lived experience. The lived experience is the kernel from which human nature cultivates both the objective and the subjective, science and poetry. They necessarily coexist and can be traced back to the same origin.

To Laßwitz, Dilthey's poetics seemed to be a confirmation of his reading of Dilthey's "Einleitung". In his review he treated both texts as a unity. This time he stressed the point that nature and art are both developed by the human mind and soul ("Gemüt"). He points out, moreover, that by exploring the lived experience we not only understand experience but also learn about reality:

> If we explore the creative energies of our self that create the world-content of our experience we thereby get to know the very modes of creating reality. We understand how everything real consists of relations of consciousness, and how on one side nature and on the other side art result from the work of the spirit and the mind.[26]

For Laßwitz this seems to be the most attractive element to be found in Dilthey's thinking. By adopting this view, the troublesome gap disappears between nature and art – analogous to the situation between science and poetry – without giving up on seeing them as distinct features of human nature. This is what Dilthey's poetics once more brings to Laßwitz's attention.

---

[25] "Fechner selbst hat den Weg zu Kant nicht gefunden. [...] Für die kritische Grundlegung des psychophysischen Parallelismus Fechners, ebenso wie für seine Abgrenzung von Wissenschaft und Glauben ist es zu bedauern, daß Fechner die mächtigen Hilfsmittel nicht ausgiebiger benutzt hat, welche er bei Kant hätte finden können" (Laßwitz 1896a, 194).

[26] "Denn wenn wir die schöpferischen Kräfte in unserem Ich erforschen, welche den Weltinhalt unserer Erfahrung erzeugen, so lernen wir eben damit die Gestaltungsarten der Wirklichkeit kennen. Wir verstehen wie alles Wirkliche selbst in Beziehungen des Bewußtseins besteht, wie auf der einen Seite Natur, auf der anderen Kunst als Resultate der Geistes- und Gemütsarbeit entspringen" (Laßwitz 1887b, 536).

When, according to Dilthey, the lived experience can be relived in the act of reading poetry, it also offers the possibility to make knowledge accessible to a broad public in an intensive and comprehensive way. Laßwitz seems to have drawn this conclusion and made use of it with his popular science writing. He decided not to teach his readers a simplified version of single scientific facts, but rather to let them relive the lived experience that contains these facts and additionally gain insight into life as a whole. He does not want to explain, but to make his readers understand.

Since the late 1880s, Laßwitz had turned to writing short texts of fiction he called "modern" or "scientific fairy tales". His first collection of scientific fairy tales "Seifenblasen" came out in 1890 (Laßwitz 1890). Wilhelm Bölsche reviewed this volume with a stress on the scientific fairy tale as a way of writing that perfectly fitted the Weltanschauung of the scientific age:

> The facile game makes sense, like the antennæ of the mind reaching out for the first time to feel their way into a territory that perhaps will be conquered on the next day: As the children's fairy tale precedes the awakening of morality, so the scientific fairy tale necessarily precedes knowledge even more than it intends to do.[27]

He states that the scientific fairy tale corresponds and reveals a stage in the intellectual and spiritual activity of man that necessarily precedes scientific knowledge, which one can easily connect to the Diltheyan concept of the lived experience. In a statement reported after his death, Laßwitz says that it was the fact that a person is able to distinguish between and to experience both science and art that made him write scientific fairy tales:

> Science and art should be separated by their methodology, but the individual should know and experience both. You cannot compose a pure mixture from impure ingredients. This is my guideline. Because of this I am able to write 'scientific' fairy tales, that is to say I treat scientific matters in a poetic way not in order to generate pieces of knowledge but in order to create works of art as best I can.[28]

When Laßwitz states that he does not want to produce pieces of knowledge with his scientific fairy tales, he refers to the detailed scientific knowledge that is gained by the method of explanation. Actually he aims at the other kind of knowledge that concerns the common origin of science and art and that results from understanding. In other words: he thinks about using poetry in order to create the necessary basis for grasping the scientific topics he wants to convey to the reader.

---

[27] "[D]as leichte Spiel gewinnt einen Sinn gleich erstem Vorstecken tastender Geistesfühler in ein Gebiet, das der nächste Tag vielleicht schon erobert sein läßt: wie das Kindermärchen der erwachenden Moral, so schreitet das Forschermärchen der Erkenntniß voraus in einem Zwange, der weit seinen Willen überragt" (Bölsche 1891, 199).

[28] "So sei Wissenschaft und Kunst streng in ihren Methoden getrennt, aber die Persönlichkeit kenne beide und erlebe sie beide; aus unreinen Elementen gibt es keine reine Mischung. Das ist mein Leitfaden. Deswegen kann ich 'wissenschaftliche' Märchen schreiben, d.h. nur wissenschaftliche Stoffe in poetischer Form behandeln, aber nicht um Erkenntnis zu erzeugen, sondern um Kunstwerke, so gut man's eben kann, zu schaffen" (Lindau 1919, 54).

## 4.4 Evolution Relived: The Example of "Homchen"

"Homchen. Ein Tiermärchen aus der oberen Kreide" is the second part of Laßwitz's second collection of modern or scientific fairy tales "Nie und Immer" (Laßwitz 1902). "Homchen" tells a story about the extinction of the dinosaurs and the rise of the mammals as a condition for the evolution of man. As in a fable, the list of characters consists of animals. These are not the typical animals talking and walking about in the classical fables or fairy tales, but animals supposed to have lived in the second half of the Cretaceous.

The hero is a koala-like marsupial named Homchen, the name being a German diminutive of the Latin word "homo". Like the other mammalian protagonists, Homchen is obviously modelled after the description that Alfred Brehm gives of the different members of the today existing family of Phalanger ("Kletterbeutler") in the second volume of his "Thierleben" (see Brehm 1865, 29–42). In the original manuscript, Laßwitz made a drawing of his hero with the words "Homchen (after a koala in Brehm's Tierleben)" written below (see Kempen 2002, 11). Marsupials supposedly were ancestral to placental animals. The oldest marsupials are represented by fossils dating from the Upper Cretaceous.

In Laßwitz's story, Homchen is depicted as a kind of exceptional creature, because unlike the other marsupials, he has been born already with fur and therefore left the pouch of his mother rather early. He often stays up during daytime, which is unusual for his nocturnal species. In addition Homchen doubts the status quo of the world he is living in. He feels that a change is going to come. The prehistoric world is inhabited by the cold-blooded saurians. Their leaders are the already warm-blooded Pretty Jaws ("Zierschnäbel") which is the popular translation of the scientific name Compsognathus. The predominance of the saurians is due to their physical strength and to the myth that the world had been created by the Red Snake, a myth made up by the Pretty Jaws. For their part the saurians also start worrying about an upcoming change. The climate is getting colder and makes them feel uneasy, whereas the marsupials do not seem to be affected by the cold. Both parties come into conflict after Homchen has killed a flying reptile. In revenge, the Pretty Jaws plan to exterminate all mammals, but the saurians become extinct themselves in a great flood. In the meantime Homchen has found out that the Red Snake does not exist and becomes aware of the creative power of the individual. For that reason he leads a part of his fellow marsupials north from their woods and tries to prepare them for the future by teaching them a new way of life, for instance to carry and to keep fire or to use weapons. But Homchen's efforts are in vain: Although the saurians are dead, evolution still has a long way to go before the rise of man.

Laßwitz's scientific fairy-tale can be read on several levels. On the one hand it is a fable with social and political implications (see Schweikert 1979, 1052f.; Kempen 2002, 10–12). On the other hand it is a way of doing popular science. In the following discussion of central passages of the text, I will focus on this second aspect by treating the question of how Laßwitz popularizes certain contemporary explanations of the process of evolution.

It is interesting to see that throughout the whole story, Laßwitz presents evolution as evolution of the brain and the mind. The physical side of this aspect comes into view when the Pretty Jaws (who apparently already have a brain) reflect upon the spinal medulla and the place in the body where it should be accumulated. They say that heaping up the marrow in the lumbar region like the saurians do only makes them follow their instincts. The only way to succeed in the struggle for life is to accumulate the marrow in the head in order to form a brain and to think.[29] Thus evolution appears to be a semi-voluntary process that to a certain extent depends on the activity of the individual. Here Laßwitz obviously puts forward a Psycholamarckian variant of the Neolamarckian interpretation of the evolutionary process.[30] On this view, evolution is not only pushed forward by the inheritance of acquired traits as proposed by Jean-Baptiste Lamarck but it is powered by something that corresponds to an original wish or will of all organisms to develop and improve (Lefèvre 2004, 24f.). This view was mainly developed and discussed by members of the Monist Movement.[31] It was the American palaeontologist Edward Drinker Cope who in 1887 proposed "the existence of an especial force which exhibits itself in the growth of organic beings, which was called growth-force" and that "development consists in the location of this energy at certain parts of the organism" (Cope 1887, VIII). The location and the activity of growth-force in the body were thought to be directed by use and effort and conserved by an organic or cellular memory.

Concerning the development of the brain which appears to be the central organ in the evolution of mammals and man, Cope applies his theory of growth-force to the so-called "principle of cephalization" that had already been suggested by the American geologist James Dwight Dana: "The increasing demands of intelligence locate growth-force round its organ, the brain, etc., while such location reacts by furnishing means of increased activity of the mind" (Cope 1887, 39). This clearly corresponds to the description of the location and evolution of the brain given by

[29]"Wir lehren die Echsen das Gebot, ihr Mark anzuhäufen, damit sie stark werden, an dem einen Ende des Rückens, aber wir lehren sie das Falsche. [...] Dadurch werden sie die Sklaven dessen, der klug ist. Sie haben den Mittelpunkt des Lebens am falschen Ende. Hätten sie diese Anhäufung des Markes am oberen Ende des Rückens, über dem Halse, in ihrem Kopfe, so würden sie klug werden, so würden sie denken können. Dort im Kopf, wo Augen und Ohren sind, da muß alle Macht des Lebens zusammenlaufen, da muß sich vereinigen, was in der Welt vorgeht. Und wenn dort die Fülle des Markes liegt, die Gehirn heißt, so sammelt sich an, was wir erfahren[.]"(Laßwitz 1902, 233).

[30]Concerning Laßwitz's position towards contemporary evolutionary theories and his preference for Neolamarckism, see Szukaj 1996, 145–155.

[31]The German Monistenbund was the most important society for promoting monist thinking. Similar organisations soon were founded in the USA. Paul Carus who in 1890 was the first editor of the American philosophical journal "The Monist" played an important role in this transfer of monism from Germany to America. Even though the Monist Movement was very much involved in the popularization of science, this aspect will not be discussed in detail in this paper. For a general overview of monism, the activities of the Monistenbund and Paul Carus, see Ziche 2000 and Weber 2000.

the Pretty Jaws in Laßwitz's "Homchen". It is not certain that Laßwitz had read Cope's writings but it nevertheless seems probable that he knew about these theories. Cope describes evolution as an energetic process. Laßwitz was very interested in the topic of energy and in Wilhelm Ostwald's idea of a new science called Energetik that should investigate the physical and chemical basis of life in general.[32] Ostwald also applied his Energetik to the evolution of organic beings. For him the concept of evolution powered by an energy that sums up all possible material and metaphysical causes is linked to the idea of the unity of human experience (see Ostwald 1902, 332–347).

In Laßwitz's story, evolution is due to a state of mind that includes rational thinking but has to have its roots in living the experience. Reasoning alone does not lead very far. This is exemplified in the characters of the Iguanodon and the marsupial hedgehog ("Beuteligel"): The Iguanodon always thinks about thinking but has a very limited view on the world. He conceives of himself as being the ideal creature. In his self-sufficient view, further development is not necessary, and the Iguanodon just does not care about other life-forms (see Laßwitz 1902, 239–242). In contrast, the marsupial hedgehog already has an idea how to push forward with his species. He tells his fellow marsupials that they have to breed the "super-marsupial" which obviously is a comical reaction to Friedrich Nietzsche's concept of the Übermensch:

> It's – how can I say – you marsupials have to outgrow yourself, you have to become something Higher – in brief: You have to breed the super-marsupial! [...] Why don't you become masterminds? Because of your pouch! [...] Away with the pouch, I tell you![33]

The hedgehog's demand asks too much from the intellectual capacities of his fellows. The marsupials do not know how to get rid of their pouch, nor do they understand why they should aim at doing so.

In a crucial scene of the story, Homchen experiences evolution and especially the evolution of "a higher and self-conscious life" (Laßwitz 1902, 263). After the eruption of a volcano, Homchen lies at its crater in deathlike condition. In this situation he has a vision of his own skull lying under a transparent cover and he hears a voice addressing him as "our ancestor":

> He saw a small bleached skull that softly bedded lay under a transparent cover. And a voice said: this is the skull of a small marsupial, one of our direct ancestors. It was found in a very strange environment, [...] in a thin layer of fossil ashes enclosed in eruptive rock.

[32] See Laßwitz's chapter on energy ("Energie") in Laßwitz ([1900] 1908, 101–113). The volume was first published in 1900, that is to say during the period in which Laßwitz had been working on "Homchen". Already in 1892 Laßwitz addressed several letters to Ostwald in order to learn more about his Energetik, see Archiv der Berlin-Brandenburgischen Akademie der Wissenschaften (ABBAW), Ostwald papers, Letter from Laßwitz to Ostwald, April 9, 1892, 1720, 73/1; Letter from Laßwitz to Ostwald, April 21, 1892, 1720, 73/2; Letter from Laßwitz to Ostwald, September 18, 1892, 1720, 73/7.

[33] "Nämlich – wie soll ich sagen – ihr Beuteltiere müßt über euch hinauswachsen, ihr müßt etwas Höheres werden – mit einem Worte: Ihr müßt den Über-Beutler züchten! [...] Warum werdet ihr keine Herrenseelen? Weil ihr den Beutel habt! [...] Fort mit dem Beutel! sag' ich" (Laßwitz 1902, 158).

> It cannot be explained how he came to be there, but he is there. What a long line of ancestors stands between him and us! But already in him the law of formation is alive, already in him there is the unity of forces that leads up to us![34]

This voice apparently belongs to the narrator. He talks like a guide in a natural history museum explaining a fossil that is exhibited in a showcase. Here Laßwitz refers to an experience that was well known to his contemporary readers. In doing so he offers a twofold (or maybe even threefold) perspective to the reader: Even though they might feel closer to the narrator who as a museum-guide definitely is a man belonging to the present time, they will tend to take the position of Homchen who at this moment is addressed by the voice of the narrator and therefore seems to be one of the visitors to whom the guide is talking. But Homchen is not only a visitor, he is at the same time the fossil in question. Thus Laßwitz also enables the reader to take the unusual perspective of a fossil being the subject for scientific studies.

The scientific topic treated in this scene quite obviously is the evolution of man. The marsupial skull in the showcase is referred to as a document for and part of the direct and long "line of ancestors" leading to man. This reminds of the way the German zoologist Ernst Haeckel presented evolution in his "Natürliche Schöpfungsgeschichte". In the second part of this work, he gives a detailed step by step description of the "Animal Line of Ancestors or Chain of Ancestors to Man".[35] The marsupials are to be found on the "24th Step of Ancestry" ("XXIV. Ahnenstufe"). Haeckel describes the prehistoric marsupials as the direct link between monotremes ("Kloakentiere") and mammals and he draws the conclusion that "among those marsupials there must have been ancestors to man".[36] This is the body of knowledge on evolution around 1900 that Laßwitz draws on in this passage of the text.

Starting from the museum situation, Homchen gets a visionary overview of the different stages or steps of evolution. It begins with the sea shown as the element of creation. The warm-blooded and furry mammals take the place of the saurians. The mammals develop into "a stronger species", i.e., human beings, and the vision ends with the arrival of "the rolling animal", i.e., a train, on which now Homchen sees himself riding (see Laßwitz 1902, 263–269).

---

[34] "Es sah einen kleinen, weiß gebleichten Schädel, der lag weich gebettet unter einem durchsichtigen Deckel. Und eine Stimme sprach: Das ist der Schädel eines kleinen Beuteltiers, eines unsrer direkten Vorfahren, gefunden unter ganz merkwürdigen Umständen, [...] in einer dünnen fossilen Aschenschicht. Rings eruptives Gestein. Es ist nicht zu erklären, wie er dahin kam, aber er ist da. Welch eine lange Ahnenreihe noch von ihm bis zu uns! Und doch in ihm schon lebendig das Gesetz der Bildung, in ihm schon die Einheit der Kräfte, die zu uns heraufführt!" (Laßwitz 1902, 262f.).

[35] "Tierische Ahnenreihe oder Vorfahrenkette des Menschen", see Haeckel ([1868] 1924, vol. 2, 360–371).

[36] "Die drei Unterklassen der Säugetiere stehen derart im Zusammenhang, daß die niederen Beuteltiere (Prodelphia) sowohl in anatomischer, als auch in ontogenetischer und phylogenetischer Beziehung den unmittelbaren Übergang zwischen den Monotremen und Plazentaltieren vermitteln. Daher müssen sich auch Vorfahren des Menschen unter jenen Beuteltieren befunden haben" (Haeckel [1868] 1924, vol. 2, 368).

The image of the train rolling through the landscape does not only stand for the technical progress of the late nineteenth century. It also alludes to the fact that many fossils have been excavated during railway constructions. In the beginning of the second part of the "Schöpfungsgeschichte", Haeckel mentions this fact and he points out that the railway constructions thus contribute to research in the history of the earth and mankind.[37]

The vision of the process of evolution commented by the narrator's voice suddenly makes Homchen understand life, and it is such a vivid impression that it brings him back to life. Homchen is aware that he is not able to understand all the details he has been told by the voice during his vision, that is to say, he is not able to follow the explanations. In contrast he understands the parts of the vision that are linked to experience and he is convinced that those were the most important parts that will enable him to do his best in the evolutionary process.[38] So it can be said that Homchen not only lives the experience but also lives because of this experience. Here Laßwitz represents a lived experience per se.

It can be seen already from the expressions quoted above that the story of "Homchen" is written in an associative style. Laßwitz uses a metaphorical and symbolic language. He avoids words like "man", "human being", "train" or "museum". Human beings are referred to as a higher or stronger species or as being descended from Homchen and his fellow marsupials. More often than not Laßwitz uses the pronouns "we" and "us" for "man" or "mankind". He is careful to bring the hero of the story closer to the reader. In the crucial scene Laßwitz even interweaves the perspective of Homchen with the perspective of the reader. The reader has to bring his preliminary knowledge into play in order to understand the text. This is how Laßwitz tries to make his readers relive and understand Homchen's experience. Like Homchen the reader surveys the process of evolution as a process of nature connected with a process of the mind. Only the experience of life as a whole leads to understanding and enables the individual to think in a rational way about the details that are part of this context of life.

Laßwitz's way of writing popular science in the form of scientific fairy tales has nothing to do with any notion of experimental poetry. On the contrary, he stays in the frame of the classical and romantic German literary tradition, i.e., the kind of poetry that Dilthey also focused on in his poetics. Laßwitz always thinks about poetry in relation to science. He considers the advantages that poetic expression has over scientific writing concerning the possibility to reach the readers and to make

---

[37] "Im ganzen ist wohl kaum der hundertste Teil der gesamten Erdoberfläche gründlich paläontologisch erforscht. wir können daher wohl hoffen, bei weiterer Ausbreitung der geologischen Untersuchungen, denen namentlich die Anlage von Eisenbahnen und Bergwerken sehr zu Hilfe kommen wird, noch einen großen Teil wichtiger Versteinerungen aufzufinden" (Haeckel [1868] 1924, vol. 2, 24).

[38] "Zu wem mochte jene Stimme gesprochen haben? Wohl schon zu den klugen, guten Tieren, die da kommen sollten, denn Homchen hatte vieles davon nicht verstanden. Aber einiges hatte es schon herausgehört, was es selbst schon erlebt und gedacht hatte; und das gab ihm Licht im Dunkel [...]" (Laßwitz 1902, 273).

knowledge accessible. The scientific fairy-tale appears to be the best solution for doing so.[39] Dilthey's writings, his "Einleitung in die Geisteswissenschaften" as well as his poetics, offered to Laßwitz the outline of the concept of the lived experience as a common origin of science and poetry, that is to say the basis of knowledge that cannot be explained but only understood. In this context of Laßwitz's reading of Dilthey, poetry appears to be a way of writing popular science. Laßwitz is inspired by Dilthey to write scientific fairy tales. Laßwitz's preferred means of popularizing science is by acquainting the reader with what the author thinks to be the common source of general and scientific knowledge. As could be seen by the example of "Homchen", Laßwitz tries not just to clothe contemporary scientific theories in poetic garments but connects them with contemporary philosophical thinking thereby also popularizing Dilthey's concept of the lived experience and his method of understanding. In doing so, Laßwitz seems to have taken seriously one of Dilthey's statements about philosophy in the beginning of his poetics: "Only insofar as philosophical thought exerts an influence does it have a right to exist" (Dilthey [1887] 1985, 44).[40]

## References

Anderson RL (2003) The debate over the geisteswissenschaften in german philosophy. In: Baldwin T (ed.) The Cambridge History of Philosophy 1870–1945. Cambridge University Press, Cambridge, pp 221–234

Archiv der Berlin-Brandenburgischen Akademie der Wissenschaften (ABBAW) Jägerstraße 22–23, 10117, Berlin

Bölsche W (1889) Selbstanzeige ‚Das Liebesleben in der Natur'. Die Zukunft 25:89–91

Bölsche W (1891) Naturwissenschaftliche Märchen. Freie Bühne 2(8):195–199

Bölsche W (1913) Wie und warum soll man Naturwissenschaft ins Volk tragen? In: Stirb und Werde. Naturwissenschaftliche und kulturelle plaudereien. Jena, Diederichs, pp 294–324

Brehm AE (1865) Illustrirtes thierleben. Eine allgemeine kunde des thierreichs, vol 2. Verlag des Bibliographischen Instituts, Hildburghausen

Cope ED (1887) The origin of the fittest. Essays on Evolution. London, MacMillan

Daum AW ([1998] 2002) Wissenschaftspopularisierung im 19. Jahrhundert: bürgerliche kultur, naturwissenschaftliche bildung und die deutsche öffentlichkeit 1848–1914. 2nd revised edn, 1st edn. Oldenbourg, München

---

[39] "Gerade je mehr der Forscher und Techniker von allem Gefühlsmäßigen abstrahieren muß, um die Natur als das Gesetzliche kühl und verstandesmäßig vor sich zu haben, umsomehr entsteht beim Publikum das Verlangen und dem Dichter die Aufgabe, die neue objektive Macht wieder im subjektiven Gefühle sich anzueignen. Es gilt, das neue Naturgefühl persönlich zu gestalten. [...] Und darin eröffnet sich ein ungeheures Feld für das wissenschaftliche Märchen, eine echt künstlerische Aufgabe, wenn anders es Aufgabe der Kunst ist, vor unseren Augen eine neue und höhere Welt entstehen zu lassen. Wenn dabei die intellektuelle Seite des Menschen zu ihrem Rechte kommt, so ist dies um so besser; denn sie ist ja an sich wesentlich und um so mächtiger, je höher der Bildungszustand eines Volkes ist" (Laßwitz [1900] 1908, 440f.).

[40] "Und nur sofern philosophisches Denken wirkt, hat es ein Recht, zu existieren" (Dilthey [1887] 1994, 116).

Dilthey W ([1883] 1989) Introduction to the human sciences. In: Makkreel RA, Rodi F (eds) Wilhelm Dilthey. Selected works, vol 1. Originally published as Einleitung in die geisteswissenschaften. Versuch einer grundlegung für das studium der geschichte der Gesellschaft. Duncker & Humblot, Leipzig. Reprint in Wilhelm Dilthey (1990) Gesammelte schriften, vol 1, 9th edn., Groethuysen B (ed.) Teubner, Stuttgart and Vandenhoeck & Ruprecht, Göttingen. Princeton University Press, Princeton, NJ

Dilthey W ([1883] 1990) Einleitung in die Geisteswissenschaften. Versuch einer Grundlegung für das Studium der Geschichte der Gesellschaft. In: Gesammelte Schriften, vol. 1, 9th edn. Groethuysen B (ed) Originally published by Duncker & Humblot, Leipzig. Teubner, Stuttgart and Vandenhoeck & Ruprecht, Göttingen

Dilthey W ([1887] 1994) Die Einbildungskraft des Dichters. Bausteine für eine Poetik. In: Gesammelte Schriften, vol 6, 7th edn. Misch G (ed) Originally published in Vischer F (ed) Philosophische aufsätze. Eduard zeller zu seinem fünfzigjährigen doctor-jubiläum gewidmet. Fues, Leipzig, pp 305–482. Teubner, Stuttgart and Vandenhoeck & Ruprecht, Göttingen, pp 103–241

Dilthey W ([1887] 1985) Poetry and experience. Makkreel RA, Rodi F (eds) Selected works, vol 5. Originally published as Die Einbildungskraft des dichters. Bausteine für eine Poetik in Vischer F (ed) Philosophische aufsätze. Eduard zeller zu seinem fünfzigjährigen doctor-jubiläum gewidmet. Fues, Leipzig, pp 305–482. Princeton University Press, Princeton, NJ

Dilthey W ([1900] 1990) Die Entstehung der Hermeneutik. In: Gesammelte Schriften, vol 5, 8th edn. Misch G (ed) Originally published in Erdmann B (ed) Philosophische abhandlungen. Christoph sigwart zu seinem 70. Geburtstage, 28. März 1900. Mohr, Tübingen, pp 185–202. Teubner, Stuttgart and Vandenhoeck & Ruprecht, Göttingen, pp 317–338

Feest U (2007) Hypotheses, everywhere only hypotheses – on some contexts of Dilthey's critique of explanatory psychology. Stud Hist Philos Biol Biomed Sci 38(1):43–62

Felt U (1996) 'Öffentliche Wissenschaft'. Zur Beziehung von Naturwissenschaften und Gesellschaft in Wien von der Jahrhundertwende bis zum Ende der Ersten Republik. ÖZG 7:45–66

Forschungsbibliothek Gotha (FB-Gotha) Schloss Friedenstein, 99867 Gotha

Haeckel E ([1868] 1924) Natürliche Schöpfungsgeschichte. In: Gemeinverständliche werke, vols 1 and 2. Schmidt H (ed) Originally published in 1 vol., Reimer, Berlin. Henschel, Leipzig and Kröner, Berlin

Kant, I (1902/1905) Kant's gesammelte Schriften. Abth. I: Werke, Vorkritische Schriften, Bd. I und II: Vorkritische Schriften. Königlich Preußische Akademie der Wissenschaften (ed.) Reimer, Berlin

Kempen B (2002) Vorwort des Herausgebers. In: Laßwitz K (author), Kempen B (ed.), Homchen. Ein Tiermärchen aus der oberen Kreide. SHAYOL, Berlin, pp 7–12

Laßwitz K (1873) Über Tropfen, welche an festen Körpern hängen und der Schwerkraft unterworfen sind. PhD dissertation, Universität von Breslau, Philosophische Fakultät

Laßwitz K (1887a) Die poetische und die wissenschaftliche Betrachtung der Natur. Nord und Süd 41(2):270–289

Laßwitz K (1887b) Wilhelm Dilthey's Grundlegung der Poetik. Die Nation 4(36):536–538

Laßwitz K (1890) Seifenblasen. Moderne märchen. Hamburg, Voss

Laßwitz K (1896a) Gustav Theodor Fechner. Frommanns, Stuttgart

Laßwitz K (1896b) Nature and the individual mind. The Monist 6(3):396–431

Laßwitz K ([1900] 1908) Wirklichkeiten. Beiträge zum weltverständnis, 3rd edn. Originally published by Felber, Berlin. B. Elischer Nachf, Leipzig

Laßwitz K (1902) Nie und immer. Neue märchen. Leipzig, Diederichs

Lefèvre W (2004) "Der Darwinismus-Streit der Evolutionsbiologen". In: Kleeberg B, Lefèvre W, Voss J (eds) Zum darwinismusstreit. Max Planck Institute for the History of Science, Berlin, Preprint, vol 272, pp 5–29

Lindau H (ed) (1919) Empfundenes und Erkanntes. Aus dem nachlasse von kurd laßwitz. B. Elischer Nachf, Leipzig

Makkreel RA (1991) Dilthey Philosoph der geisteswissenschaften. Suhrkamp, Frankfurt am Main

Mill JS ([1843] 1996) A system of logic ratiocinative and inductive. Being a connected view of the principles of evidence and the methods of scientific investigation. In: Collected Works of John Stuart Mill, vol 8. Robson JM (ed) Originally published by John Parker, London. Routledge, London

Ostwald W (1902) Vorlesungen über naturphilosophie, gehalten im sommer 1901 an der universität leipzig. Veit, Leipzig

Sarasin P (1995) La Science en Famille: Populäre Wissenschaft im 19. Jahrhundert als bürgerliche Kultur – und als Gegenstand einer Sozialgeschichte des Wissens. In: Gyr U (ed.) Soll und haben. Alltag und lebensformen bürgerlicher kultur. Offizin, Zürich, pp 97–109

Schweikert R (1979) Von geraden und schiefen Gedanken. Kurd Laßwitz – Gelehrter und Poet dazu. In: Laßwitz K (author), Schweikert R (ed.) Auf zwei planeten. Roman in zwei büchern. Zweitausendeins, Frankfurt am Main, pp 978–1074

Shinn T, Whitley R (eds) (1985) Expository Science: Forms and Functions of Popularisation. Reidel, Dordrecht

Siemens W ([1886] 1891) Das naturwissenschaftliche Zeitalter. In: Wissenschaftliche und technische Arbeiten, vol 2, 2nd edn. Originally published in Tageblatt der 59. Versammlung deutscher Naturforscher und Ärzte in Berlin 59(3):92–96. Justus Springer, Berlin, pp 491–499

Szukaj H (1996) Empfundenes und Erkanntes: Kurd Laßwitz als Wissenschaftspopularisator 1848–1910. Ph.D. dissertation. Westfälische Wilhelms-Universität, Münster, Medizinische Fakultät

Weber H (2000) Monistische und antimonistische weltanschauung. Eine auswahlbibliographie. Verlag für Wissenschaft und Bildung, Berlin

Whitley R (1985) Knowledge producers and knowledge acquirers. Popularisation as a relation between scientific fields and their publics. In: Shinn T, Whitley R (eds) Expository Science: Forms and Functions of Popularisation. Reidel, Dordrecht, pp 3–28

Willmann F (2002) Kurd Laßwitz' Popularisierungswerk: Wissenschaft im Märchen. In: Maillard C, Titzmann M (eds) Literatur und wissen(schaften) 1890–1935. Metzler, Stuttgart/Weimar, pp 97–109

Ziche P (ed) (2000) Monismus um 1900. Wissenschaftskultur und weltanschauung. Verlag für Wissenschaft und Bildung, Berlin

# Chapter 5
# Explaining History. Hippolyte Taine's Philosophy of Historical Science

**Philipp Müller**

Historians of European historiography have often characterized Hippolyte Taine (1828–1893) as an adherent of the positivist school of thought, typical for the development of a scientific culture in Western Europe that differed from its German counterpart.[1] In accordance with that view, Wilhelm Dilthey grouped him together with other scholars like John Stuart Mill and Herbert Spencer against who Dilthey tried to develop his conception of the human sciences based on the notion of "verstehen" (see Dilthey [1924] 1957, 139ff.). Dilthey understood Taine as proposing to analyze the human mind by identifying its individual components and then explaining their meaning by laws of their relation. He argued that such an approach might be adequate for the natural sciences, but neglected the fact that an analysis of the mind had to start from a given psychological connection that was prior to any definition of particular phenomena. From Dilthey's point of view, applying Taine's theory to historical studies only made them look more objective while actually Taine was unaware of just following the prevailing convictions of his time (idem, 191f.).

Without doubt Taine has given some ground to that interpretation of his theory by provocative statements that were meant to advance a more scientific approach in the moral sciences (as they were called in France). In the introduction to his 'Histoire de la littérature anglaise' he insisted on the idea of cause and explanation as universal scientific principles that could be used to gain knowledge of the natural as of the moral world: "Que les faits soient physiques ou moraux, il n'importe, ils ont toujours des causes; il y en a pour l'ambition, pour le courage, pour la véracité, comme pour la digestion, pour le mouvement musculaire, pour la chaleur animale. Le vice et la vertu sont des produits comme le vitriol et le sucre" (Taine 1863, XV). In contrast to Dilthey, Taine did not feel the need to draw a line between the natural and the human sciences – rather he was trying to bridge possible gaps between them. In his eyes, a science of history could be developed when it was based on psychology, modeling its approach on the other sciences' dependence upon their

P. Müller (✉)
Institut für Geschichtswissenschaften,Unter den Linden 6, 10099 Berlin, Germany
e-mail: muellerp@cms.hu-berlin.de

[1] See among others von Srbik (1951, 222f.), Breisach (1983, 275f.). Against this view, see Charlton (1959, 106f.).

U. Feest (ed.), *Historical Perspectives on Erklären and Verstehen*, Archimedes, Vol. 21,
DOI 10.1007/978-90-481-3540-0_5, © Springer Science+Business Media B.V. 2010

explanatory instruments. "De même qu'au fond, l'astronomie est un problème de mécanique et la physiologie un problème de chimie, de même l'histoire est un *problème de psychologie*" (idem, Taine's emphasis). The only differences between scientific disciplines Taine could spot ran between abstract sciences like mathematics and geometry and the sciences of experience and they were not designed to make different kinds of scientific methods necessary (see below). Still, just like Dilthey Taine wanted to found historical studies on a psychology that left metaphysical notions behind. In basic agreement with Dilthey (but contrary to Dilthey's sketch of his theory) Taine did not derive mental content from particular sense data but conceived it as being produced by psychological connections that preceded individual instances of experience. At the same time, like Dilthey, he did not understand the connections of the mind as innate psychological structures but considered them to result from an interplay of inner mental and outer historical forces. Finally, Taine held that through art one could discover the psychological foundations of the mind most easily, an argument that was employed similarly by Dilthey (see Dilthey [1927] 1992, 207).[2] Relating Taine's philosophy of historical science to his contemporary context I want to argue that Taine's idea of founding history on psychology was much closer to Dilthey's conception of the human sciences than traditional views have suggested.

In the following, I will first explain Taine's idea of historical science within the context of his discussions with the literary critic Charles-Augustin de Sainte-Beuve and the historian Ernest Renan and their common opposition against the academic philosophy of the French university. Then I will discuss whether Taine's effort to discover historical laws by turning history into historical psychology separates his conception from an approach that reflects the time-bound dimension of scientific reasoning. I am arguing that the idea of specific historical structures of the mental made Taine aware of his own outlook as subject to historical change. His 'Voyage en Italie', therefore, rests on a reflexive form of scientific understanding that can be demonstrated if the technique of his historical writing is taken into account. Considered from this point of view, Taine and Dilthey do not represent two different scientific cultures, irreconcilable with each other.

## 5.1 Historical Psychology and Its Context

After a brilliant student career at the *École Normale* Taine had originally planned to become a professor of philosophy. But first, he was unable to pass the examination for teaching that would have allowed him to get a professorship at a *lycée* in Paris because the commission of the *agrégation* doubted Taine's aptness to defend Christian convictions. Then, Taine's proposal on writing a thesis about the nature of the human mind met with resistance at the Sorbonne, which motivated him to postpone the subject to the time after his dissertation. When Taine finally received his doctorate for a treatise on the French novelist Lafontaine the Dean told him that – considering his

[2] On Dilthey's conception of aesthetics as a model for the 'Geisteswissenschaften,' see Makkreel (1991, 117f.).

reputation as a naturalist *révolté* – it would be useless to give a career at the university any further thought (Léger 1993, 97). Taine found himself forced to make a living by lecturing as a private tutor and delivering articles to journals. As a consequence, Taine's philosophy of historical science developed within a context that was opposed to the views upheld by the academic scholarship in France at that time.

Taine soon made contact with other intellectuals concerned with questions that were close to his own interests. Among his friends were eminent figures of the cultural life of Paris during the Second Empire such as the literary critic Sainte-Beuve, the novelists Gustave Flaubert and the brothers Goncourt and the historian Ernest Renan. They met regularly at Magny's, a restaurant on the left bank in Paris discussing recent literary and historical writing as well as contemporary social and moral questions (Dewald 2003, 1,012f.). Their intellectual ties were strengthened by frequent encounters at the salon of Napoleon III's cousin, the Princess Mathilde, who sympathized with liberal artists and writers. The discussions of the group demonstrated the compatibility of different intellectual interests and literary genres outside of academic disciplinary boundaries. At the same time, at the beginning of the 1860s they all believed in one way or another in the 'Future of science' as a famous book by Renan was called that he had written a couple of years before.[3] The members of the group used images and expressions taken from biology and medicine to present their works to the public and to each other. For example, Sainte-Beuve described Flaubert in a now well-known expression as someone who was handling a pen like a surgical knife; similarly, the brothers Goncourt wanted their novel about the mental troubles of a young woman to be understood as a clinical study (Dufour 1998, 109f.). An important subject of the intellectual exchanges within the group was the question of a connection between the mind and its natural and historical environment as it was denied by academic philosophy at that time.

Academic philosophy was dominated by supporters of Victor Cousin who had used the means at his disposal during the July monarchy to institutionalize his views throughout the French educational system.[4] Cousin thought that philosophy had to be based on what he called the "méthode psychologique", a method that was directed against attempts to understand the human soul as depending on empirical matter. According to Cousin, introspection showed that the fundamental principles of all understanding were purely mental and could not be derived from the senses or other non-spiritual factors. Rather, for him, the mind consisted of autonomous faculties that were psychological in substance and primordially given to man (Schmauss 2004, 60–68). In Cousin's eyes, their existence proved the presence of God within human reason: "La vérité est en quelque sorte prêtée à la raison humaine, mais elle appartient à une toute autre raison, à savoir cette raison suprême, éternelle, incréée, qui est Dieu même" (Cousin [1853] 1854, 103). The participants of the meetings at Magny's shared Cousin's interest in psychology but were not

[3]Renan had written 'L'avenir de la science' in 1848. It was only to be published in 1890. See Renan ([1890]1995), préface.

[4]Besides dominating the most important commissions that decided about the admission of candidates to universities and academic degrees Cousin developed a syllabus that was mandatory for the school teaching of philosophy. See Brooks (1998, 36ff.).

willing to accept his spiritualist conception and opposed his premises at various occasions. A brief survey of some of their writings shows that Taine developed his philosophy of science within an intellectual environment that discussed the nature of the mind as related to its historical surroundings (Dewald 2003, 1023–1025).

Instead of believing in a divine origin of poetic imagination Sainte-Beuve stressed the conditions that formed an author's character. In his literary history of the Jansenist monastery 'Port Royal' he delivered an intellectual history of France during the seventeenth and eighteenth century that considered its writers as representatives of the currents of thought of their time. Sainte-Beuve explained their literary and theological convictions as the consequence of a general spiritual crisis that expressed itself throughout the social and cultural life before the Revolution. He presented the book as the result of impartial observation that studied the poetic spirit not as a supranatural capacity but as a scientific object and its changes under the influence of society and its doctrines. "Il m'a semblé qu' [...] il n'y a point d'emploi plus légitime et plus honorable de l'esprit que de voir les choses et les hommes comme ils sont [...], de décrire autour de soi, en serviteur de la science, les variétés de l'espèce, les diverses formes de l'organisation humaine, étrangement modifiée au moral dans la société et dans le dédale artificiel des doctrines" (Sainte-Beuve [1859] 1955, 674). Throughout his work Sainte-Beuve insisted on the necessity of taking the details of a writer's biography into account in order to be able to fully grasp the significance of his artistic production. Sainte-Beuve – who had studied medicine before he became a professional writer – hoped to contribute to the development of a "méthode naturelle en littérature" (Sainte-Beuve [1862] 1870, 32). His aim was to turn the study of literature into a science of man that would observe the particular facts of the past and its influence on novelists and poets.

Likewise, Ernest Renan promoted a new science of humanity that could live up to the methodological progress that the natural sciences demonstrated. He considered the mental as the essence of human nature and, therefore, argued to study psychological activities with scientific rigor – "une science exacte des choses de l'esprit" (Petit 1995, 26f.). Renan's work at that time was close to Sainte-Beuve's approach in its conception of the mind as a malleable form that was influenced by changing social and intellectual circumstances.[5] Renan focused on the history of religion that he conceived as offering especially promising material for his research. In his 'Origines du christianisme' Renan treated the texts of the Bible as documents for a history of the human consciousness. He argued, for example, that the story of the conversion of St. Paul could be explained naturally if one considered the spontaneous and unreflected state of mind of people in ancient times. In combination with St. Paul's exhaustion that was caused by the persecutions and his journey, Renan explained, he interpreted a hallucination as a divine vision (see especially Pfeiffer 1990, 325–329).

[5] Accordingly, he explained the origins of Christianity by a state of mind that was formed by the world of the ancients: "C'est le mouvement intérieur de ces fortes natures qui a préparé le coup de tonnerre; mais le coup de tonnerre a été déterminé par une cause extérieure. Tous ces phénomènes se rapportent, du reste, à un état moral qui n'est plus le nôtre" (Renan [1863]1949, 29).

The way Renan unmasked Christian traditions as legends enraged his Catholic readers and provoked fierce reactions by the spiritualists. In the eyes of Cousin, Renan represented the incarnation of atheism and materialism.[6] On the other hand, Renan and Sainte-Beuve were strongly supported by Taine who wrote an entire book to attack and ridicule Cousin's philosophical convictions. In his 'Les philosophes français du XIXe siècle' Taine contended that Cousin was not able to prove the divine presence within human reason but rather substituted arguments for figures of speech. He explained Cousin's theory of the human mind as the result of rhetorical needs in salon conversation (see Taine 1857, 192ff.).

Taine delivered a more detailed explanation of his own ideas about the relation between the human mind and its environment in his 'Histoire de la littérature anglaise'. He pointed out three basic notions to disclose a relation between an inner realm of primitive tendencies and an outer realm of its conditions by isolating different factors that contributed to the specific shape of their connection. With the notion of "la race" Taine aimed at an inner mold of a people that changed its form in the course of history. The term "le milieu" summed up all the external conditions that shaped the inner dispositions. The circumstances Taine wanted to include in this respect were as much cultural as social, economic and political. The notion of "le moment" gave the interdependence of race and environment a temporal dimension. Each relationship between the inner mold and its external conditions existed only because it was built on an earlier state of their relation – the interplay of the two realms was influenced by its own history.

The three factors should not be understood independently of each other. For example, the category of environment cannot be entirely separated from the category of time; also, what Taine called "race" cannot be defined without recourse to its environment. More important than the question of his terminological exactitude is the main thrust of his theory: Taine was trying to analyze a relation between inner dispositions and their historical conditions that dismissed the spiritualist insistance on the autonomy of the human mind, replacing it with a psychology that could form the basis for a new method in historical studies. In agreement with Sainte-Beuve und Renan who both conceived the mind as a historically moldable form, Taine believed that history should treat the details of the past as embodying a hidden psychological content.[7]

[6]Because of persistent efforts of the catholic church (and Cousin) who accused him of having undermined the Christian faith when giving his inaugural lecture Renan was suspended as a professor at the *Collège de France* in 1862. Sainte-Beuve protested unsuccessfully against the Imperial decree in the senate. See Gaigalas (1972, 34f.).

[7]According to Taine, Sainte-Beuve and Renan had proved how the understanding of the actions of historical individuals and groups could benefit from psychological studies. See Taine ([1864]1878, 21). In his 'Histoire de la littérature anglaise' Taine characterized the renewal of methods within the moral sciences as Sainte-Beuve's merit. "Personne ne l'a fait plus juste et aussi grand que Sainte-Beuve; à cet égard nous sommes tous ses éléves; sa méthode renouvelle aujourd'hui dans les livres et jusque dans les journaux toute la critique littéraire, philosophique et religieuse" (Taine 1863, XIV). In a similar fashion Taine called Sainte-Beuve his intellectual mentor in their correspondence. "Si je fais de la physiologie morale, c'est grâce à vous." Taine, To Sainte-Beuve June 15th 1867, in Taine (1904, 339).

Even though Taine's psychological approach was supposed to lead to scientific explanations of the past that have been perceived as positivistic, Taine's ideas resemble much of what Dilthey developed within his theory of the human sciences. Like Dilthey, Taine was especially interested in the actual functioning of human thinking and elaborated his views much further than Sainte-Beuve and Renan. An important goal of his argument was to show that it was only a superficial view to conceive sensations as the constituents of human thinking. Taine pointed out that ideas and categories of the mind could not be derived from atomized sense data but had to be explained on the basis of acquired connections of experiences. In his long prepared study on the subject Taine argued that cognition was made possible by a mental function that substituted sensual perception for an image. He described the origin of that image as the result of an accommodation of inner psychological drives to continued corrections. The basic form of the mental image was created early in life by a primitive instinct of the mind that formed an abstract category out of a succession of similar experiences by extracting one of their common features. This category achieved its final content only by its connections with the abstracted form of other experiences. As a consequence, Taine held that thought activity did not rely on sense data but rather on a system of mental representations where one group of images limited the scope of their neighboring ones. According to his theory, Taine believed that for the image of an experience to reoccur under given circumstances it was not necessary that the object of the original experience was actually present; rather, its presence rose into consciousness when its substitutional image was not suppressed but stimulated by the system of the surrounding mental images it was associated with (see Taine [1870] 1878, vol. 1, 49f.). The correspondence of each individual's images with its contemporaries was achieved by a process of adaption to the physical and social environment (idem, 123ff.).

Taine conceived singular thoughts as resulting from an acquired structure of the mind. The interplay of external and internal forces produced a system of mental images mutually depending on each other. According to Taine, the thinking substance that the spiritualists had conceived as an autonomous self should, therefore, be re-described as a harmony of images that conditioned each other to form an organized whole. "De même que le corps vivant est un polypier de cellules mutuellement dépendantes, de même l'esprit agissant est un polypier d'images mutuellement dépendantes, et l'unité, dans l'un comme dans l'autre, n'est qu'une harmonie et un effet" (idem, 124). Taine believed that his theory of the mind could function as a starting point for a more scientific approach in historical studies. In his eyes, it was the task of the moral sciences to disclose the network of mental images working behind the external appearances of human action and behavior. This network of mental images was not primarily to be seen as the psychological structure of an individual, but as being shaped by general currents of thought which were representative for a historical time period. Since, following Taine's conception, the structure of the mind was influenced by its social setting it was capable of reflecting the spirit of an entire society, expressing the moral sense, religious beliefs, and political institutions (see Taine 1863, XVIII–XX). In Taine's eyes, historical events were only commonly perceived as particular phenomena – behind their individual appearance they were really based on general connections of a mental system of which they were an expression. "À proprement parler les faits [...] n'existent pas; ils n'existent qu'au regard de notre esprit; au fond il n'existe que des choses générales [...] nous n'apercevions les choses qu'à l'envers" (see Taine, "To Éduard de Suckau July 24th 1862", in Taine 1904, 258).

In agreement with Dilthey, Taine was convinced that the historical dimension of the mental was best reflected in works of art. For him, changes within the organization of mental representations changed the way of thinking and acting entirely. They could lead to a mental form that was inclined to perceive the world as an object of creative practices or rather to a more contemplating outlook on reality. From the range of its possible expressions it was the realm of literature and art that Taine considered to articulate the mind's composition in the most representative form. He believed that the spirit of an artist could serve as a model to understand psychological functions of a certain time. Explaining his preferences of study Taine declared: "Tout peintre, poëte, romancier d'une lucidité exceptionelle devrait être questionné et observé à fond par un ami psychologue. On apprendrait de lui la façon dont les figures se forment dans son esprit, sa manière de voir mentalement les objets imaginaires, l'ordre dans lequel ils lui apparaissent [...] etc." (Taine [1870] 1878, vol. 1, 14). Taine claimed that a historian equipped with adequate interpretive skills learned more about the past studying a piece of literature or a work of art than working with other sources (see Haskell 1993, 347–353). Its value as historical documents stemmed from the function Taine ascribed to works of art: He thought that novelists and painters had a specific capacity to concentrate the spirit of their time because they were assembling the sentiments prevailing within their society. A writer or a painter, therefore, gave a clear expression of the mental structure of his contemporaries. As a consequence, poetry and art in return enabled the historian to grasp the hidden mental forces of the past and to use them when explaining historical developments. Taine conceived art as an irreplaceable instrument that could partly compensate for the lack of other measuring tools in historical scholarship. "[L'art] ressemble à ces appareils admirable, d'une sensibilité extraordinaire, au moyen desquels les physiciens démêlent et mesurent les changements les plus intimes et les plus délicats d'un corps" (Taine 1863, XLVI).

## 5.2 Taine's Concept of Mental Laws in Historical Studies

Even though Taine's approach bears resemblance to Dilthey's ideas, an important difference seems to remain concerning the explanatory role of the concepts which stood at the center of Taine's theory. For Taine, the goal of historical studies was to cast mental structures into psychological laws. By contrast, not only Dilthey but also Sainte-Beuve and Renan believed that the realm of the human world could not be properly understood applying the concept of law. In order to understand Taine's reflections about the nature of the mind in this respect it is necessary to take his idea of scientific practice into account.

According to Taine, the explanatory categories of historical science relied on the abstraction of a mental structure that was (in principle) indicated by all the possible cultural expressions within a society. First, the historian could analyze the works of philosophy, theology and art of a certain age and then complete the intellectual side of that picture by detecting how state and family were organized, and how trade and agriculture were carried out. The impression gained during the first reading of a novelist, for example, would be rounded off by the information collected by studying the form of government etc. Having completed the first step the historian would notice – according to Taine – that the characteristic elements of each part depended mutually on the others and that they were all related to common needs and tendencies. Starting from what seemed to be isolated facts the historian would develop an idea of how the facts could be related to a reigning system of inclinations and attitudes. From Taine's point of view, that system should be understood as a mental structure that caused the outer appearances of the past to appear as they did. Taine argued that, as a result of the procedure, the historian possessed a general term of a psychological law that designated the moral temperament and explained the actions, modes of perception and behavior within the society of a certain time.[8]

Just like Taine, Sainte-Beuve und Renan were trying to render the historical studies more scientific by focusing on mental structures that were supposed to underlie the events of the past. But when discussing their work with each other the general agreement also proved to include a complex of questions where their approaches did not concur. While Taine agreed with Sainte-Beuve and Renan about the necessity to naturalize the concept of the human mind by renouncing the philosophy of the spiritualists they differed on the question whether it was possible or desirable to describe mental structures as abstract psychological laws.

Sainte-Beuve recognized that Taine was heading towards a natural explanation of the mind similar to his own and welcomed his approach early in Taine's career. In his review of Taine's first works on literary history Sainte-Beuve noted: "M. Taine excelle à situer les auteurs qu'il étudie, dans leur époque et dans leur moment social, à les encadrer, à les y enfermer, à les en déduire: ce n'est pas seulement chez lui une inclination et une pente; c'est un résultat de méthode et une conséquence qui a force d'une loi" (Sainte-Beuve [1857] 1858, 210).[9] While Sainte-Beuve agreed with Taine's way of interpreting authors within their particular environment, he also noticed that Taine tried to arrive at general terms that were supposed to express strict mental structures. Even if in his works on Livy, La Fontaine or the English literature Taine did not explicitly present his analysis that way, Sainte-Beuve still suspected that, within Taine's methodological guidelines, an

[8] See especially Taine's exposition of his research procedures in Taine (1866, I–XXVII).

[9] In another review Sainte-Beuve expresses their mutual agreement on the idea of a natural method of literature and the human mind: "Nous tous, partisans de la méthode naturelle en littérature [...], nous tous, artisans et serviteurs d'une même science que nous cherchons à rendre aussi exacte que possible, sans nous payer de notions vagues et de vains mots, continuons donc d'oberserver sans relâche, d'étudier et de pénétrer les conditions des œuvres diversement remarquables [...]" (Sainte-Beuve [1864]1867, 87f.).

author had to write like he did, forced by his own psychological law. If carried to the extreme he felt that Taine's view was in danger of replacing historical agents by an abstract formula that functioned blindly. "Que le savant, chez lui [Taine], ne domine pas trop le littérateur, c'est là le seul conseil général qu'on doive lui donner" (Sainte-Beuve [1864] 1867, 232).

Similar to Sainte-Beuve, Renan argued against Taine that the historian should try to come to understand the actual character of historical minds rather than deduce general notions from their behavior. When he read an extract of his 'Vie de Jésus' to Taine they could not agree on Renan's methodological procedure.[10] Above all, in Taine's eyes, Renan lacked a clear concept of the psychological structure he purported to explain. In a personal summary of their conversation Taine wrote: "Renan est parfaitement incapable de formules précises, il ne va pas d'une vérité à une autre. Il tâte, palpe. Il a des impressions, [...] les généralités ne sont pour lui que le retentissement, l'écho des choses en lui. Il n'a pas de système mais des aperçus, des sensations" (Taine 1904, 242f.). Renan, on the contrary, argued that accepting Taine's criticism would lead to a dry treatise that could not convey historical knowledge because it did not reach the experiences of the past.[11]

Judging from Sainte-Beuve's and Renan's critical remarks Taine's scientific endeavor seemed to be directed towards a reduction of the human mind to strict relations of cause and effect. But even if Taine thought that the goal of analysis within the humanities was the discovery of psychological laws, thereby stressing the correspondence between the natural and the moral sciences, it is not easy to see what he exactly meant by that term. Since psychological laws could only be analyzed indirectly by way of their historical expressions Taine did not believe that they could be fully grasped. Consequently, in his historical writing he did not present them as depersonalized categories of historical explanation but rather as depending on the researcher's point of view.

Taine wanted to use a general category of mental structures to explain the past but he had difficulties to develop a clear and convincing definition of its status (see Nordmann 1990, 122f.). While the concept of a law seems to be associated with a strict sequence of cause and effect independent from an individual perspective Taine's use of the concept within his historical studies does not correspond to that idea. First of all, it is quite clear that Taine could not aim at anything like a mathematical law that was supposed to be valid at all times and under all circumstances. Rather, the search for a relation between the three factors of race, environment and time was designed to lead to a connection that was bound to the period of history under scrutiny. Accordingly, instead of conceiving psychological structures

[10] Taine criticized Renan's efforts to reproduce the outlook of the ancients without relying on source material because thereby he would replace the legends of the Bible with a speculative hypothesis created through literary means. "En vain Berthelot et moi nous lui disons c'est mettre un roman à la place de la légende; qu'il gâte les parties certaines par un mélange d'hypothèses" (Taine 1904, 245).

[11] Taine described Renan's reaction to his suggestions: "Il [Renan] n'entend rien, ne voit que son idée, dit que nous ne sommes pas artistes, qu'un traité simplement et dogmatique ne rendrait pas la vie" (Taine 1904, 245).

as universal Taine pointed out that the historian had to be aware of the fact that the world of the past was different from his own because of the different mental systems. In his history of the Italian Renaissance Taine wrote: "[S]urtout il faut se dire et se redire qu'alors l'âme du spectateur n'était pas la même qu'aujourd'hui" (Taine [1866] 1965, vol. 1, 159). By explaining that the way of thinking in the past was guided through different mental forms Taine wanted to make clear that one could not use modern judgments to understand the beauty of Renaissance art but had to reconstruct and adopt the position of its culture.[12] Consequently, whenever Taine claimed to advance the knowledge of laws through historical psychology he could only refer to the knowledge of a fixed mental structure that prevailed in a certain period of time. In his eyes, limiting the scope of historical concepts in that way did not run the risk of reintroducing a methodological distinction between the moral and the natural sciences because the categories that were used in biology or zoology could not claim to be applicable without circumstantial restrictions either.[13]

If psychological laws were not meant to have the same status as universal laws because they changed over time, the problem remained how they could be defined as temporary structures. If the structures of the mind developed different forms through the course of history it became difficult to see how they were to be determined as specific entities at all. Even though Taine remained hesitant on this question he was eventually forced to accept that the scientific categories of historical explanation could not represent more than an ideal type constructed from the historian's point of view.

In many theoretical expositions of his theory Taine argued that the meaning of scientific concepts could be derived from their correspondence to reality. For example, in the last chapter of his 'Les philosophes français' he held the view that the psychological state discovered by scientific observation represented a substantial part of the real world (see Taine 1857, 357). Similarly, in his small book on the philosophy of John Stuart Mill Taine criticized Mill's conception that only the ideas of facts were connected, not the facts themselves. In Taine's view, the goal of science was to arrive at the actual causes for the way things had happened. "Nous découvrons des couples, c'est-à-dire des composés réels et des liaisons réelles. Nous passons de l'accidentiel au nécessaire, du relatif à l'absolu, de l'apparence à la vérité" (Taine [1864] 1878, 142). But faced with the contemporary state of the empirical sciences Taine explained that historical psychology could not yet pass from the accidental to the necessary and absolute and that it was doubtful whether it would ever be able to do so. "[À] une certaine limite notre explication s'arrête, et, quoique, de siècle en siècle nous la poussions plus avant, il est possible qu'elle

[12] The relation between the historical state of mind and its specific context as it was disclosed by the historical account should help the reader to change his point of view and look at the carnal scences of early modern painting with different eyes. "Avec de la réflexion, des lectures et de l'habitude, on réussit par degrés à reproduire en soi-même des sentiments auxquels d'abord on était étranger; nous voyons qu'un autre homme, dans un autre temps, a dû sentir autrement que nous-mêmes" (Taine [1866]1965, vol. 1, 17).

[13] "L'histoire n'est pas une science analogue à la géométrie, mais à la physiologie et à la géologie." Taine, To Ernest Havet April 19th 1864, in Taine (1904, 229).

vienne toujours s'arrêter devant une limite" (Taine [1870] 1878, vol. 2, 428). In Taine's eyes, the difficulty of historical explanations resulted only from the lack of knowledge concerning the general relations behind the facts of the past. But as a consequence, the explanatory tool the historian developed through research had to be derived from the converging effects of a cause that could never be observed by itself. Although Taine was ambiguous about the evaluation of his endeavor he still recognized that relying on the emergence of impressions during research failed to deliver a definite criterion for mental structures as specific entities. Explaining the status of general terms in historical psychology Taine declared that they differed from geometric and mathematical categories because questions concerning the human mind could not be measured like the corresponding values of the artificial sciences ("sciences de construction"). Since it was impossible to make sure that the primitive elements of the mental system as they were detected through the historian's impressions while reading the sources actually were the decisive elements in question, the general notion of a mental structure could only represent an explanatory instrument of relative value.[14] Rather than discovering the "rigid designator" (Saul Kripke) of psychological laws that were attached to some main feature of past reality, Taine conceived the result of historical research as a description that was not rigid but literary. "Nous ne pouvons la fixer dans une formule exacte ou approximative; nous ne pouvons donner, à propos d'elle, qu'une impression littéraire" (Taine 1863, XXXI). The meaning of the historical categories could only be thought of as forming an ideal type cast in a literary description that was assumed by the historian for the sake of scientific explanation. It could never be proved independently of the elements that research had produced to construct it and, therefore, it was designed to be modified or revised whenever the picture of the psychological structure of the past changed. Taine did not elaborate the revisable character of scientific notions within his philosophy of historical psychology much further; rather, its consequences are reflected in his historical writing.

## 5.3 The Display of Explanation in Historical Writing

According to Taine, history as a science should be concerned with discovering psychological structures that were supposed to be the foundation of the events of the past. They could not be studied directly but only through their impact on the outer

[14]Similarly – according to Taine – the explanatory categories of the natural sciences were relative to the detected elements from which the general cause of their appearance was deduced. "La grandeur est toujours relative; rien n'empêche que nos molécules différentes, aussi petites par rapport à elles qu'elle le sont elles-même par rapport à une planète, et ainsi de suite, sans trêve ni fin." (Taine [1870]1878, vol. 2, 429) "De même qu'il y a des rapports fixes mais non mesurables quantitativement, entre les organes et les fonctions d'un corps vivant, de même il y a des rapports précis, mais non susceptibles d'évaluation numérique entre les groupes des faits qui composent la vie sociale et morale." Taine, To Ernest Havet April 19th 1864, in Taine (1904, 299f.).

appearances of historical life. Under these circumstances Taine believed that the historian could never reach the structure of past mentalities themselves. In order to be able to understand the meaning of historical facts he had to develop an impression about their psychological ground which – as it did not represent the definite explanation – was defined as a literary description. A closer look at Taine's historical writing reveals narrative techniques that reflect the circular relation between the knowledge of a psychological structure necessary to understand the past and its construction by the historian.

Taine's historiography relied on discussions between members of the Magny group about the relation of fictional objectives and scientific knowledge. From their point of view, literature and science were not to be looked at as exclusive spheres but rather as two ways of achieving one goal (see Dufour 1998, 109f.). Accordingly, Flaubert characterized the development of contemporary fictional writing by comparisons that modeled literature on science. "La littérature prendra, de plus en plus, les allures de la science. – Le grand Art est scientifique" (Flaubert quoted in Thiher 2001, 79). Renan used literary devices in his historical writing and defended the connection of science and art explicitly when reviewing the historiography of his time: "L'histoire [...] est un art autant qu'une science, la perfection de la forme y est essentielle, et toute critique qui ne tient compte, dans l'appréciation des œuvres historiques, que des recherches spéciales est par là même défectueuse" (Renan [1857] 1948, 103). Taine pleaded for a conjunction of science and art, declaring that history demanded a spirit that was capable of analytical abstraction and poetic imagination: "L'histoire est un art [...] mais elle est aussi une science; elle demande à l'écrivain l'inspiration, mais elle lui demande aussi la réflexion; si elle a pour ouvrière l'imagination créatrice, elle a pour instrument la critique prudente et la généralisation circonspecte" (Taine 1858, 323). None of them wanted to reduce literature to the procedures of scientific demonstration or reveal the fictional character of scientific truth claims;[15] rather, they thought that the insights conveyed by literary devices were compatible with scientific knowledge. In this respect, they were following authors like Balzac who had announced in a famous and often quoted statement that he wanted to liken literature to science and study by literary means the way of life in the modern French society like Buffon and Saint-Hilaire had studied the organization of animals and plants (de Balzac [1842] 1956, 78ff.). Endorsing a similar reorientation of literature towards empirical matter Stendhal opposed the tradition of idealized plots by stressing the necessity of basing figures and narrative on the real circumstances of time and place. He wanted to transform literary writing into a mirror that would reflect nothing but the true details of reality (see Auerbach [1946] 2001, 424–426; Blin 1954, 60). Stendhal was especially praised by Taine who took his novels as the first efforts to use literature to turn the study of psychology into a scientific discipline. For Taine, Stendhal demonstrated how literary devices could capture the complicated functions of the human soul. "[O]n n'a pas vu que

[15]It is therefore unnecessary to rehearse the discussions around the view – most often attributed to Hayden White's 'Metahistory' – that historical writing cannot amount to more than verbal fictions. See Müller (2008, 16–29).

sous des apparences de causeur et d'homme du monde, il [Stendhal] expliquait les plus compliqués des mécanismes internes [...] qu'il importait dans l'histoire du cœur les procédés scientifiques" (Taine 1863, XLIV). But the novels of authors like Stendhal could not only be read as related efforts to study the human mind within its social setting, their literature could also offer a procedure to deal with the literary character of the scientific categories used in historical explanation.[16]

Even if they purported to be based on the observation of reality, Stendhal's and Flaubert's novels did not restrict themselves to display empirical descriptions but also reflected the structure of observation itself. An adequate literary technique to represent the purely analytical and neutral position of an author could seem to be withdrawing his organizing function from the surface of the text. In literary history the change in narrative perspective that results from this move is known as telling a story by fusing the behavior of the narrator with the behavior of a character in the story. The opening scene of Flaubert's 'Madame Bovary' may demonstrate the procedure to some extent. At the same time it reveals the relativity of any attempt to reach an impartial point of view (Kablitz 2003, 99).

In the first scene of Flaubert's novel a new student, Charles Bovary, who enters a classroom is described from the point of view of a group of other students denoted by a "We". This "We" is the narrating subject that watches Charles Bovary as a stranger. With this technique the narrator is established as a figure within the story who discovers an unknown character gradually by observing his external appearances. At the same time the portrait of the young boy that is conveyed is nothing but an enumeration of shortcomings which show that the narrator is not an objective or benevolent observer. The way the observation is conducted creates a certain distance to the "We" that is telling the scene. As a result the standpoint of observation is doubled. The group is observing Charles Bovary and is by itself observed from the outside while observing (Kablitz 2003, 107–109).

The sliding in and out of another observer, into and out of the character's perspective and the mutual observation that follows the procedure is considered an important feature of Flaubert's writing. In its structure it resembles Stendhal's narrative technique to develop his literary realism. Stendhal thought that reality could not be adequately described by achieving an omniscient standpoint; rather, he stressed the reality of perception that he conceived to be by its very nature limited to a perspective. Accordingly, his novels do not rely on a narrator that is established as an independent and distanced observer; rather, they follow and show the limited scope of the character's subjective observation of the world. At the same time, this perspective is doubled when the limits of these perspectives are themselves revealed (see Blin 1954, 120ff., 142ff.).

The overlap and the distance between the perspectives of the character and the narrator can also be found in Taine's historical writing and they fulfill a similar narrative function. Applied to historical writing, the doubling of perspectives makes the reader aware of the fact that the narrator's point of view is bound to a character's

[16] See for the relation between Taine and Flaubert (Donatelli 2001, 75–87); for Taine and Stendhal see among others Seys (2000, 13–31).

perception of reality that is part of an ongoing story. Taine's 'Voyage en Italie' offers an example of how his historical writing reflects the revisable status of its explanatory categories by employing literary techniques.

The book is written as a description of a journey that uses the literary devices of a traveler's account to analyze the history of Italy focusing on the art of the Renaissance. After publishing the last volume, Taine explained to one of his critics that the title of his book should have indicated that it was a historian's trip through the history of painting. "Mon livre devrait s'appeler: Voyage d'un historien à travers la Peinture" (Taine, "To Paul de Saint-Victor", in: Taine 1904, 334). But actually, the title of his study was carefully chosen because it prepared the reader for a mode of writing that did not correspond to a historical narrative that purported to portray its subject from a distanced point of view. Taine employed his theory of race, environment and time to study the psychological structure of the Italian people of the fifteenth and sixteenth century but he expressed his results in a literary form that was adopted from books of travels.

According to Taine, a proper understanding of Early Modern painting had to situate its artists within their cultural milieu. He characterized Renaissance Italy as a historical setting where social and judicial restrictions had been abolished and replaced with the possibility to follow individual desires and needs. In his 'Voyage en Italie' the streets of Rome were dominated by rivaling families and their vendettas. Festivities did not show the fulfillment of a perfect civilization but rather the absence of moral conventions. Even the popes did not shy away from committing crimes to realize their plans. According to Taine's account, the culture of the Renaissance had liberated the individual from religious and moral constraints and thereby set free an untamed life of the senses. "Les instincts corporels étalent encore toute leur nudité à la lumière, et ni le raffinement du monde, ni les convenances de l'habit ne sont venus tempérer ou déguiser la fougue intacte des sens déchaînés" (Taine [1866] 1965, vol. 1, 189). The lack of regulating institutions did not only produce a situation of violence and insecurity but also entailed the rise of new forms of art. Creative energy that had been channeled by social traditions before could now develop independently of the control of the church and the state and led to different aesthetic ideals.[17] Taine was especially struck by the dominating role the representation of the human body and its strength played in Italian painting. He argued that in a time of frequent armed conflicts and threats the knowledge about body motion and function had to acquire new importance. Also, he thought that people who lived through an age of persisting nervous strains could only recognize themselves in powerful images (see Taine [1866] 1965, vol. 2, 91).[18]

[17] "Ces mœurs étaient la température vivifiante qui de toutes parts faisait germer et fleurir la grande peinture" (Taine [1866] 1965, vol. 1, 194).

[18] "Il y a un germe dont le reste n'est que le développement, c'est le beau corps bien portant, solidement et simplement peint dans une attitude qui manifeste la force et la perfection de sa structure; c'est cela seul qu'il faut chercher; les autres parties de l'art sont subordonnées" (Taine [1866] 1965, vol. 1, 168).

Taine interpreted the interdependence of Renaissance culture and new ways of creating and conceiving artistic representation as an expression of a specific mental structure that was formed by the historical circumstances. Rather than perceiving the world within the terms of faith or reason Taine explained that the way of thought of the Renaissance was oriented by forms. Inducing the fundamental psychological layer of the Italian history from his observations and describing it as figural thinking Taine declared about Raffael: "Il pensera par des formes comme nous pensons par des phrases" (Taine [1866] 1965, vol. 1, 162).[19] The psychological structure of Early Modern Italy Taine claimed to have discovered was supposed to explain the main features of its society and culture. Even though his historical explanation was also related to aesthetic ideals that pointed towards the idea of representing an Italian nature his historical writing does not follow these aims.[20] Taine presented his findings from an observer's point of view that was by itself observed.

Taine declared at the beginning of his 'Voyage en Italy' that it had to be read as the result of personal experience, following from his individual impressions and ideas that he wanted to publish as they were, without adapting them afterwards to other ends.[21] Of course, Taine had not only prepared his trip by extensive reading about Italian history and art, he also spent long hours in the libraries after his return before turning his journal notes into a book. Still, he maintained the style of immediate descriptions and delivered his historical explanations as the instantaneous reflections he had had during his visits of cities, museums and churches. By characterizing and presenting his study as untouched thought he followed the genre of travel descriptions (Guentner 1997, 14–26). At the same time, it corresponded to the restrictions Taine thought the categories of the experimental sciences were bound to. Taine tried to portray the general structure of the Italian mind by presenting particular facts since historical psychology could not rely on displaying universal laws. Given the belief that the expressions of the moral sciences were too vague to transmit precise analytical results Taine believed that historical knowledge could only be conveyed by means of small details which were to be understood as meaningful specimen. "À mon sens, les seuls moyens de transmission sont [...] les petits faits, anecdotes, citations, spécimens expressifs et significatifs [...] qui sont des fragments intacts, extraits de la réalité" (Taine, "To Frantz Funk-Brentano March 13th 1891", in Taine 1907, 320). The representation of the process of seeing and

[19] After his last volume was published Taine pointed out repeatedly that figural thinking was in his eyes the founding mental structure of the Renaissance: "[J]e me persuade de plus en plus, chez le sculpteur et le peintre de la Renaissance, le moule de la pensée était autre que chez nous. Ils pensaient par des formes colorées et nous par des mots abstraits." Taine, "To Paul Saint-Victor", in (Taine 1904, 334).

[20] For contradictions within Taine's aesthetics see Gerhardi (1989, 50–53).

[21] "Prends ceci comme un Journal [...] de plus tout personnel. [...] Ce que chacun sent lui est propre et particulier comme sa nature; ce que j'éprouverai dépendra de ce que je suis" (Taine [1866] 1965, vol. 1, 4).

writing was supposed to lead to an image of the reality of the past and especially of its mental foundation that emerged from the relation between the facts as they were continuously noted. The state of mind of the Renaissance was to be introduced by the order of the extracts of reality that the historian arranged. In this respect, Taine conceived historical science as an art that had to paint its objects in order to make their structure known. "La science devient art. Elle ne prend point pour cela un habillement étranger et extérieur. Elle ne reçoit que sa forme naturelle et définitive. [...] L'artiste dans l'historien n'est pas séparé du savant" (Taine [1856] 1994, 145; see also Kahn 1970, 50f.).

At the same time the narrative form of personal notes marked of the perspective of the historian as the source of historical descriptions. In his preface Taine advised the reader explicitly to bear in mind that the following account was produced by the perceptive faculties of a modern individual who had developed his instruments under specific circumstances.[22] By addressing the reader Taine introduced a point of view that was designed to observe the observing subject. He called attention to the fact that it was the historian's point of view that was producing an image of the past and that this way of perceiving was itself part of history. Taine adopted this perspective at various occasions throughout the book, especially when he declared his judgments as modern rational reconstructions that could not live up to the way the world was looked at during the Renaissance (see Taine [1866] 1965, vol. 2, 43, 220f.). By relating the image that organized the observed specimen into a whole to a specific point of view the narrative procedure doubled the perspectives of historical understanding. The method of observation was introduced as a means to collect aspects of the mentality of the Italian Renaissance; yet the image of the mental structure it delivered was not presented as taken from a neutral stance but rather as created by the faculties of the time-bound observing subject. In Taine's eyes, the effort to reconstruct the point of view of Early Modern Italy could serve to develop a sense of the relativity of contemporary forms of judgment and therefore reflected the value of the categories of historical science. "Voyager en critique, les yeux fixées sur l'histoire, analyser, raisonner, distinguer, [...] qu'est-ce autre chose qu'une manie de lettré et une habitude d'anatomiste?" (idem, vol. 2, 221) Accordingly, the doubling of the narrative perspective served as a device to reveal the dependence of historical knowledge on a point of view that was itself a part of historical continuity. Once Taine's historical writing is taken into account his effort to turn historical studies into a science does not seem to contradict an outlook on historical psychology that stresses the conceptual relativity of explanatory structures of the mental. Taine's historiography of the Italian Renaissance rather overtly conveyed the history of the human consciousness as a contemporary reconstruction.

---

[22] After explaining his intellectual preferences and procedures as the decisive influences on his way of understanding Taine wrote: "C'est cet instrument que j'emporte aujourd'hui en Italie; voilà la couleur de ses verres; tiens compte de cette teinte dans les descriptions qu'il produira" (Taine [1866]1965, vol. 1, 17).

## 5.4 Conclusion

Taine's belief that the particular facts of the human world should be understood as appertaining to a mental system that structured life in general does not contradict Dilthey's conception of the *Geisteswissenschaften*; rather, it corresponds with many of its aspects. Taine developed his philosophy of historical psychology in opposition to the spiritualist conception of the human mind that was dominating academic philosophy at that time. Rather than conceiving the soul as independent from empirical matter Taine wanted to study the dependence of ways of thought on specific historical circumstances thereby turning history into a science. In his opinion, the mind was made up of connected images that gave individual instances of experience its meaning. He applied his theory to the study of history arguing that historical events should be treated as embodying a mental content. Taine defined the notions of race, environment and time to disclose a relation between the mental form and its conditions and aimed at discovering psychological laws. But his terminology has misled many of his contemporary critics and later interpreters. Taine was not looking for universal laws but sought to detect psychological structures prevailing at a certain time. These structures could not be studied by themselves but only by conceiving the appearances of historical life as expressions of an inner mental form. Like Dilthey, Taine argued that it was necessary to presuppose a mental whole that gave the particular details of history its meaning even though the whole could only be reached by studying the particular. Consequently, Taine conceived the categories of historical explanation as instruments that were closer to literary descriptions than to precise measuring tools. His 'Voyage en Italy' employed a narrative form that presented historical insights as the product of an observing subject that was itself a part of history and, therefore, unable to deliver results independently from its own culture. But even if Taine's philosophy of historical science corresponded with Dilthey's thinking on various issues Taine was never led to the idea of separating the human from the natural sciences. Rather than distinguishing between 'Erklären' and 'Verstehen' Taine felt that his theory was able to prove the benefits of a single method of explanation.

## References

Auerbach E ([1946] 2001) Mimesis. Dargestellte wirklichkeit in der abendländischen literatur, 10th edn. A. Francke, Tübingen

Balzac H de ([1842] 1956) Avant-propos pour La Comédie Humaine. In: Oeuvres complètes de balzac, vol 1. Le Prat, Paris, pp 77–90

Blin G (1954) Stendhal et les problèmes du roman. Corti, Paris

Breisach E (1983) Historiography. Ancient, Medieval and Modern. University of Chicago Press, Chicago, IL

Brooks JI (1998) The eclectic legacy. Academic Philosophy and the Human Sciences in Nineteenth Century France. University of Delaware Press, Newark

Charlton DG (1959) Positivist Thought in France During the Second Empire 1852–1870. Clarendon, Oxford
Cousin V ([1853] 1854) Du vrai du beau et du bien, 2nd edn. Didier, Paris
Dewald J (2003) 'A la Table de Magny'. Nineteenth-century men of letters and the sources of modern historical thought. Am Hist Rev 108:1009–1033
Dilthey W ([1924] 1957) Ideen über eine beschreibende und zergliedernde Psychologie. In: Gesammelte Schriften, vol 5, 2nd edn. Teubner, Leipzig, pp 139–240
Dilthey W ([1927] 1992) Plan der Fortsetzung zum Aufbau der geschichtlichen Welt in den Geisteswissenschaften. In: Gesammelte schriften, vol 7, 8th edn. Teubner, Stuttgart
Donatelli B (2001) Taine lecteur de Flaubert. Quand L'histoire rencontre la littérature. Romantisme 111:75–87
Dufour P (1998) Le réalisme, de balzac à proust. Presse Universitaire de France, Paris
Gaigalas VV (1972) Ernest Renan and His French Catholic Critics. Christopher Pub, North Quincy
Gerhardi GC (1989) La philosophie de l'art de Taine: Étude des métaphores sur la nostalgie des origines. In: Drost W, Leroy G (eds) La lettre et la figure. La littérature et les arts visuels à l'époque moderne. Winter, Heidelberg, pp 45–56
Guentner W (1997) Esquisses Littéraires. Rhétorique du spontané et récit de voyage au XIXe siècle. Saint-Genouph, Nizet
Haskell F (1993) History and Its Images. Art and the Interpretation of the Past. Yale University Press, New Haven, CT
Kablitz A (2003) Realism as a poetics of observation. The function of narrative perspective in the classic French novel: Flaubert – Stendhal – Balzac. In: Kindt T, Müller H-H (eds) What Is Narratology? Questions and Answers Regarding the Status of a Theory. De Gruyter, Berlin, pp 99–135
Kahn SJ (1970) Science and Aesthetic Judgement. A Study in Taine's Critical Method. Greenwood Press, Westport
Léger F (1993) M. Taine. Paris, Critérion
Makkreel RA (1991) Dilthey. Philosoph der geisteswissenschaften. Translated by Barbara M. Kehm. Frankfurt am Main, Suhrkamp
Müller P (2008) Erkenntnis und erzählung. Ästhetische geschichtsdeutung in der historiographie von ranke, burckhardt und Taine. Böhlau, Köln
Nordmann J-T (1990) Taine et la critique scientifique. Presse Universitaire de France, Paris
Petit A (1995) Introduction. In: Ernest Renan, L'avenir de la science edited by A Petit. Flammarion, Paris, pp 7–45
Pfeiffer H (1990) Die Gegenwart der Spätantike. Christentum und Antike in Ernest Renans 'Histoire des origines du christianisme'. Poetica 22:323–354
Renan E ([1857] 1948) M. Augustin Thierry. In: Essais de morale et de critique. Œuvres complètes, vol 2. Calmann-Lévy, Paris, pp 86–108
Renan E ([1863] 1949) Vie de jésus. In: Essais de morale et de critique. Œuvres complètes, vol 4. Calmann-Lévy, Paris
Renan E ([1890] 1995) L'avenir de la science. Edited by A Petit. Flammarion, Paris
Sainte-Beuve A de ([1857] 1858) Divers écrits de M. H. Taine (9 mars 1857). In: Causeries du lundi, vol 13. Garnier Frères, Paris, pp 204–232
Sainte-Beuve A de ([1862] 1870) Chateaubriand. Jugé par un ami intime en 1803 (28 Juillet 1862). In: Nouveaux lundis, vol 3, 2nd edn. Calmann-Lévy, Paris 15–33
Sainte-Beuve A de ([1864] 1867) Histoire de la littérature anglaise par M. Taine (30 Mai 1864). In: Nouveaux lundis, vol 8. Calmann-Lévy, Paris, pp 66–89
Sainte-Beuve A de ([1859] 1955) Port-royal, vol 3. Edited and annotated by Leroy M Gallimard, Paris
Schmauss W (2004) Rethinking Durkheim and His Tradition. Cambridge University Press, Cambridge
Seys P (2000) Taine, inventeur de Stendhal. Lettres Romanes 64:13–31

von Srbik HR (1951) Geist und geschichte vom deutschen humanismus bis zur gegenwart, vol 2. Bruckmann, München
Taine H ([1856] 1994) Essaie sur tite-live. Economica, Paris
Taine H (1857) Les philosophes français du XIXe siècle. Hachette, Paris
Taine H (1858) Michelet. In: Essais d'histoire et de critique. Paris, Hachette, pp 311–360
Taine H (1863) Histoire de la littérature anglaise, vol 1. Hachette, Paris
Taine H ([1864] 1878) Le positivisme anglais. 2nd ed. Germer Baillière, Paris
Taine H (1866) Préface. In: ibid., Essais de critique et d'histoire, 2nd edn. Hachette, Paris, pp I–XXVII
Taine H ([1866] 1965) Voyage en Italie. 2 vols. Juillard, Paris
Taine H ([1870] 1878) De l'intelligence, corrigée et augmentée. 2 vols. Hachette, Paris
Taine H (1904) Vie et correspondance, vol 2. Hachette, Paris
Taine H (1907) Vie et correspondance, vol 4. Hachette, Paris
Thiher A (2001) Fiction Rivals Science. The French Novel from Balzac to Proust. University of Missouri Press, Columbia

# Chapter 6
# Understanding and Explanation in France: From Maine de Biran's Méthode Psychologique to Durkheim's Les Formes Élémentaires de la vie Religieuse

**Warren Schmaus**

## 6.1 Introduction

My task here is to compare the ways in which the relations between the human and the natural sciences were conceived in late nineteenth and early twentieth century France and Germany. Historical generalization may be a mug's game. But if I had to generalize, I would say that the French distinguished the human or cultural sciences from the natural sciences only in terms of their subject matters, while the Germans were more likely to try to distinguish them in terms of their goals, methods, foundations, and normative content as well. Although we may be able to find many philosophical positions among the French that resemble certain aspects of the thought of Wilhelm Dilthey, Wilhelm Windelband, or Heinrich Rickert, no one in France held exactly the same combination of philosophical views concerning the human sciences as that held by any of these German thinkers. In particular, no one in France tried to distinguish the human from the natural sciences in terms of understanding versus explanation in the way that Dilthey did. Thus, although there were other disputes in France in regard to the human sciences, such as that between Émile Durkheim and Gabriel Tarde over the role of psychology in sociological explanation, or that between sociologists and philosophers over the methods of ethics, there was no controversy analogous to the conflict among Dilthey, Windelband, and Rickert over the best way to distinguish the human from the natural sciences.

At least through the 1890s, Dilthey defended a program for a descriptive, as opposed to an explanatory, psychology that bears some resemblance to the introspective psychology of the French spiritualist philosophers. Unlike Dilthey, however, the French did not regard this sort of psychology as providing a foundation for the human sciences that distinguished them from the natural sciences. In addition, although the French may have distinguished the search for general laws in science from an emphasis on particular descriptions in history in ways that bring to mind Windelband and Rickert, they did not perceive this distinction as entailing any methodological

W. Schmaus (✉)
Lewis Department of Humanities, Illinois Institute of Technology, Chicago, Illinois, USA
e-mail: schmaus@iit.edu

U. Feest (ed.), *Historical Perspectives on Erklären and Verstehen*, Archimedes, Vol. 21, DOI 10.1007/978-90-481-3540-0_6, © Springer Science+Business Media B.V. 2010

distinctions in the way that Windelband at least claimed it did. The French recognized methodological distinctions among the various academic and scientific disciplines, but these cut across distinctions in terms of goals and subject matter, and distinctions among goals did not coincide with distinctions in subject matter.

We can also find parallels between Dilthey's later hermeneutic methodology and Durkheim's methods in *The Elementary Forms of Religious Life* (Durkheim [1912] 1985). Both thinkers appear to have increasingly emphasized the interpretive character of the sciences of humanity. However, where Dilthey saw this as entailing a fundamental methodological distinction between the *Geistes* and the *Naturwissenschaften*, Durkheim did not. For Durkheim, the method of interpretation was but a species of the general methods of induction and hypothesis testing that the social shared with the natural sciences. In *The Elementary Forms*, Durkheim set out to understand primitive religion in order to explain contemporary religion. Thus, far from explanation and understanding being conflicting goals, achieving one was conducive to achieving the other. Indeed, it is this very way of melding explanation with understanding that distinguishes Durkheim's sociology of religion and knowledge from that of Lucien Lévy-Bruhl. Lévy-Bruhl shared with Durkheim the notion of modeling the social sciences on the natural sciences. However, throughout most of his career Lévy-Bruhl held that there was a radical break between so-called primitive and modern thought, such that we could only *explain* the things that primitives say and do but not hope to *understand* them. Durkheim, on the other hand, held that primitive was continuous with modern thought in a way that would make the interpretation of primitive thought possible.

## 6.2 The Méthode Psychologique and Dilthey's First Views on the Method of Verstehen

One thing that distinguished the French philosophical community from the German during this period was its greater institutional unity. Philosophy was dominated by professors who were trained in Paris. They all received a similar training not only at the university level, but even earlier, at the *lycée* as well. All students in France were required to take a standard 1-year course in philosophy during their final year at the *lycée. Agrégés* in philosophy typically began their careers by teaching this course, helping students to prepare for their *baccalauréat* and thus to obtain access to higher education. The curriculum for this course was set by a committee in Paris and published by the Ministry of Public Instruction. This course took a systematic approach to philosophy, beginning with a philosophical psychology and then proceeding through logic, which included methodology, and then ethics and metaphysics. This systematic philosophy was laid out for students in textbooks, typically written by members of the curriculum committee. By perusing the chapters on method in these textbooks, one can obtain a fairly good idea of what philosophers in the Third Republic were able to take for granted.

I will draw on Paul Janet's *Traité élémentaire de philosophie à l'usage des classes*, first published in 1879 but reissued in many editions, right up until the early 1950s. Although perhaps less well-known than his famous nephew Pierre Janet, Paul Janet was an important and powerful figure in French philosophy. Early in his career he had been the amanuensis of Victor Cousin. His philosophy continued to bear the mark of this early influence as he rose to the chair in the history of philosophy at the *Faculté des lettres*. He was on the committee that drafted the *lycée* philosophy syllabus. He also served on the dissertation examination committees for many doctoral candidates in philosophy.[1]

Janet, following John Stuart Mill, divided what he called the "moral sciences" from the natural sciences in terms of subject matter.[2] Janet then subdivided the moral sciences into philosophy, which included philosophical psychology, logic and methodology, metaphysics, and moral philosophy; the social sciences, which included politics, jurisprudence, and political economy; the philological sciences; and the historical sciences. However, he did not see the distinction between the moral and the natural sciences as coinciding with any methodological distinctions. The moral sciences employ a mixture of deductive or demonstrative and inductive, experimental methods, in varying proportions according to the nature of the sciences (Janet 1883, 487ff.). Thus the methods of the moral sciences overlap with those of the natural sciences. According to a more detailed textbook, *Leçons de Philosophie*, by Élie Rabier, the Director of Secondary Education in France, the important methodological distinction was not between the natural and the moral sciences but between the empirical sciences, including physics, psychology, and history, and the analytic sciences, including geometry, moral philosophy, and politics (Rabier [1886] 1909, vol. II, 316).

Through much of the nineteenth century, both empirical and analytic methods were also distinguished from a method peculiar to philosophy, the "*méthode psychologique*." For Cousin, this term had two meanings: it referred both to a method of direct, unmediated perception of the self and to the grounding of all philosophy in this self-perception (Schmaus 2004, 61) The French referred to the first sense of this method variously as "*réflexion*," "*aperception*," or "*sens intime*." This method had its source in the philosophy of Pierre Maine de Biran, for whom it was a kind of self-perception that is fundamentally different than the perception of external objects in space:

> By the internal apperception or the first act of reflection, the subject distinguishes itself from the sensation or the affective or intuitive element localized in space, and it is this very distinction that constitutes the fact of consciousness, personal existence. (Maine de Biran 1982, vol. II, 324)

[1]For more details on Janet's philosophy and career, see Schmaus (2004) and Brooks (1998).

[2]Mill's *A System of Logic* was translated into French by Jean Louis Hippolyte Peisse and first published in 1866 in Paris by Ladrange. Although Janet did not cite Mill in his classification of the sciences, he did cite the French translation of Mill's *Logic* elsewhere in the *Traité*, for example, in his account of Mill's methods of eliminative induction.

Although some of Biran's terminology brings Kant to mind, there are important differences between Biran's philosophy and anything we can find in the *Critique of Pure Reason*. Biran's internal apperception is neither Kant's empirical apperception, through which one knows oneself only as internal appearance that can be subsumed under the categories, nor his transcendental apperception, through which one is conscious of oneself neither as appearance nor as noumenal self, but simply that one is. It is more like the intellectual faculty in Kant's 1770 Latin inaugural dissertation, *On the Form and Principles of the Sensible and Intelligible World*, in that it is supposed to reveal the self as it really is in itself, and not merely as phenomenon or appearance. Biran did not read German, and drew most of his knowledge of Kant's philosophy from either this dissertation, which is not representative of Kant's later critical philosophy, or the secondary literature written in French.

There are also Cartesian, Leibnizian, and Condillacian elements in Biran's concept of apperception. Like Descartes, Biran thought that this method of internal reflection revealed the self as thinking substance, but perhaps more like Leibniz, he thought of it as more fundamentally revealing the self as active cause. This sense of the self as active cause derives from reflecting upon the experience of willed effort, which Biran, like Condillac, held to be the most fundamental experience. Biran grounded the category of causality in this experience of willed effort. Indeed, he held that all the categories that are fundamental to our knowledge of the external world are derived from internal experience (Schmaus 2004, Chapter 3). This psychological method was adopted and promulgated by Cousin, who compared this foundational philosophical psychology to the rational psychology of the Germans (Cousin 1860, 34). Janet (1883, 493) also compared it to the methods of Descartes, Plato, Malebranche, and Hegel.

There are parallels between Biranian apperception and Dilthey's appeal to lived experience in his descriptive psychology. In his 1894 essay, *Ideas concerning an analytical and descriptive psychology* (*Ideen über eine beschreibende und zergliedernde Psychologie*) (Dilthey [1894] 1977), Dilthey drew a distinction between an explanatory (*erklärende*) psychology following the methods of the natural sciences and a descriptive and analytic (*beschreibende und zergliedernde*) psychology of lived experience. This psychology of lived experience was to serve as the foundation of the *Geisteswissenschaften* and to employ a method of *Verstehen* that is completely different than the method of *Erklären* in the natural sciences. In the natural sciences, we are able to perceive only coexistence and succession. Causal connections among the phenomena must be introduced through hypotheses that take us beyond what we can experience. In the human sciences, on the other hand, the connections among our lived experiences can be directly experienced, making hypotheses unnecessary. For Dilthey, this methodological difference between the natural and the human sciences is due to a difference in their objects of study:

> The human studies are distinguished from the sciences of nature first of all in that the latter have for their objects facts which are presented to consciousness as from outside, as phenomena and given in isolation, while the objects of the former are given originaliter from within as real and as a living continuum [*Zusammmenhang*]. (Dilthey [1894] 1977, 143, in transl. of 1977, 27)

In his review of Dilthey's *Ideen*, Hermann Ebbinghaus argued that Dilthey had not managed to avoid the use of hypotheses in psychology, but relied on them in relating part to whole (Harrington 2000, 440). Dilthey's reply to these charges is contained in a "Remark" that is appended to later editions of the *Ideen*.[3] Here he did not so much deny any use of hypotheses in psychology as argue that they were unnecessary to reveal the "structural nexus" of lived experience. He had no objection to supplementing the results of an analytic and descriptive psychology with "hypotheses of independent units of sensation, of psycho-physical parallelism, of determinism, of unconscious representations" (Dilthey [1894] 1977, 238, transl. of 1977, 118). For Dilthey, it seems, it was methodologically appropriate to make use of hypotheses concerning postulated entities that exist outside experience. But they were unnecessary for understanding the relationships that held among the parts within our experience, since these could be immediately known.

Dilthey held throughout his career the notion that the connections among lived experience are known directly. In a later essay, the *Formation of the historical world in the human sciences* (*Der Aufbau der geschichtlichen Welt in den Geisteswissenschaften*) (Dilthey [1910] 1961), he said:

> The way in which a lived experience is there for me is completely different from the way in which images stand before me. The consciousness of a lived experience is one with its nature, its being-there-for-me and what in it is there for me are one. The lived experience does not stand over against an observer as an object, but its existence for me is indistinguishable from what in it is there for me. ((Dilthey [1910] 1961, 139), quoted in Makkreel (2003, 156))

Dilthey's notion that the structure or interconnectedness of one's lived experience can be known directly brings Biran's apperception to mind. That is, both Biran and Dilthey believed that one's inner self or life is known in some way that is not mediated by appearances, phenomena, intuitions, images, or any other such mental entity. However, where Biran's position reflects an incomplete knowledge of Kant's critical philosophy, Dilthey's attests to his rejection of much of the Kantian program. He distanced himself from Kant even to the point of arguing that epistemology could be grounded in the psychology of lived experience (Dilthey [1894] 1977, 150, transl. of 1977, 34).

Dilthey's interpretive psychology aims at a description and an analysis of the structure or interconnectedness (*Zusammenhang*) of lived experience. What gives experience this interconnectedness for Dilthey appear to be intentions or purposes. These intentions must be understood within the context of the person's life as a whole. One begins by understanding how one's own experiences hang together through intentionality, and then proceeds by analogy to understand how someone else's experiences hang together. For Dilthey, this inference by analogy differs from inductive inference in that it proceeds from particular to particular instead of from particular to universal, and is only an implicit, not a conscious, explicit inference (Harrington 2000; Makkreel 2000).

---

[3]Dilthey's reply to Ebbinghaus was first published in 1896 as part of his *Beiträge zum Studium der Individualität* (Makkreel 1975, 207).

Janet also defended something resembling Dilthey's method of *Verstehen,* combining introspection with drawing analogies between oneself and others. Much as Dilthey had distinguished a descriptive from an explanatory psychology, Janet distinguished what he called a "subjective" from an "objective psychology." Janet's subjective psychology resembles Dilthey's descriptive psychology, insofar as it employs a method of internal observation. This method could penetrate beyond internal phenomena and reach the very subject of experience itself. It is thus different than observation in the natural sciences, in which the phenomena are observed only *du dehors*, as effects of unknown causes, and for which we cannot penetrate their essence. Janet's distinction between subjective and objective psychology is not so much an epistemological distinction as a distinction in terms of subject matter and method. Subjective psychology was no less certain than objective psychology, as it was not simply the results of introspection by a single individual but of many philosophers practicing introspection who could correct one another's results. However, Janet thought an introspective psychology was incomplete by itself and must be joined with an objective psychology, which includes empirical studies of animal, physiological, pathological, and ethnological psychology.[4] On the other hand, this objective psychology would be unintelligible and even impossible if it were not founded upon subjective psychology, since it is only through analogy with ourselves that we are able to understand what is happening in the minds of others (Janet 1883, 488–490).

There are additional parallels between the French spiritualists and Dilthey besides the idea that we have direct, unmediated experience of our inner lives. Like Biran, Dilthey thought that our reflexive awareness of the will is more fundamental than our experience of external objects (Makkreel 1993, 432), and that our notion of causality derives from this source and not from the external world (ibid., 436). For Dilthey, just as for the French spiritualists, this inner experience revealed the categories that structure internal as well as external experience.

> If I start with inner experience, then I find the whole external world to be given in my consciousness and all the laws of nature to be subject to the conditions of my consciousness and, therefore, dependent on them. (Dilthey 1883, Einleitung in die Geisteswissenschaften, quoted by Makkreel 1993, 425)

As he explained in the *Ideen*, "the concepts of the unity of a diversity, of parts of a whole, of causal relations" are formulated on the basis of lived experience. These concepts allow us to understand nature when we apply them under conditions of coexistence and succession (Dilthey [1894] 1977, transl. of 1977, 53). Similarly, the concept of a functional system is taken from our inner lives and applied to botany and zoology (ibid., 56). Dilthey's notion that we can directly experience the causality of the will may even have had its source in Biran. He was certainly aware of Biran's

[4]Janet took the need for a physiological psychology sufficiently seriously that he encouraged his nephew, Pierre Janet, to go to medical school after completing his philosophical studies at the École normale supérieure, to enable him to pursue research in this area.

work. In several places, Dilthey distinguished three major world views in philosophy: the materialist, naturalist, or positivist, which runs from Lucretius up through Comte and Avenarius; objective idealism, which runs from Heracleitus through Hegel; and subjective idealism or the idealism of freedom, which runs from Plato through Kant and the French spiritualist tradition of Maine de Biran, Cousin, and Henri Bergson.[5]

In spite of these parallels between Dilthey's thought and French spiritualism, and although later spiritualists, such as Bergson, continued to seek some sort of basis for philosophy in inner experience, no one in France seems to have thought that this method of internal observation provided a distinct grounding for the historical or human sciences in the way that Dilthey thought that it did. Bergson himself regarded the social and cultural sciences as parts of biology (Gayon 2005). Although we can find methodological views about psychology in France that resemble Dilthey's in various ways, they never seem to have been a part of any debates over the goals, methods, or foundations of the cultural or human sciences analogous to the debates that arose in Germany among Dilthey, Windelband, and Rickert.

To be sure, Tarde held that sociology as well as psychology used a method of introspection. Durkheim quoted Tarde as follows:

> In sociology, we have, by a singular privilege, the intimate knowledge of the element that is our individual consciousness (*conscience*) as well as the compound that is the assemblage of these consciousnesses (Tarde, 1894, La Sociologie élémentaire. Annales de l'Institut internationale de sociologie 1, 222; quoted in Durkheim [1897] 1988, 350f.).

Durkheim found Tarde's claim to be

> a bold negation of all of contemporary psychology. It is generally understood today that psychic life, far from being able to be known by an immediate vision, has, on the contrary, profound depths where the internal sense (*le sens intime*) is not able to penetrate and that we can attain only bit by bit by roundabout and complex procedures, analogous to those used by the sciences of the external world. (Durkheim [1897] 1988, 351)

Similarly, as Durkheim explained in a letter to his colleague Celestin Bouglé, "Society is not in any individual, but in all of the individuals associated under a determinate form. It is thus not through the analysis of the individual consciousness that one will be able to do sociology" (6 July 1897 [(Durkheim 1975, vol. II, 400)]). Years later, in a discussion at a meeting of the *Société française de philosophie* of the role of the unconscious in historical explanation, Durkheim claimed that psychologists had given up on introspection because it could reveal only facts and not the causes of these facts, hidden in the unconscious. If psychological facts

[5] Dilthey, *Das Wesen der Philosophie* (1907), *Gesammelte Schriften* V, 402–404, excerpted and translated in Hodges ([1944] 1969, 152ff.); see also *Die Drei Grundformen der Systeme in der Ersten Hälfte des 19 Jahrhunderts* (1900?), *Gesammelte Schriften* IV, 528–554, cited in Hodges (1952, 90); *Die Typen der Weltanschauung und ihre Ausbildung in den metaphysischen Systemen* (1911), *Gesammelte Schriften* VIII, 75–118, trans. in Dilthey (1976, 133–154); see also Ermarth (1978, 329ff).

cannot be studied through introspection, he argued, it must be even truer that social facts cannot be studied this way, since their causes escape the consciousness of the individual even more (Durkheim et al. [1908] 1975, 230).

But although Tarde, like Dilthey, defended the use of introspection in the human sciences, there were other ways in which Tarde's position was quite different than Dilthey's. The goal of Tarde's introspective method was not understanding in anything like Dilthey's sense, but rather explanation. Tarde, like Durkheim, took the natural sciences as his model for the human sciences, seeking general explanatory laws. But where Tarde believed that social phenomena could be explained by the psychological laws of imitation, Durkheim held that sociological explanations were irreducible to psychology (Durkheim [1895] 1927, 128). Tarde regarded Durkheim's anti-reductionist position as "pure ontology," and compared their dispute to the scholastic debates between nominalism and realism, with Tarde defending the nominalist position (Durkheim and Tarde [1904] 1975, 87).

## 6.3 Normativity and the Cultural Sciences

Although the polemics between them appear to have heated up only after the turn of the twentieth century, Windelband had already laid out a position in his 1894 Strasbourg rectorial address to which Dilthey felt compelled to reply in a later edition of his *Einleitung in die Geisteswissenschaften* (Oakes 1980). According to Windelband, Dilthey's distinction between the *Geisteswissenschaften* and *Naturwissenschaften* is only a substantive difference or a difference in subject matter, and does not entail a difference in method. Windelband regarded psychology for instance as a *Geisteswissenschaft* from the point of view of subject matter, but a *Naturwissenschaft* from the point of view of method. As a Kantian, Windelband rejected the notion of a psychology based on direct experience of the activity of the will. Psychology, he thought, was restricted to the methods of the natural sciences. However, Windelband did not think that the natural sciences shared a single method that set them apart from humanistic inquiries. The objects of each of the natural sciences call for specialized empirical methods unique to each. From this point of view, he said, psychology is as different from chemistry as mechanics is from biology (Windelband [1894] 1980, 173f.). Rickert also maintained that the natural sciences do not all pursue a single method (Adair-Toteff 2003, 39).

The difference between the natural and the cultural sciences for Windelband is perhaps more accurately described as having to do with goals, rather than methods. The cultural sciences, in particular history, are more interested in providing exhaustive descriptions of particular events than in achieving knowledge of general laws. Thus the cultural sciences were typically what he called "idiographic" rather than "nomothetic" sciences. The nomothetic sciences seek general laws of the phenomena, pursuing "general, apodictic judgments" or "necessary connections," or "what is invariably the case." The idiographic sciences, on the other hand, seek a description of a single process or event, resting content with a mere "assertoric proposition" describing "what was once the case" (Windelband [1894] 1980, 174f.). Similarly, Rickert, drawing on Windelband's nomothethic/idiographic distinction, distinguished "generalizing" from "individualizing" sciences (Outhwaite 1975, 38ff.). However, Windelband emphasized that this was not a distinction in terms of subject matter, as one could pursue either idiographic or nomothetic goals in the same field of inquiry. This is because the distinction between what is invariable and what is unique or particular is relative to one's time frame. Windelband provided the example of the study of a language. Certain forms may appear invariable for a certain time-span, but in the overall history of human languages, this particular language along with its "laws" is only a unique and transitory phenomenon. Windelband added that one could make similar points about natural sciences such as physiology, geology, and even astronomy. For example, in biology, systematic taxonomy has a nomothetic character, while evolutionary history is an idiographic discipline. Windelband referred to the distinction between idiographic and nomothetic sciences as a "methodological" distinction, as he seems to have thought that this distinction in goals entailed a difference in method as well (Windelband [1894] 1980, 175f.). That is, he appears to have thought that since the goal of the nomothetic sciences was to achieve general laws, they must proceed by a method of inductive generalization, which would be an inappropriate method for the idiographic sciences. But of course, to use Windelband's own examples, one could seek both laws of taxonomy and particular reconstructions of evolutionary histories through the same method of hypothesis and test. Thus the distinction between nomothetic and idiographic is a distinction regarding only epistemic goals and neither method nor subject matter.

Rickert recognized that the distinction between generalizing and individualizing did not suffice to separate the cultural from the natural sciences and proposed an alternative. According to Rickert, what sets these sciences apart is that human values determine what are significant topics for inquiry in the cultural sciences, but not in the natural sciences (Makkreel 1999, 559; Outhwaite 1975, 38ff.). This distinction is difficult to maintain in the light of recent work in the history, philosophy, and sociology of science. Philip Kitcher, for example, argues that there is no way to define epistemic significance independently of human concerns, even in the natural sciences (Kitcher 2001, Chapter 6). Windelband ([1894] 1980, 182) also drew a connection between the cultural sciences and human values, as he held that our sense of values is grounded in the particular and the unique. Thus it appears that the idiographic sciences are appropriate for investigating questions concerning human values. However, Windelband's way of drawing this connection appears dangerously close to committing the naturalistic fallacy, the attempt to establish normative judgments on an empirical basis.

Perhaps a more defensible way to draw the distinction that these German philosophers were seeking might be to point to the role that values play in explanation. In the humanities, the actions of individual agents are often explained in terms of their beliefs and desires. These beliefs may include such things as moral and cultural values as well as factual beliefs. There is a second role that values play in humanistic explanation, as well. When we explain a person's actions in terms of her beliefs and desires, we are saying that those actions were *rational* in the light of these beliefs and desires. The ascription of rationality to the agent is itself a value judgment, one that is not relevant to naturalistic explanations. Even in an explanation as simple as saying "She reached for the glass of water because she was thirsty," we are making at least an implicit value judgment, assuming the agent is not acting irrationally. However, if we draw the distinction between the natural and the cultural sciences in this way, we are making a distinction in terms of content, not method. The very ascription of beliefs, desires, and rationality to an agent involves the making of hypotheses that are subject to revision in the light of additional evidence.

Based on a search I conducted through the *Revue Philosophique*, the *Revue de métaphysique et de morale*, and *L'Année Sociologique* for the relevant years, it appears that the debates among Dilthey, Windelband, and Rickert were largely ignored in France, at least at the time they were taking place in Germany. Dilthey and Rickert are only briefly mentioned in the *Revue Philosophique,* which provided one or two paragraph summaries of their articles in German journals such as *Philosophische Studien*, *Vierteljahrsschrift der wissenschaftlichen Philosophie*, *Archiv für Geschichte der Philosophie*, *Archives de Psychologie*, and *Zeitschrift für Psychologie.* I found only a single article concerning understanding *(comprendre)* and explanation *(explication)*, an article that appeared in 1914 in the *Revue philosophique*, in which the author, René Paucot, argues that they cannot even be equated in the natural sciences (Paucot 1914). I found nothing in the *Revue de métaphysique et de morale. L'Année sociologique* appears to have entirely ignored the philosophical debates going on in Germany over the methods of the social and cultural sciences, although they reviewed plenty of other German works, dealing with criminology, political economy, ethnography, geography, and other social sciences.

Nevertheless, we also find in French thought a distinction resembling Windelband and Rickert's between the nomothetic, generalizing and the idiographic, individualizing sciences. However, there were two important differences: First, where Windelband saw the nomothetic sciences as aiming at necessary truths, in France, due to the work of philosophers of science such as Claude Bernard, Émile Boutroux, Henri Poincaré, and Pierre Duhem, there was a greater sense of the contingent and hypothetical character of scientific laws. Second, where Windelband characterized this as a "methodological" distinction, the French regarded it as a distinction concerning only the goals of inquiry, and did not see it as entailing any fundamental methodological distinction.

For instance, Janet regarded history as an inductive science, grounded in testimony. History used a special sort of induction, proceeding not from particular facts to general laws, but from particular facts to other particular facts. Specifically, one infers from one sort of class of facts, testimony, to another, events. According to Janet, "The historical method … can be reduced to the inductive method" (Janet 1883, 508). However, this special sort of induction does not divide the human from the natural sciences for Janet, as he saw geology as using the same method. According to Janet, this method of induction was a method of interpretation: "In a word, if induction, as Bacon says, is the interpretation of nature, the historical method is the interpretation of human testimony" (Janet 1883, 509). Durkheim defended a similar position early in his career as a philosophy instructor at the *Lycée de Sens.* According to student notes from his course, Durkheim held that while the sciences study laws in abstraction, history shows their working out in particular times and places. Nevertheless, historical reconstruction, just like the search for general laws, proceeds by methods of induction and imaginative hypotheses (Durkheim [1883–1884] 2004, 218–220).

The French during this time also debated the relationship between human values and the sciences. However, whereas for Windelband and Rickert the cultural sciences in general are concerned with values, the French debated whether values are the special province of philosophy, distinct from history and the social sciences. Consider the controversy stirred up by Durkheim's *De la division du travail social* (Durkheim [1893] 1986). Durkheim had originally submitted this work as his philosophy dissertation.[6] In it, he made use of empirical methods in order to answer the question whether specialization was desirable from a moral point of view. Drawing on evidence such as ancient and medieval codes of law, he showed that a new, organic form of social solidarity was replacing an older, mechanical form of social solidarity. The philosophers on his examination committee had no problem with this conclusion. What they objected to was Durkheim's attempt to draw moral implications from it. Henri Marion told Durkheim that his thesis did not "reach morality," but was only a "physics of mores (*physique des moeurs*)." Similarly, Paul Janet argued that Durkheim's thesis could in no way establish that people in contemporary society actually had a moral duty to specialize. In short, the committee had no philosophical objection to a science of moral facts modeled on the methods of the natural sciences, as long as one did not try to draw ethical conclusions from it (Muhlfeld [1893] 1975, 441f.).

Although they did not use the term, these philosophers in effect criticized Durkheim for committing the naturalistic fallacy. However, to say that the naturalistic fallacy is indeed a fallacy is to presuppose that naturalists are trying to do what previous moral theorists have tried to do, which is to provide some basis from which one can logically derive moral rights and duties. It assumes that the naturalist is simply substituting empirical premises for premises grounded in reason or intuition in such a derivation. But it is not at all clear that Durkheim was seeking this sort of empirical foundation for ethics. Horner (1997) points out that there were French philosophers during this period such as Lucien Lévy-Bruhl and Frédéric Rauh who were arguing for a non-foundationalist ethics. According to Durkheim's favorable review of Lévy-Bruhl's *La Morale et la science des moeurs* (1903), Lévy-Bruhl argued that the very idea of an abstract, theoretical science of morals is incoherent. What passes for moral philosophy simply reflects in abstract language

[6] To be precise, it was his French thesis. Candidates for the *doctorat d'état* were also required to submit a Latin thesis. Durkheim's examination committee raised only minor objections to his Latin thesis, which was about Montesquieu (Muhlfeld [1893] 1975, 440).

the moral ideas of the day. There can be no absolute basis for morality. However, there can be an empirical science of the moral phenomena of various societies that describes, compares, and provides causes. This can be used as a guide for making practical moral decisions, by analogy with the way in which medicine uses physiology. Durkheim found Lévy-Bruhl's approach to practical decision-making no worse than a Kantian or utilitarian approach. He argued that any time one attempts to derive a specific practical conclusion from one of these philosophies, one is taking a risk, and at best reaches only uncertain and approximate conclusions (Durkheim [1904] 1969). Throughout his career, Durkheim similarly sought to replace philosophical ethics with an empirical, sociological approach to morality. He was pursuing this goal towards the very end of his life in a book to be called *La Morale*. However, he never finished this work. All that survives is the first, introductory chapter (Durkheim [1920] 1975).

## 6.4 Hermeneutics and the Method of Interpretation

According to scholars such as Rudolf Makkreel, Dilthey appears to have modified his original position in the last decade of his life. He did not abandon the psychology of lived experience, but he no longer thought it sufficient for achieving understanding. He realized that there could be no such thing as a foundation for the human sciences in descriptive psychology, as all psychological claims are always open to re-interpretation. Thus Dilthey came to the conclusion that our inquiries should begin not with a psychological description of inner experience, but with outward forms of expression that are public and available to sense experience. Dilthey began to place new importance on the ways in which an individual's expressions of his or her thoughts are permeated with cultural commonalities. These expressions had to be interpreted, and this called for the hermeneutical method (Makkreel 1977, 12; 1999, 566; 2003, 157).

Lanier Anderson characterizes Dilthey's hermeneutic method in terms of the making of hypotheses about the whole, then using them to interpret the parts of the whole, and then using these latter interpretations to revise the hypothesis about the whole, in an endless cycle:

> The hermeneutic method approaches holistic cultural meanings by its famous circular procedure: first, the interpreter projects a hypothesis about the meaning of the whole, which she uses as background for understanding each part in turn; but the initial hypothesis is only tentative, and the interpreter allows her gradual discoveries about the meanings of the parts to influence her hypothesis about the whole, revisions of which, in turn, once again affect the way she sees the parts. Hermeneutic procedure consists in this repeated mutual adjustment, aiming at interpretive equilibrium. (Anderson 2003, 228f.)

For example, Dilthey spoke in the *Aufbau* about how memos, speeches, decisions, and other outward forms of expression can be used in the interpretation of political life, and how the more different situations from which such documents are drawn the better our understanding will be. But no matter how many we acquire, we will still achieve only a partial view of the entire political landscape (Dilthey [1910] 1961, 319–322, transl. of 1961, 77). Thus our historical interpretations resemble nothing so much as our theories in the natural sciences. The more evidence we have, the better our interpretations or theories, but no matter how much evidence we acquire, we never achieve more than a kind of tentative knowledge, that is subject to revision in the light of further evidence.

The major difference between Dilthey's historical hermeneutics and the method of hypothesis in the natural sciences is that the hermeneutical method involves using evidence to test and revise a particular interpretation of an historical event, whereas the natural sciences use evidence to test and revise putative general laws. To be sure, Dilthey also saw the natural sciences as making use of a method of interpretation. However, he thought that the objects of interpretation in the natural sciences are the theoretical constructs of previous scientists, whereas in the humanities, they are the constructs of ordinary members of society (Harrington 2001, 325). However, this is a distinction only in subject matter, and not a methodological distinction.

## 6.5 Durkheim's The Elementary Forms of Religious Life

To the extent that Dilthey's method of hermeneutical interpretation can be characterized in terms of making tentative hypotheses about the whole that are refined through an analysis of the parts, it appears to be not all that different than the method that Durkheim employed in *The Elementary Forms.* On the face of it, no two social thinkers would appear to be more different than Durkheim and Dilthey. Durkheim rejected those aspects of French thought that bore the closest resemblance to Dilthey's philosophy of the *Geisteswissenschaften*, refusing to ground the social sciences in any sort of psychology and dismissing from his sociology explanations in terms of the intentions of individual agents (Durkheim [1895] 1927, 111–113). For Durkheim, there was to be no radical separation between the human and the natural sciences. Both pursued the epistemic goal of explanation and both made use of the method of hypothesis and test. Indeed, the motivation behind his sociology was his desire to transcend the traditional academic divide between the sciences and the humanities, or at least philosophy. Not only did he seek to provide a sociological ethics, but also, in *The Elementary Forms*, he pursued an empirical, sociological inquiry into the formerly metaphysical question concerning the nature of god and the epistemological question concerning the nature of the fundamental categories of thought.

Nevertheless, in spite of his differences from Dilthey, Durkheim bequeathed to us a highly interpretive social science. We find a similar emphasis on cultural commonalities in both thinkers, especially in their later works. For Durkheim, these cultural commonalities were to be explained in terms of his hypothesis of collective representations. This term had a double meaning for him, referring both to inner mental states shared by the members of the same society and to public expressions of these mental states, ranging from flags and emblems to the written and spoken word. Through the analysis of these public expressions, the social scientist comes to understand a culture's shared beliefs.

For Durkheim, the epistemic goals of understanding and explanation are not distinct but complementary. There are two explanatory goals specific to *The Elementary Forms*: to analyze and explain the simplest known religion and to explain the genesis of the most fundamental categories of thought. Sociology pursues these goals not as ends in themselves, as one would in history or ethnography, but in order to understand the present (Durkheim [1912] 1985, 1f.). By showing that our present modes of thought have their origin in primitive thought, Durkheim was establishing a continuity between them, making it possible for us to interpret and understand primitive thought.

Durkheim thought that the only way to understand present religion is to resolve it into its simplest elements by tracing it back to its origins (Durkheim [1912] 1985, 4). He analogized this method of historical analysis to the use of experiments in the natural sciences. The physicist designs his experiments in such a way as to simplify the phenomena in which he is interested in order to discover the laws that relate them. Early religions, being simpler than contemporary religions, are like naturally occurring experiments at the dawn of history (Durkheim [1912] 1985, 11). Durkheim characterized his study of indigenous Australian religions as a "well-made" (Durkheim [1912] 1985, 135) or "well-defined" experiment (Durkheim [1912] 1985, 594). He explained this concept by contrasting it with what he perceived to be James Frazer's method of simply collecting as many facts as possible without care for the social environment from which they come. What mattered for Durkheim is not the number, but the value of the facts. Facts cannot be compared just because they resemble each other; one must take care that they are from societies with the same type of social organization (Durkheim [1912] 1985, 132–135). An induction based on a well-defined experiment, Durkheim maintained, "is less risky than summary generalizations that try to attain at one blow the essence of religion without resting on the analysis of any particular religion" (Durkheim [1912] 1985, 594).

Although Durkheim took the natural sciences as his model for the social sciences, he was no positivist. That is, he did not conceive scientific explanation as restricted to providing the laws governing the phenomena, in the way that philosophers such as Hume and Comte did. To give an explanation of a phenomenon is to give the cause of it. And a cause is not simply the antecedent of an empirical generalization, but more like an underlying essence that could be known only through offering hypotheses and successively refining them in the light of empirical evidence.[7]

Durkheim's method for uncovering the underlying essence of religion bears a very close resemblance to the hermeneutic method of successively refining one's hypotheses about the whole of some phenomenon in light of one's hypotheses about its parts, and vice versa. The first chapter of *The Elementary Forms* begins by

[7]I defend this interpretation of Durkheim in great detail in Schmaus (1994).

seeking a provisional definition of religion, since according to Durkheim one cannot arrive at the "profound and truly explicative characteristics of religion" except at the end of one's research (Durkheim [1912] 1985, 31f.). This provisional definition should indicate a number of easily perceptible external signs that would allow us to recognize religious phenomena wherever they are met with and not confuse them with others (Durkheim [1912] 1985, 32). Durkheim proceeded by analyzing the whole of religion into its two principal parts: beliefs, which consist in collective representations, and rites, which are modes of action. Rites can be distinguished from other prescribed modes of action, such as moral rules, only by their object. Since the object of a rite is expressed in a belief, he argued, one can define the rite only after having defined the belief. Now according to Durkheim, all known religious beliefs presuppose a classification of things into two opposed classes: the sacred and the profane (Durkheim [1912] 1985, 49f.). Thus the first perceptible sign we have for the recognition of religious phenomena is the radical separation of everything into these two categories, with sacred things isolated and protected by interdictions from profane things. Religious rites are rules of conduct dictating how one is to comport oneself around sacred objects. Religious beliefs are the representations that express the nature of sacred things and their relation to each other and to profane things (Durkheim [1912] 1985, 56). Then, in order to distinguish religion from magic, Durkheim modified this preliminary definition to say that religion is a unified system of beliefs, regarding the sacred, that unifies a community (Durkheim [1912] 1985, 65).

This provisional definition then sets the explanatory goal for the sociology of religion: how to account for the origin of the distinction between the sacred and the profane (Durkheim [1912] 1985, 122). Durkheim argued that the hypothesis that totemism is the oldest form of religion explains this dichotomy better than either Frazer and Edward B. Tylor's hypothesis that it originated in animism or Max Müller's that it originated in the worship of natural forces. To understand [*entendre*] totemism, he said, one must know the beliefs on which it rests. The most important beliefs are those relating to the totem (Durkheim [1912] 1985, 141f.). Relying principally on Australian totemism for his evidence, he then described the ways in which totemic emblems are used in religious ceremonies, pointing out, for instance, how women and uninitiated young men are not allowed to touch it or even see it (Durkheim [1912] 1985, 169). Durkheim then argued that the hierarchy of totems can be explained by the tribal organization of the clans these totems represent. Society, according to Durkheim, provides the "canvas" for logical thought (Durkheim [1912] 1985, 211). Ultimately, he argued that the totem is the symbol of both the clan and the god of the clan, since the god of the clan is nothing other than the clan itself hypostasized as the animal or plant species that serves as the totem (Durkheim [1912] 1985, 295). The totemic emblem chosen as the focal point for the expression of social unity at the same time serves as a constituent element in the formation of this sentiment. Individual consciousnesses are closed to one another and can communicate only by means of external signs that translate their internal states. In order for their communication to result in the expression of a common sentiment, it is necessary that they express the same sign: the same cry, the same gesture, or a dance that imitates the motions of the totemic species. In this way they become aware of their moral unity. This external sign is thus the outward expression of a collective representation. The totemic emblem can keep this sentiment alive even when the group is not present (Durkheim [1912] 1985, 329–331).

Durkheim then derived his explanatory definition of religion from this account of totemism. The impression that we are in relation to two sorts of realities, sacred and profane, he said, arises from the dichotomy between individual and collective representations (Durkheim [1912] 1985, 304). The collective representation carries with it the authority that society has over the individual (Durkheim [1912] 1985, 295ff.). This power of society over us is represented as something divine (Durkheim [1912] 1985, 298ff.). Hence, the believer is not deceived when he thinks there is a superior moral power over him: it is society. For Durkheim, what the primitive experiences as spiritual force or power is actually nothing but the obscure or confused perception of the power of society over him or her (Durkheim [1912] 1985, 322). Having reached this conclusion, Durkheim used it to explain contemporary religion as well. Religious feelings, according to Durkheim, are simply the result of sentiments stirred up during religious gatherings. When the believer thinks she is having a religious experience, she is actually experiencing the effects of social forces.

A fundamental assumption running through his entire work is that the mind of the primitive is continuous with ours. Consider, for instance, his reasons for rejecting Tylor's hypothesis that animism is the oldest form of religion. According to Tylor, the most fundamental religious belief is the belief in the soul. This idea, he proposed, was suggested by dreams. The primitive thought that when one dreamt, one's soul left one's body and went to the place one dreamt about (Durkheim [1912] 1985, 70). Durkheim argued that this interpretation of one's dream would have been easily refuted on waking and talking with others, and suggested some alternative, simpler ways in which the primitive may have explained his dreams (Durkheim [1912] 1985, 79f.). According to Durkheim, Tylor's hypothesis explains the origin of religion in terms of "hallucinatory representations" with no basis in objective reality, attributing to the primitive a confusion between waking and dreaming, death and sleeping, animate and animate. If this were the origin of religion, Durkheim reasoned, one would not be able to explain how religion could give rise to law, morality, and science (Durkheim [1912] 1985, 97f.). He also rejected Frazer's explanation of totems as a later development of animism, according to which they were originally conceived as a place to hide one's soul in war time. For Durkheim, this is only to attribute an "absurd" belief to primitives: Why would they think their souls were safer that way? How could one's soul be safer in a totemic animal if there are hunters about who would kill and eat the animal (Durkheim [1912] 1985, 248–250)? Durkheim raised a similar objection to the hypothesis that religion has its origin in an attitude of fear and wonder with regard to natural forces, which were conceived under the form of personal agents like human beings. He argued that if the purpose of religion were to serve as a guide to the natural world, its failure would have been quickly noticed and it would have been rejected (Durkheim [1912] 1985, 109–113). In all of these arguments, Durkheim assumed that the primitive thinks in much the same way that we do: rejecting putative explanations when they are contradicted by the facts or lead to false predictions, making careful distinctions concerning the value of various experiences, and choosing the simplest among alternative explanations.

Durkheim's emphasis on making primitive thought and religion understandable set his work apart from Lévy-Bruhl's. Beginning with *Les fonctions mentales dans les sociétés inférieures*, Lévy-Bruhl took what could be regarded as the *Erklären* approach to so-called primitive thought. He tried to explain the strange things he read in ethnographic reports by postulating a "prelogical" mentality that does not recognize contradictions but instead follows an alternative "logic," governed by the "law of participation" (Lévy-Bruhl 1910, 79). For example, a primitive could work witchcraft on his neighbor through obtaining some of his hair, nails, bodily fluids, clothing, or utensils, since the primitive thought of all these things as somehow "participating" in his intended victim. People in so-called primitive societies are thus held accountable for harm to others in ways that seem irrational to us. As Durkheim recognized, Lévy-Bruhl's approach to explaining primitive thought and behavior does not make it comprehensible to us, but merely presents it to us as something completely foreign. Durkheim, on the other hand, thought that there was no need to postulate the existence of an alternative logic among primitives, and that what appear to be contradictions could be more simply explained once we recognize that primitive cultures have different classificatory systems than we do. Totemic systems of classification function the way scientific theories do in our culture, according to Durkheim, providing a set of relations or connections, including causal connections, among things in the natural world. So for example for an Australian totemist to say a man is a kangaroo is no more a contradiction than for us to say that heat is the motion of molecules. Once humanity acquires the sense that there are connections among things, philosophy and science become possible: "Our logic was born from this logic," he said (Durkheim [1912] 1985, 340). There is no "abyss" between primitive religious and contemporary scientific thought as there was for Lévy-Bruhl (Durkheim [1912] 1985, 342). By thus stressing the continuity of their thought with ours, Durkheim endeavored to make their thought understandable to us. However, the difference between Durkheim and Lévy-Bruhl is not so much a methodological difference as a theoretical difference or a difference in explanatory goals and principles. These two French thinkers were largely in agreement in questions of method, conceiving the social sciences as proceeding through methods of induction, hypothesis, and test.[8]

According to Boudon (1995), the idea of taking apparently irrational acts or beliefs and trying to make sense of them is characteristic not just of Durkheim's later work on primitive thought, but of *Suicide* (Durkheim [1897] 1988) as well. For instance, in this work Durkheim was concerned to explain the fact that the social rate of suicide increases when financial markets are going up as well as

---

[8]For a more detailed discussion of the Durkheim and Lévy-Bruhl's respective approaches to primitive thought, see Schmaus (1996).

down. On the face of it, rising markets would seem like good news. Durkheim tried to explain this surprising fact on the grounds that economic booms as well as busts could be indicative of rapid social change leading to anomie.

## 6.6 Conclusion

Although we can find many of the same distinctions regarding the goals and subject matters of the natural and human sciences in France and Germany, these did not lead the French to draw any fundamental methodological distinctions between them. Perhaps what needs to be explained is not the fact that there was no French parallel to the debates among Dilthey, Windelband, and Rickert, but the fact that these debates occurred in Germany in the first place. The methodological distinctions drawn by Dilthey, Windelband, and Rickert do not bear up under close scrutiny. Either they turn out to be distinctions in terms of content or epistemic goals instead of method, or they underestimate the hypothetical or provisional character of all knowledge. The distinction between the epistemic goals of *Erklären* and *Verstehen* does not entail any methodological distinction, since both goals, as Durkheim saw, can be approached through a method of hypothesis and test.

Starting with Comte, the French recognized that the sciences pursued a variety of methods of inquiry specific to the subject matter of each. As Durkheim put it, each discipline "has its own object, its own method, its own spirit" (Durkheim [1893] 1986, 2). However, these methods may all be considered as specific instances of one and the same general method of hypothesis and test. Any attempt to distinguish these specialized methods would distinguish the natural sciences from each other as much as from the social sciences, as Windelband and Rickert came very close to recognizing. There is nothing that all and only the methods of the natural sciences have in common that distinguishes them from the methods of the social sciences, and there is nothing that all and only the methods of the social sciences have in common that distinguishes them from the methods of the natural sciences. If the French drew a methodological distinction anywhere, it was between philosophy and the empirical sciences, which included both the natural and the social sciences. The important question for the French was not whether interpretation or understanding called for different methods than explanation did. Rather, it was whether normative philosophical inquiries such as ethics and epistemology could be grounded in empirical social science, or required a different foundation.

## References

Adair-Toteff C (2003) Neo-Kantianism: The German idealism movement. In: Thomas B (ed) The Cambridge History of Philosophy 1870–1945. Cambridge University Press, Cambridge, pp 27–42

Anderson RL (2003) The debate over the Geisteswissenschaften in German philosophy. In: Thomas B (ed.) The Cambridge History of Philosophy 1870–1945. Cambridge University Press, Cambridge, pp 221–234

Boudon R (1995) Weber and Durkheim: beyond the differences a common important paradigm? Revue Internationale de Philosophie 49(192):221–239
Brooks JI (1998) The eclectic legacy. University of Delaware Press, Newark, DE
Cousin V (1860) Du Vrai, de beau et du bien. Didier et Cie, Libraires-Éditeurs, Paris
Dilthey W ([1894] 1977) Ideas concerning a descriptive and analytic psychology. In: ibid., Descriptive Psychology and Historical Understanding. Translated by Richard M. Zaner. Originally published as "Ideen über eine beschreibende und zergliedernde Psychologie. In: Gesammelte Schriften, vol. V. Teubner, Leipzig. Martinus Nijhoff, Den Haag
Dilthey W ([1910] 1961) Pattern and Meaning in History: Thoughts on History & Society. Edited, introduced and translated by Hans Peter Rickman. Contains selected passages from Wilhelm Dilthey (1910) Der Aufbau der geschichtlichen Welt in den Geisteswissenschaften. In: ibid., Gesammelte Schriften, vol. VII. Teubner, Leipzig. Harper Torchbooks, New York
Dilthey W (1976) W. Dilthey, Selected Writings. Edited, translated, and introduced by H. P. Rickman. Cambridge University Press, Cambridge
Durkheim É (1975) Textes. Edited by Victor Karady, Paris, Les Éditions de Minuit
Durkheim É ([1883–1884] 2004), Durkheim's philosophy lectures: Notes from the Lycée de Sens Course, 1883–1884. Edited and translated by Neil Gross and Robert Alun Jones. Cambridge University Press, Cambridge
Durkheim É ([1893] 1986) De la division du travail social. Reprint edition of the 2nd 1902 ed. Originally published as ibid., Alcan, Paris
Durkheim É ([1895] 1927) Les règles de la méthode sociologique. Reprint edition of the 2nd 1901 ed. Originally published as ibid. Alcan, Paris
Durkheim É ([1897] 1988) Le suicide; étude de sociologie. Reprint. Originally published as ibid. Presses Universitaires de France, Paris
Durkheim É ([1904] 1969) Review of L. Lévy-Bruhl, La morale et la science des moeurs. In: Duvignaud J (ed) Journal sociologique. Reprint. Originally published in L'Année sociologique 7:380–384. Presses Universitaires de France, Paris, 467–470
Durkheim É ([1912] 1985) Les formes élémentaires de la vie religieuse. Reprint. Originally published as ibid. Alcan, Paris
Durkheim É ([1920] 1975) Introduction à la morale. In: Karady V (ed) Textes, vol II. Reprint. Originally published in Revue philosophique 89:79–87. Les Éditions de Minuit, Paris 313–331
Durkheim É, Tarde G ([1904] 1975) La sociologie et les sciences sociales. Résumé d'une conférence faite au début de l'année scolaire 1903–1904 à l'école des hautes études sociales à Paris. In: Karady V (ed) Textes, vol I. Reprint. Originally published in Revue international de sociologie 12:83–87. Les Éditions de Minuit, Paris, pp 160–165
Durkheim E, et al. ([1908] 1975) Débat sur l'explication en histoire et en sociologie. In: Karady V (ed) Textes, vol I. Reprint. Originally published in Bulletin de la société française de philosophie 8:229–245. Les Éditions de Minuit, Paris, pp 199–217
Ermarth M (1978) Wilhelm Dilthey: The critique of Historical Reason. University of Chicago Press, Chicago, IL
Gayon J (2005) Bergson's spiritualist metaphysics and the sciences. In: Gutting G (ed) Continental Philosophy of Science. Blackwell Publishing, Malden, MA, pp 43–58
Harrington A (2000) In Defence of Verstehen and Erklären. Theor Psychol 10(4):435–451
Harrington A (2001) Dilthey, Empathy, and Verstehen: A contemporary reappraisal. Eur J Soc Theor 4(3):311–329
Hodges HA ([1944] 1969) Wilhelm Dilthey: An Introduction. Reprint. Originally published as ibid. Routledge & Kegan Paul, London
Hodges HA (1952) The Philosophy of Wilhelm Dilthey. London: Routledge & Kegan Paul. Reprint, Greenwood Press, Westport, CT, 1974
Horner R (1997) A Pragmatist in Paris: Frédéric Rauh's 'Task of Dissolution'. J Hist Ideas 58(2):289–308
Janet P (1883) Traité élémentaire de philosophie à l'usage des classes. Librairie Ch. Delagrave, Paris
Kant I ([1770] 1900ff.) De mundi sensibilis atque intelligibilis forma et principiis. In: Kant's gesammelte schriften. Akademie-Ausgabe, Bd. II. Berlin, pp 385–419
Kitcher P (2001) Science, Truth, and Democracy. Oxford University Press, Oxford

Lévy-Bruhl L (1910) Les fonctions mentales dans les sociétés inférieures. Alcan, Paris
Maine de Biran and François-Pierre-Gonthier (1982) Oeuvres complètes, 14 vols. Tisserand P, Gouhier H (eds). Slatkine, Geneva/Paris
Makkreel RA (1975) Dilthey: Philosopher of the Human Sciences. Princeton University Press, Princeton, NJ
Makkreel RA (1977) Introduction. In: Dilthey W, Zaner RA (eds) Descriptive Psychology and Historical Understanding. Martinus Nijhoff, Den Haag, pp 3–20
Makkreel RA (1993) The underlying conception of science in Dilthey's introduction to the human sciences. In: Blosser P et al (eds) Japanese and Western Phenomenology. Kluwer, Dordrecht, pp 423–439
Makkreel RA (1999) The problem of values in the late nineteenth century. In: Popkin RH (ed) The Columbia History of Western Philosophy. Columbia University Press, New York, pp 556–567
Makkreel RA (2000) From simulation to structural transposition: A Diltheyan Critique of Empathy and defense of Verstehen. In: Kögler HH, Stueber KR (eds) Empathy and Agency: The Problem of Understanding in the Human Sciences. Westview Press, Boulder, CO, pp 181–193
Makkreel RA (2003) The cognition-knowledge distinction in Kant and Dilthey and the implications for psychology and self-understanding. Stud Hist Philos Sci 34:149–164
Muhlfeld L ([1893] 1975) Débat sur les types de solidarité et la division social du travail. In: Karady V (ed) Textes, vol II. Reprint. Originally published in Revue universitaire 2(1): 440–443. Les Éditions de Minuit, Paris, pp 288–291
Oakes G (1980) History and natural science; Wilhelm Windelband; Note. Hist Theor 19(2):165–168
Outhwaite W (1975) Understanding Social Life: The Method Called Verstehen. George Allen & Unwin, London
Paucot R (1914) Expliquer et Comprendre. Revue Philosophique 78:285–295
Rabier É ([1886] 1909) Leçons de philosophie. II. Logique, 6th edn. Librairie Hachette et Cie, Paris
Schmaus W (1994) Durkheim's Philosophy of Science and the Sociology of Knowledge. University of Chicago Press, Chicago, IL
Schmaus W (1996) Lévy-Bruhl, Durkheim, and the positivist roots of the sociology of knowledge. J Hist Behav Sci 32(4):424–440
Schmaus W (2004) Rethinking Durkheim and His Tradition. Cambridge University Press, Cambridge
Windelband W ([1894] 1980) Geschichte und Naturwissenschaft. Reprinted as "History and Natural Science", translated by Guy Oakes in Hist Theor 19(2):165–185. Heitz, Strassburg

# Chapter 7
# Instead of Erklären and Verstehen: William James on Human Understanding

**David E. Leary**

## 7.1 Introduction

Perhaps more than any other American psychologist and philosopher, William James (1842–1910) was intimately familiar with contemporary European thought and debate, including the discussion of *Erklären* and *Verstehen* advanced by Wilhelm Dilthey (1833–1911) and others around the turn of the twentieth century. Even before this discussion was initiated, James had been dealing with related issues, pondering alternative solutions, and formulating his own original views on human understanding. These views coalesced in a distinctive approach to cognition. Fundamental to this approach was a belief in possibility and probability as innate features of the physical as well as mental manifestations of the universe. Also fundamental was a conviction that understanding is understanding, regardless of its viewpoint, object, or label as either "descriptive" or "explanatory."

A review of James's approach, preceded by a discussion of its multiple contexts, may help us comprehend why James held these premises and never participated in the debate about *Erklären* and *Verstehen*, despite a personal connection to Dilthey and striking similarities between their views and concerns.

## 7.2 The Scientific and Philosophical Context of James's Thought

Well before the turn of the twentieth century, old verities about nature and cognition gave way to uncertainty and even to disbelief in the possibility of certainty regarding ultimate matters. *Ignoramus et ignorabimus* was a common cry (Croce 1995; Mandelbaum 1971), and chance as well as change were common catchwords (Peirce [1893] 1992, 358f.).

In this context, the scientific advances and philosophical consequences associated with the names of Quételet, Buckle, Clausius, Krönig, Maxwell, and Darwin were

D.E. Leary (✉)
Ryland Hall 320, University of Richmond, Richmond, VA 23173, USA
e-mail: dleary@richmond.edu

U. Feest (ed.), *Historical Perspectives on Erklären and Verstehen*, Archimedes, Vol. 21,
DOI 10.1007/978-90-481-3540-0_7, © Springer Science+Business Media B.V. 2010

discussed by the members of a small "Metaphysical Club" that met with some regularity in Cambridge, Massachusetts, in the early 1870s (Fisch 1964; Menand 2001; Wiener 1949). Revolving around Chauncey Wright, this small group included Peirce (later America's greatest logician), Oliver Wendell Holmes, Jr. (later an important Supreme Court justice), and James (later a leading contributor to the transformation of both psychology and philosophy in the United States and elsewhere). It was in the crucible of this group that much of James's later thinking took initial shape, largely in response to the "nihilistic" dicta of Wright's positivistic way of thinking. Somewhat older, more intellectually mature, and quite skilled at intellectual boxing (as reported by (Peirce [1907] 1998, 399)), Wright was a thoroughgoing empiricist, who refused to conjecture beyond available sensory data.

Nature or phenomenal reality, Wright ([1873] 1877, 205–229) argued, is simply what is given in experience: so many data, or facts, experienced at particular moments in time. That is all that we can be sure there is, and hence all that we can know. Anything beyond that is mere speculation, and there is no warrant – and no need – to engage in speculation. In his view, there is no reason to concern ourselves about the way our experience *comes* to be ordered, as if the ordering is the result of some other process or reality. Experience simply *is* ordered, and we can formulate and test hypotheses about that order staying entirely within the realm of experience; indeed, that is all we can do, Wright asserted, if we wish to advance the cause of knowledge. For no matter how frequently our hypotheses may be confirmed by experience, we can never arrive at rational, apodictic rules that will guarantee what *must* happen next or what *can not* happen next. To claim that we can do so would be to assume that we can attain knowledge of underlying realities and causes that is simply beyond our reach.

When Wright died unexpectedly in 1875, James published a notice that characterized him as "a worker on the path opened by [David] Hume" and suggested that if Wright had lived to complete his treatise on psychology, it "would probably have been the last and most accomplished utterance of what he liked to call the British school." With this utterance, James said, "he would have brought the work of [John Stuart] Mill and [Alexander] Bain for the present to a conclusion" (James [1875] 1987, 16). This was a worthy tribute to his esteemed friend and interlocutor, especially since James himself was spending a great deal of time studying the work of Mill and Bain. I emphasize this because James was to become famous for his own version of "radical empiricism" that bears some resemblance to that of Wright. For instance, it starts with and maintains an emphasis on the continuity of experience, and it places a heavy burden on description at the expense of explanation. But in the end it is clearly *not* the kind of aseptic empiricism that Wright advocated, and to understand why it isn't we need to explore the experiences that James had before participating in the Metaphysical Club.

## 7.3 The Personal and Literary Context of James's Thought

William James had moved to Cambridge in 1861, when he entered the Lawrence Scientific School at Harvard University in order to study chemistry and eventually physiology, after giving up an earlier commitment to become a painter.

This "suppressing" or "murdering" of a potential self (his painter-self), as he put it (James [1890] 1981, vol. 1, 295), led to frequent periods of depression between 1861 and 1878, the year he was married (Leary 1992). He experienced particularly difficult times between 1865 and 1874. During this time he pondered the meaning of life in general and the direction of his own life in particular. Especially salient in light of his increasing familiarity with advances in the natural sciences (including Darwin's *Origin of Species*, 1859) was his growing fear that "we are Nature through and through, that we are *wholly* conditioned, that not a wiggle of our will happens save as the result of physical laws" (James 1992–2004, vol. 4, 370).[1] This fear was at odds with his deeply felt desire to "make my *nick*, however small a one in the raw stuff the race has got to shape, and so assert my reality" (ibid., 250).

Repeatedly over years from 1861 to 1878, as he reflected on the conflict between his fear and desire, James had recourse to literature – most notably to the works of Shakespeare, Goethe, Wordsworth, Robert Browning, and Emerson. As he ruminated on these works, he continued to study the empiricist accounts of mental life offered by Locke, Hume, Bain, Wright, and others as well as the growing literature on physiology, neurology, and psychiatry. The combined impact of this diverse reading and consideration was the development of a distinctive view of human cognition.

Although Shakespeare came to serve as the primary example of James's view of human understanding (James [1890] 1981, vol. 1, 984–986), James was also drawn to the works of Goethe, at first as a means of improving his German, but after a period of incomprehension and even distaste, out of an increasing sense that Goethe had much else to offer. Taking a vital clue from Emerson, James came to see Goethe as "living through every pore of his skin" (Emerson [1850] 1983, 750), and as he read his various works, he remarked more and more frequently in his diary, notebooks, and letters that Goethe had achieved what he himself could not yet achieve: he put aside his own individual ego, with its surfeit of selfish concerns, in lieu of striking a balance between the subjective and objective that allowed him to describe human experience with an accuracy, specificity, and wealth of insight that few other individuals had ever attained. In short, James became convinced that Goethe, by checking his ego, setting aside his prejudices, and applying his perceptual and cognitive sensitivities without hindrance from subjective factors, had managed to understand and convey a richness of detail and significance that transcended typical, prosaic description of the phenomena that he studied. For Goethe, each individual experience was unique and worthy in its own right; yet in focusing upon each experience with sufficiently unfettered attention, he was able to discern patterns ("archetypes") that elucidated the general nature as well as particular distinctiveness of that experience. Goethe's method of pure, unhindered observation inspired James to assume what we would now call a phenomenological approach to the analysis of experience: an approach that goes beyond the initial, surface manifestations of reality, shaped to fit the preexisting, subjective, conceptual categories of the observer.

[1]I have discussed James's personal crises as a background for understanding his theories of self and personality in Leary (1990b).

James also turned to the works of Wordsworth, Browning, and others, as he intermittently pursued his studies and recuperation. For instance, at the same time that he was studying the work of Helmholtz, Wundt, Hume, and Bain, and exchanging ideas with Wright and Peirce, he was reading and re-reading Wordsworth's long poem "The Excursion" (Wordsworth [1814] 1977), especially the section on "Despondency Corrected." He left that work – as Mill and Darwin had – with an enhanced sense that human cognition entails what Wordsworth called a "marriage" of mind and matter: a union of the subjective and objective brought about by the mind's "excursive power" to "walk around" phenomena, viewing them this way and that, from one vantage point and another, trying first this metaphorical key and then that, all the while noting that some keys open special doors of perception, while others do not. He also spent time with Browning's works, especially but not only *The Ring and the Book* (Browning [1868–1869] 1971), which schooled him in Browning's groundbreaking explorations of internal monologue, raising his awareness of ongoing streams of consciousness, the very topic he would soon be analyzing with sensitivity and insight, and which would provide the foundation for his psychology and philosophy.[2]

Thus, by the early to mid-1870s, James's sense of "experience" was considerably different from Chauncey Wright's. Goethe in particular, but also Wordsworth had taught James something that Emerson, too, had argued: that nature is richer than typical positivists realize and that more can be revealed in experience if one sets aside presuppositions and relies on one's own unfettered sensibilities. And Browning helped him realize the inextricable relations between experience and consciousness. James's growing conviction in these regards was apparent in a variety of ways, including his choice of reading materials throughout this crucial period of his life. And all of this – all of James's unique education – started coming to fruition in 1878, less than 3 full years after Wright's death and in the same year that James was finally married, at the age of 36. This fruition was apparent in two public lecture series, his plans for a textbook, and the publication over 2 years of six substantive articles, which served as preliminary drafts for different parts of his proposed textbook. It was also apparent in the courses he was teaching at Harvard. Clearly, James's life and thought were turning away from personal turmoil and indecision – away also from simple empiricism – toward the distinctive intellectual achievements for which he is now so well known.

## 7.4 James's Vision of a World of Possibility and Probability

Even before 1878, James had hinted at the ongoing development of his thinking. In "The Teaching of Philosophy in Our Colleges" (James [1876] 1978), he had granted a place for Wright's positivism, but asserted the need for other philosophical views:

[2] The preceding analysis of James's reading of literature draws upon my current research, which is based on a wide variety of untapped and in some instances previously unknown resources.

> However sceptical one may be of the attainment of universal truths. . .one may never deny that philosophic study means the habit of always seeing an alternative, of not taking the usual for granted, of making conventionalities fluid again, of imagining foreign states of mind. In a word, it means the possession of mental perspective (James [1876] 1978, 4).

By 1878, James had clarified his own alternative perspective, which was based on an innovative melding of insights and concerns from his personal, scientific, artistic, and literary backgrounds. Specifically, it drew out the implications of Darwin's theory of natural selection into a full-scale cosmology that highlighted the role of possibility and probability over necessity and certainty. It was a bold and compelling vision that made room for ideals, hope, and the belief that individuals can make a difference. To some extent it was theory-driven, but at the center of the theory was an hypothesis – of selectivity – that made sense of a wide range of observable phenomena pertaining to sensation, perception, cognition, emotion, motivation, and action (within psychology) and to epistemology, aesthetics, or morality (within philosophy). Working out this vision in something close to its entirety took the rest of his life, but he had established his basic viewpoint by the late 1870s.

The first explicit indication of James's new way of thinking was given in a lecture series on "The Senses and the Brain and Their Relation in Thought," delivered at Johns Hopkins University in February 1878. In those lectures, James asserted that "evolution accounts for mind as a product" and that "first nervous systems appeared, then consciousness" (James [1878a] 1988, 4). The implication was clear: the mind depends upon matter. And beyond that, James proposed an "important mental principle," namely, that the senses are "organs of *selection*" (ibid., 4). Even if there is "no fulcrum of certainty" for this proposition, he said, this thesis is "not only possible but probable" (ibid., 5). Thus, while granting the positivist's claim of ultimate uncertainty, he argued – with Darwin – that this need not stand in the way of reasonable, empirically grounded conclusions that have a high degree of probability.

Important corollaries followed from James's conviction regarding the selectivity of the human mind: that experience – and the reality represented in experience – is "overabundant," as Darwin put it; that there is always *something else* that could have been selected or could be selected in the future; that the world offers more than is minimally or logically necessary, thus allowing for unforeseen eventualities. As a result, strange and unpredictable things can and do happen, can and do come to be. What is possible in the future is never limited to what has happened or what has been in the past. *Contra* Chauncey Wright, the world of our experience is not simply "given"; it is constituted by our "selections" among givens. And these selections occur, James argued, at all levels of functioning: From a wide spectrum of potential sensations, our sense organs select a very limited number; from these sensory impulses, our perceptual processes select a small portion for attention; from these perceptions, an even narrower range are selected for conceptualization; and from our concepts, even fewer are selected for additional cognitive processing, aesthetic appreciation, and moral consideration. Selection is the key all up and down the line. At each stage, potential phenomena "rise as possibilities into consciousness," with only the "most attractive" becoming "plenary" (ibid., 16).

James clarified what he meant by "most attractive" in subsequent writings. More typically, the word he used was "interesting" rather than "attractive." Every organism, he noted, has "subjective interests" that direct the organism's attention to different aspects of the "objective world." These interests "form a true spontaneity" on the part of the organism, thus allowing the organism to "cooperate" with the environment in "molding" what it notices and knows – much as Wordsworth indicated when he wrote of the "marriage" of mind and nature (James 1992–2004, vol. 5, 24f.). Interests are "the real *a priori* element in cognition"; they are "an all-essential factor which no writer pretending to give an account of mental evolution has a right to neglect" (James [1878] 1978, 11). In fact, "interests which we bring with us, and simply posit or take our stand upon, are the very flour out of which our mental dough is kneaded," and they are equivalent to the "spontaneous variations ... which form the ultimate *data* for Darwin's theory" (ibid., 18f.). But they actually represent, James argued, a much wider range of values and ideals than Darwin proposed. "Survival is only one out of many interests. ... The story-teller, the musician, the theologian, the actor, or even the mere charming fellow, have never lacked means of support" for the simple reason that people are interested in what they have to offer (ibid., 12f.). There are, then, a variety of interests for which people live, and by directing their attention and shaping their behavior according to those interests, they contribute to the variation of human experience that affords human and social evolution its primary motive force (James [1890] 1981, vol. 1, 380f.).

James pointed out the moral significance of this analysis in the other lecture series that he gave in 1878 – his Lowell Lectures on "The Brain and the Mind." After discussing the role of "caprice" in the working of the nervous system, and noting how the lability of the brain provides opportunities for the "possible," James argued against a rigid division between the world as science portrays it, and the world as "Philosophy, Metaphysics, Religion, Poetry, [and] Sentiment" deal with it (James [1878b] 1988, 30). There is no need for a chasm between science and "all that makes life worth living," he suggested, because science itself has revealed that the "possible" often comes about only through the selective thoughts and efforts of interested parties, whatever their interests may be. "I trust," he said,

> that you too after the evidence of this evening will go away strengthened in the natural faith that your delights & sorrows, your loves & hates, your aspirations & efforts are real combatants in life's arena, & not impotent, paralytic spectators of the game (ibid., 30).

In other words, if no one is interested in attending to phenomena in certain ways, science will not be advanced; and if no one is interested in attending to them in other ways, the arts or philosophy or ethics or whatever else will not be advanced. Always it is a matter of selective attention and action, with different expectations and results proceeding from each type of attention and action. And no one can say, beforehand, which approach will be more useful in novel situations. Life is an open-ended game with guidelines, perhaps, but no inviolable rules. The world is full of possibilities and probabilities waiting to occur, with or without human assistance.

## 7.5 James's Views on Human Understanding

In several well known essays, Ralph Waldo Emerson wrote of experience or consciousness as a "stream" or "flowing river" (e.g., Emerson [1841] 1983, 385). When James attended closely to his own experience, as Goethe had taught him to do, he realized that Emerson had selected a fitting metaphor. Experience flows; it is continuous – it is "a teeming multiplicity of objects and relations" (James [1890] 1981, vol. 1, 219).

Those final words, "*and relations*," constituted an important addition to the usual empiricist analysis of consciousness, as presented in the works of Locke, Hume, and Mill. We do not simply have "objects" in our consciousness – be they "sensations" or "ideas" or "thoughts." Rather, all of these potentially discriminable "things" are embedded – and firmly related – within the ongoing stream of our experience. And the relations, which he called "transitive parts" of consciousness, are just as experientially real as the "substantive parts" (ibid., 236). In other words, the feelings of "and" and "if" and "but" are just as real and present to our consciousness, if we sufficiently attend to it, as are the feelings of "hot" and "cold" (ibid., 238).

This discovery was at the root of James's later articulation, as a philosopher, of "radical empiricism" (James [1909] 1977, 126f.; James [1912] 1976). If we are truly and thoroughly empirical, James argued, we will notice that we do not need any "laws of association" to combine what is present to our mind. It all comes to us related; it is only subsequently that we abstract, or *select out*, certain features of the stream, which we may then associate with other features. In the beginning, we encounter the-table-as-hard-and-brown; later, we separate out the notions of "hard" and "brown." And we do so, James argued, according to our interests and previous experience, which are themselves frequently related.

James gave some graphic illustrations of the selective nature of consciousness in his *Principles of Psychology* (James [1890] 1981). For instance:

> Let four men make a tour in Europe. One will bring home only picturesque impressions – costumes and colors, parks and views and works of architecture, pictures and statues. To another all this will be non-existent; and distances and prices, populations and drainage-arrangements, door- and window-fastenings, and other useful statistics will take their place. A third will give a rich account of the theatres, restaurants, and public balls, and naught beside; whilst the fourth will perhaps have been so wrapped in his own subjective broodings as to tell little more than a few names of places through which he passed. Each has selected, out of the same mass of presented objects, those which suited his private interest and has made his experience thereby (James [1890] 1981, vol. 1, 275f.).

Besides reflecting variations in interest, differences in consciousness may result from what someone has learned in past experience:

> Men have no eyes but for those aspects of things which they have already been taught to discern. Any one of us can notice a phenomenon after it has been pointed out, which not one in ten thousand could ever have discovered for himself . . . .*The only things which we commonly see are those which we preperceive* [those for which we are on the lookout], and the only things which we preperceive are those which have been labeled for us, and the labels stamped into our mind (ibid., 420).

This insight about the stamping done by labels was behind one of the major methodological cautions that James raised: a caution that has significance for comprehending his views on the limitations of human understanding. It concerned "the Misleading Influence of Speech" (ibid., 193–196). In psychology as in all other fields, we have inherited our basic vocabulary from previous times and peoples. And having been given a word for a particular experience, we are "prone," on the one hand, "to assume a substantive entity existing beyond the phenomena, of which the word shall be the name" (ibid., 194). Hence, we take for granted that there are such "things" as "ideas" and "thoughts" separate from the steam of consciousness from which we have conceptually extricated them. On the other hand,

> the *lack* of a word quite as often leads to the directly opposite error. We are then prone to suppose that no entity can be there; and so we come to overlook phenomena whose existence would be patent to us all, had we only grown up to hear it familiarly recognized in speech. It is hard to focus our attention on the nameless, and so there results a certain vacuousness in the descriptive parts of most psychologies (ibid., 194).

Besides our vocabulary, our previous experience – and what we have learned from it – may get in the way of our attending closely and accurately to new experiences, as Goethe had pointed out and as James knew from his experiences as a painter:

> We shall see how inveterate is our habit of not attending to sensations as subjective facts, but of simply using them as stepping-stones to pass over to the recognition of the realities whose presence they reveal. The grass out of the window now looks to me of the same green in the sun as in the shade, and yet a painter would have to paint one part of it dark brown, another part bright yellow, to give its real sensational effect. We take no heed, as a rule, of the different way in which the same things look and sound and smell at different distances and under different circumstances. The sameness of the *things* is what we are concerned to ascertain; and any sensations that assure us of that will probably be considered in a rough way to be the same with each other (ibid., 225f.).

This tendency toward stereotyping our experience, and ignoring any unexpected qualities, is an important concern of James's since in actuality "our state of mind is never precisely the same. Every thought we have of a given fact is, strictly speaking, unique, and only bears a resemblance of kind with our other thoughts of the same fact" (ibid., 227).

What James believed to be appropriate, of course, was an openness to experience and an awareness of the ever ongoing changes in it. In essence, he wanted everyone to be like the good philosophers he had described, who are in "the habit of always seeing an alternative, of not taking the usual for granted, of making conventionalities fluid again" (James [1876] 1978, 4). This openness to variation is healthy and good, he believed; it prepares us to deal with life's unexpected eventualities. By continuing to develop discrimination, individuals become more expert at sizing up and responding to situations in quick and appropriate ways.

But how do we arrive at better understanding *beyond* simply making finer discriminations? How does *qualitatively* better understanding develop over time? These questions lie at the heart of James's views on human understanding. The answer, for him, has to do with what someone does *after* discriminating, or selecting out, relevant properties from among the "well-spring of properties" that adhere to any given "phenomenon" or "fact" (James [1878] 1983, 17; James [1890] 1981, vol. 2, 959). The person who significantly advances our knowledge does so, generally, by associating this or that property – as no one has before – with a similar property in something else. Using the language of his time, James referred to this process as "association by similarity," and he noted that "geniuses are, by common consent, considered to differ from ordinary minds by an unusual development of association by similarity" (James [1890] 1981, 972). In more contemporary terms, this means that knowledge tends to be advanced by analogical thinking, which I have referred to elsewhere as "comparative thinking" (Leary 1990a). As we shall see, James made a further distinction between insights reached by analogical thinking alone, and insights, originally based on an analogy, that have subsequently been worked out with greater specificity through empirical study and/or with more generality through abstract reasoning. Still, it all starts with comparison and the fact that

> some people are far more sensitive to resemblances, and far more ready to point out wherein they consist, than others are. They are the wits, the poets, the inventors, the scientific men, the practical geniuses. A native talent for perceiving analogies is. . .the leading fact in genius of every order (James [1890] 1981, vol. 1, 500, italics removed).

At root, then, creative or innovative thinking is based upon "the rapid alteration in consciousness" that takes place as one compares two or more phenomena with regard to their "points of difference or agreement" (James [1890] 1981, vol. 2, 971). The poet who makes an apt, new comparison provides fresh insight into human experience, and accordingly may change the way we think about ourselves and perhaps even how we feel and behave. Indeed, James said, Shakespeare may well be the best exemplar of those who "astonish" us by the "unexpectedness" no less than the "fitness" of their comparisons (ibid., 985). Quoting Thomas Carlyle's assessment that "Shakespeare possessed more *intellectual power* than anyone else that ever lived," as demonstrated by his ability to make novel yet meaningful connections," James went on to conjecture that "Shakespeare himself could very likely not say why [he had made any given association]; for his invention, though rational, was not ratiocinative" (ibid., 985f.). Yet "the dry critic who comes after [him]," though incapable of having *created* the connections that Shakespeare made, can after the fact "point out the subtle bonds of identity that guided Shakespeare's pen" (ibid., 986).

Most humans – even many of the other higher animals – can make "intuitive" connections of this kind, even if the connections are not as inspired as Shakespeare's. Some humans, however, can and do go further, though James was quick to note that "it would be absurd in any absolute way to say that a given analytic mind was superior to any intuitional one" (ibid., 986). Nonetheless, it is true, he said, that "the former *represents* the higher stage. Men, taken historically, reason by analogy long before they have learned to reason by abstract characters" (ibid., 986). The difference in philosophy – or in science or in any other field of abstract knowledge – is that one is able to "give a clear reason" for one's views, and in many instances, at least, to give a greater number of relevant instances, which might qualify those views.

In sum, no matter what type of understanding he was considering (the more concrete, intuitive form of understanding or the more abstract, reasoned form of understanding), James argued against the then-common notion, held by Chauncey Wright among others, that the mind is *and should be* passive, simply recording sense-impressions and their cumulative consequences. Instead, James's view was that the mind is *and should be* active; that the knower and what is known are *and should be* closely related; that, as Goethe had exemplified, there is *and should be* no such thing as absolute subjectivity or absolute objectivity. In fact, the pertinent challenge, for James as for Goethe, was to balance the polar tension in experience in order to avoid the distortions resulting from extreme subjectivity and the simplifications resulting from extreme objectivity (James [1878] 1978, 21).

## 7.6 Description and Explanation in James's Thought

James provided a summary of his view of human understanding in *The Principles of Psychology* (James [1890] 1981), fittingly expressed in metaphorical terms:

> The mind is at every stage a theater of simultaneous possibilities. Consciousness consists in the comparison of these with each other, the selection of some, and the suppression of the rest by the reinforcing and inhibiting agency of attention. ... The mind, in short, works on the data it receives very much as a sculptor works on his block of stone. In a sense the statue stood there from eternity. But there were a thousand different ones besides it [within the same block of stone], and the sculptor alone is to thank for having extricated this one from the rest. Just so the world of each of us, howsoever different our several views of it may be, all lay embedded in the primordial chaos of sensations, which gave the mere *matter* to the thought of all of us indifferently. . . .Other sculptors, other statues from the same stone! Other minds, other worlds from the same monotonous and inexpressive chaos! My world is but one in a million alike embedded, alike real to those who may abstract them (ibid., vol. 1, 277).

It should be clear from this passage that James was a realist. There is a very tangible "block of stone" that all humans experience and deal with, albeit from different sides or perspectives. But James was also a constructivist, to use a more recent term. Every person has to *make* and then *live with* his or her own partial version of reality, or else accept that of someone else. Beyond that, complicating the sculptor-and-stone metaphor, we "*add*, both to the subject and to the predicate part of reality" (James [1907] 1975, 123). That is, we are not simply knowers standing *beside* and hence *outside* that "block of stone"; we are *part* of the "subject" itself – part of the reality we are trying to comprehend with the "predicates" we are using to get in touch with it. And since our thoughts, emotions, and actions are just as real as any mountain or chair, each new thought, feeling, and act assures that "the world is still in process of making" (James [1909] 1975, 123). Each literally *adds to* and thus *changes* the world that we are trying to describe and possibly to explain.[3]

[3]On his application of this perspective to the question of "truth," see James ([1909] 1975).

How we try to describe or explain this open-ended, changing world is largely a matter of temperament, James believed – or if you wish, it is a matter of personal interest or preference (James [1907] 1975, 11). Some people gravitate toward one mode of description or explanation, others to another. Empiricism and rationalism represent two ends of a spectrum. "Reduced to their most pregnant difference, empiricism means the habit of explaining wholes by parts, and rationalism means the habit of explaining parts by wholes" (James [1909] 1975, 9, italics removed).[4]

Bringing his psychological acumen to bear on the matter, James noted that "different men find their minds more at home in very different fragments of the world" (ibid., 10). The result, James said, is that any philosophy, or science, provides "but a summary sketch, a picture of the world in abridgment," from one of many possible "perspective[s] of events" (ibid., 9). Switching back to a sculpturing metaphor, "we *carve out* order by leaving the disorderly parts out" (ibid., 10). Which parts we find disorderly depends on our experience and preference, but we cannot carve just anything or any way we want. The "block of stone" yields more readily to some potential statues and not at all to others. James ([1909] 1977) called himself a pluralist – someone who believed deeply that there is always *another* way to see, understand, and deal with reality, but he did not think that *anything* goes. Our formulations and actions must pay off, must bear fruit, in some way. Hence his famous notion of pragmatic truth (James [1907] 1975), according to which, as in Darwin's work, tangible *consequences* are what matter.[5] Stated obversely, reality for James is that which *resists* some of our formulations and actions (ibid., 99).

As an example of what he meant by individual preference or style (James [1912] 1976) pointed out that "in some men, theory is a passion" that leads them to "systematize and classify and schematize and make synoptical tables and invent ideal objects for the pure love of unifying" (ibid., 136). In distinction, James admitted that he had a strong bias toward empiricism, with its tendency toward "irrationalism" (ibid., 137), by which he meant to indicate an aversion to the kind of "intellectualism" that would reduce everything to a single set of terms and rules (James [1909] 1977, 32). In defense of his own preference, he (James [1912] 1976) simply asked, "Can anything prevent Faust from changing 'Im Anfang war das Wort' [a belief in the priority of reason] into 'Im Anfang war die That [a belief in the priority of experience]?'" His response: "Nothing in earth or heaven" (ibid., 140).[6]

---

[4]James's point is nicely illustrated by the fact that the British tended to understand and develop Newton's work in an empirical way while the French tended to do it in a rational way (Guerlac 1977). On the role of "style" in science, see Hacking (2002).

[5]As James ([1890] 1981) put it, "every scientific conception is in the first instance a 'spontaneous variation' in someone's brain. For one that proves useful and applicable there are a thousand that perish through their worthlessness" (vol. 2, 1,232). James's "Darwinian epistemology" has been developed by Campbell (1960) and Richards (1977).

[6]Reflecting his artistic background, James frequently contrasted the "classic" and "romantic" styles of intellectual work (James [1907] 1975, 9–26; James [1909] 1977, 7–23). His friend Friedrich Wilhelm Ostwald (1909), the German Nobel laureate in physical chemistry, elaborated the same distinction with regard to natural scientists, using temperament (as discussed by his Leipzig colleague Wundt) as the crucial variable.

As for the issue of description vs. explanation, one might ask if (James [1909] 1977) was offering a description or an explanation when he noted that "different men find their minds more at home in very different fragments of the world" (ibid., 10). James's own criterion for whether or not this is a meaningful question would be: What difference does it make? What meaningful consequences result from using one of these terms rather than the other? The bulk of James's work suggests that there is no single, significant, definitive distinction to be made between description and explanation, especially in scientific work, and that the use of either term is generally a matter of individual preference. What counts as one person's description will be another person's explanation, and vice versa. For instance, one person's psychological explanation will be taken by someone else as a description to be explained by physiological factors, which in turn will be taken by yet another person as a description to be explained neurologically or genetically, and so on.

Nonetheless, James made it clear that "all attempts to *explain* our phenomenally given thoughts … are metaphysical" and that as a scientist he was committed, instead, to a "strictly positivistic point of view." Having underscored this point, he admitted that the kinds of empirical descriptions in his *Principles of Psychology* would keep "running out into queries which only a metaphysics alive to the weight of her task can hope successfully to deal with" (ibid., 6). So, even though he added that such successful dealing "will perhaps be centuries hence" and that "meanwhile the best mark of health that a science can show is this unfinished-seeming front" (ibid., 6f.), he seemed to leave the door cracked open to metaphysics and hence, presumably, to explanation in a sense clearly distinct from description. This crack might be seen as widening further when he defined psychology as "the Science of Mental Life, both of its phenomena *and of their conditions*" (ibid., 15, italics added). However, James's use of "conditions" signaled, not causality in a metaphysical sense, but causation in the positivistic sense discussed by John Stuart Mill in *A System of Logic* (James [1872] 1973). According to Mill, "cause" is simply a term for the "assemblage" of observable "conditions" that seem invariably to accompany a phenomenon (ibid., 327–342). The use of the term should not, Mill insisted (and James now implied), be understood as a metaphysical claim about ultimate, ontological causality (ibid., 326).[7]

We could leave the matter here, except for James's response to George Trumball Ladd's review of *Principles*. In that review, this philosopher-psychologist (1892) averred that "as descriptive science, the work is admirable," but "as explanatory science, without metaphysics, it is, at best, *no science at all*" (ibid., 24, 38). These comments reveal Ladd's very different conception of what science should be, and that would be the end of the matter, except that James ([1892] 1984) changed his definition of psychology *to the one that Ladd had suggested* when he published an abbreviated version of his textbook in 1892:

[7]James owned and carefully annotated the eighth edition of Mill's *Logic* and taught courses on Mill's logic in 1881–1882 and 1886–1887 (James [1881–1882] 1988, 490).

> The definition of Psychology may be best given in the words of Professor Ladd, as the *description and explanation of states of consciousness as such*. By states of consciousness are meant such things as sensations, desires, emotions, cognitions, reasonings, decisions, volitions, and the like. The 'explanation' must of course include the study of their causes, conditions, and immediate consequences, so far as these can be ascertained (James [1892] 1984, 9).

In assessing this change, we should note that James mentions both "causes" and "conditions," as if they are separate things, and that he put "explanation" in quotation marks. This may provide a key to a final understanding of his views on description vs. explanation. It seems that by explanation James meant an *ultimate* description of how things fit, in absolute rather than relative terms, in some definitive, rational account of the way things are. (This accords with his later, more formally philosophical statements on the matter.) But, as we have seen, James did not believe that such an account was actually possible – and certainly not within the foreseeable future. We need to keep that in mind as we read this additional statement from his introduction to his abbreviated textbook:

> Most thinkers have a faith that at bottom there is but one Science of all things, and that until all is known, no one thing can be completely known. Such a science, if realized, would be Philosophy. Meanwhile it is far from being realized; and instead of it, we have a lot of beginnings of knowledge made in different places, and kept separate from each other merely for practical convenience' sake, until with later growth they may run into one body of Truth. These provisional beginnings of learning we call 'the Sciences' in the plural (ibid., 9).

Having underscored that no individual science offers a final, definitively "true" explanation of its subject matter, James went on to write that "a *provisional* body of propositions about states of mind, and about the cognitions which they enjoy, is what I mean by Psychology considered as a natural science" (ibid., 10, italics added).

So James's new way of talking about psychology as science, though it included the word "explanation," did not signal a new way of thinking about it, and any ambiguities vested in "conditions" and the implications of his use of "cause" can now be resolved. He still believed what he wrote in the final chapter of *Principles* (James [1890] 1981, vol. 2): "Reality *exists* as a *plenum*. . . . But we can neither experience nor think this *plenum*" (ibid., 1,231). Instead we select and focus on "collateral contemporaneity" (ibid., 1,232): what he – and Mill – had meant when speaking of "conditions." As for "causes," James takes an agnostic perspective:

> We have no definite idea of what we mean by cause, or of what causality consists in. But the principle expresses a demand for *some* deeper sort of inward connection between phenomena than their merely habitual time-sequence seems to us to be. The word 'cause' is, in short, an altar to an unknown god; an empty pedestal still marking the place of a hoped-for statue (ibid., 1,264).

In short, it is a promissory note: a hunch that some "conditions" are more significant than others, but what we have in psychology, as in other sciences, nonetheless remains "a mass of descriptive details" (James [1890] 1981, vol. 1, 7) for which our "causal explanations" are, in fact, only hypotheses (James [1912] 1976, 143). Whether these disciplines have assumed logical or narrative form, none has achieved the status of an idealized, apodictic science. The latter might be a good heuristic goal, but the open-ended nature of reality, the partiality of human viewpoints, and the novelty of human experiences, including innovative ways of associating all those "descriptive details," make its realization unlikely. Nonetheless, the provisional understanding that we have reached through science and other forms of knowledge can be useful in the course of human life: so useful that it is well worth pursuing.[8]

## 7.7 The Personal and Intellectual Relations Between James and Dilthey

In 1867, about 5 years before the Metaphysical Club was established, a remarkable event took place in Berlin. William James, in Europe on one of his periodic recuperative trips, had presented a letter of introduction from Emerson, a family friend, to Herman Friedrich Grimm, who was translating Emerson's works into German. Grimm responded by inviting the 25-year-old James to supper on at least three prior occasions. Then, on October 16, James arrived for another meal and found another visitor, a professor (he wrote home) "whose name I cd. not catch," who was "a man of a type I have never met before. He is writing now a life of Schleiermacher of wh. one vol. is pub[d]" (James 1992–2004, vol. 4, 213). This professor

> was overflowing with information with regard to every thing knowable & unknowable. He is the first man I have ever met of a class wh. must be common here, of men to whom learning has become as natural as breathing. . . .He talked and laughed incessantly at table, related the whole history of Buddhism to Mrs. Grimm, and I know not what other points of religious history. After dinner. . .while G. & the Prof. engaged in a hot controversy about the natural primitive forms of religion.

Then, after a brief nap, the visitor

> rose refreshed like a giant and proceeded to fight with Grimm about the identity of Homer. . . .From there through a discussion about the madness of Hamlet. . . . the sun waned low & I took my leave in co. with the Prof. We parted at the corner, *without* the Prof. telling me. . .

[8]James's point, as Perry (1935) summarized it, is that "the problem of causation has remained unsolved. . .because philosophers have oscillated between the two impossible views that cause and effect are identical and that they are external and mutually exclusive" (ibid., vol. 2, 665). As James ([1911] 1979) wrote, "'enlightened opinion' about cause is confused and unsatisfactory," leading at best to a view of the universe as "consisting of nothing but elements with functional relations between them" (ibid., 104). James was not disappointed by this state of affairs since, as he admits, if *causa aequat effectum* were shown to be true, it would mean that "no *real* growth and no *real* novelty could effect an entrance into life. . . .Such an interpretation of nature, would of course relegate variety, activity and novelty to the limbo of illusions, as far as it succeeded in making its static concepts cancel living facts" (ibid., 103). James had built his entire psychology and philosophy on the premise that variety, activity and novelty – or if you wish, pluralism, freedom and possibility – were real factors in the universe.

> that he would be happy to see me at his domicile, so that I know not whether I shall be able to continue acquainted with a man I wd. fain know more of (ibid., 213f.).

This professor was, of course, Wilhelm Dilthey, 9 years older than James and on holiday from his university post at Basel – a good reason for not giving James his home address! (Dilthey was called from Basel to Kiel in the next year and then to Berlin in 1882, to replace Hermann Lotze in the chair that once belonged to Hegel.) So far as I know, the two men never met again, and whether they ever realized, later, that they had met, I do not know. But they came to know each other's work when James's good friend and Dilthey's Berlin colleague, Carl Stumpf, brought James's *Principles* to Dilthey's attention in November of 1890, calling it "an excellent work, the best we have, which makes Wundt look sorry by comparison" (Ermarth 1978, 212). Four years later, Dilthey approvingly cited several passages from James's *Principles* in his "Ideas concerning a Descriptive and Analytic Psychology" (Dilthey [1894] 1977, 50, 60), in which Dilthey elaborated his distinction between explanatory and descriptive psychology. Subsequently, after Hermann Ebbinghaus's (1895) critique of Dilthey's article, Dilthey turned to James's work to help formulate his response (Ermarth 1978, 212f.).

In 1900, Dilthey nominated James, successfully, for election to the Prussian Academy of Sciences (James 1992–2004, vol. 9, 592, 598), and 2 years later, he reacted positively to James's *Varieties of Religious Experience* (James [1902] 1985), which he judged to be "the great American contribution to the psychological understanding of religion" (Hughes 1961, 197). He also referred to James as a "psychological genius" and mentioned James occasionally in materials now deposited in the Dilthey archives, including materials associated with his move *away from* his 1894 position and *toward* his later hermeneutical position (Ermarth 1978, 212, passim). In this latter regard, Dilthey's comments sometimes had a negative tinge, as when he chastised James – and himself – for a residual constructivist (explanatory) bias (ibid., 185).

We do not have a similar record of James's knowledge of Dilthey's works, probably indicating that James did not read them, even though he read widely in the German philosophical literature, including works by Wilhelm Windelband (whose work he liked, though he disagreed with much of it) and Heinrich Rickert (whose work prompted him to significant marginal note-making, despite – or because of – significant disagreement).

However, even without specific documentation, it seems almost certain that James at least knew about Dilthey's article on "Ideas concerning a Descriptive and Analytic Psychology" (Dilthey [1894] 1977). Beside the fact that it dealt with issues that concerned him, especially in the wake of Ladd's criticism of his psychology, James would have heard about the article and the subsequent controversy from Stumpf, among others. In addition, it is hard to imagine how James could have invited Dilthey to the Third International Congress of Psychology, held in Munich in August of 1896, without being aware of Dilthey's article. (The invitation was extended on behalf of both James and Theodor Lipps.) As things turned out, Dilthey declined the invitation and gave up lecturing on psychology, almost certainly in response to the controversy aroused by his article on explanatory vs. descriptive psychology (Ermarth 1978, 185).

There the record of personal and intellectual encounters between James and Dilthey seems to have come to an end.[9] What is striking is how asymmetric they were, with Dilthey drawing upon James's work, but not vice versa.

## 7.8 Conclusion

Despite the asymmetry in their relations, James and Dilthey had much in common. Their general views and concerns were remarkably similar. Both wanted a psychology that would be true to what they conceived to be the essential characteristics of humanity. Both emphasized the importance and continuity of lived experience. Both developed ideas and approaches that led, through different paths, to twentieth-century phenomenology. Both recognized that beliefs lie at the foundation of knowledge. Both had an abiding interest in religion and the ways that people seek, or lose, meaning when religion no longer attracts allegiance. Both believed that humans are fundamentally historical creatures. Both believed that basic perspectives, or worldviews, define the circumference of human knowledge.[10] Both were concerned about ethics as well as aesthetics. And both had a deep and abiding connection to literature, which in significant ways – perhaps most evidently in their shared admiration for and probing reflection upon Goethe and his various works – helped to set the tone and agenda for their scholarly careers.

Yet James proposed that humans have a single, capacious way of understanding. Whether thought to be descriptive or explanatory, all knowledge in the Jamesian scheme comes from within, so to say, as situated, interested minds select and then compare what they have taken from the ongoing flow of experience. With hypotheses derived from this process of abstraction and comparison – whether of "physical" or "mental" phenomena – they proceed to act and draw further inferences, thereby testing their understandings by their consequences. Dilthey, meanwhile, as discussed more fully in other chapters in this volume, came to believe that there are two modes of comprehension, one indicated by the German word *Erklären*, signifying causal explanation based upon inductive and synthetic construction of disparate elements inferred from experience, and the other by the German word *Verstehen*, standing for understanding based upon deductive analysis of immediately given and originally coherent experience.

---

[9] The only other communication between James and Dilthey, of which I am aware, was a recommendation from James that the French philosopher Charles Renouvier be admitted to membership in the Prussian Academy, but this communication was sent to Dilthey through Stumpf (James 1992–2004, vol. 9, 599).

[10] Interestingly, James ([1909] 1977) illustrated his belief that "a man's vision is the great fact about him" and that "a philosophy is the expression of a man's intimate character," not with a reference to Dilthey's publications on *Weltanschauungen*, but with comments on a book in which Dilthey, among others (including Wundt, Ostwald, and Ebbinghaus) presented their various "idiosyncratic" philosophies. Turning from one chapter to another, James said, was like "turning over a photograph album" (ibid., 14).

There is some irony here. James, the resolute pluralist, concluded that there is a fundamental unity among the various forms of cognition, whereas Dilthey, the would-be reconciler of idealism and realism, was convinced that there are, in fact, two different modes of cognition. These modes were not, of course, coextensive with "physical" and "mental" phenomena, since the explanatory mode could be – and had been – applied to psychology (e.g., by "associationists" who treated sensations, ideas, and affects as so many elements to be related by explanatory laws). A better way to express the difference, according to Dilthey, is to distinguish between explaining-from-outside and understanding-from-inside. But to James, as suggested above, all cognition develops from inside, whether aimed at what is conceived to be "inside" or "outside" the person who knows or understands.

The fact that, in this one regard, James maintained a kind of monism and Dilthey espoused a form of dualism should not blind us to the related fact that both James and Dilthey opposed an *explanatory* monism, whether based on idealist or materialist premises. And more specifically to the point, both opposed the encroachment of positivistic scientism into the domain of the spirit. Still, what separated them was not so much their views on psychology as their views on science. In the end it seems that James's greater familiarity with, and more thorough criticism of, contemporary natural science was the fulcrum of their differences with regard to *Erklären* and *Verstehen*. Although Dilthey had been diligent in studying the history of science, as is apparent in his epoch-making *Introduction to the Human Sciences* (Dilthey [1883] 1989), his focus had been on physical science of the more distant past. James, however, was more knowledgeable than Dilthey regarding contemporary as opposed to historical developments in the physical sciences, and *much* more knowledgeable regarding current advances in the biological sciences, about which Dilthey's knowledge seems to have been relatively rudimentary, judging from his published discussions of science (Dilthey [1883] 1989, 192–206). In fact, Dilthey made *no* reference at all to Darwin in his treatment of science in his *Introduction to the Human Sciences*, concentrating exclusively on highly abstract features of traditional physical science.[11] Since Darwin was so central to James's conceptualization of human life and experience – indeed, of the world in general – this can only be seen as relevant to their differing positions regarding the natural and the human sciences, and regarding *Erklären* and *Verstehen*. Dilthey clearly contrasted *mechanistic physics* with the understanding of human experience, whereas James used *evolutionary biology* to elucidate it. In doing so, as we have seen, James brought evolutionary thought into conformity with his personal search for a meaningful human existence, one that left room for freedom, possibility, and hope. Meanwhile, Dilthey, who was not trained in science as James was, described physical science

[11] There is *one* mention of Darwin in the drafts and plans, never realized, for a second volume of Dilthey's *Introduction*, but this reference has to do with the general notion of *adaptation* rather than Darwin's specific theory of evolution. Indeed, Dilthey ([1883] 1989) wrote that this "great law of all life," that living creatures adapt to their environment, "is the same whether one applies the explanations of Aristotle or Cuvier, Goethe, Lamarck or Darwin to it" (ibid., 468).

*as he understood it* in a way that restricted belief in freedom, possibility, and hope to the realm of subjectivity, in stark contrast to the objectivity supposedly mandated by the scientific worldview.

This may be enough of an explanation – or is it a description? – of the differences between these two otherwise similar individuals. But there may also have been a relevant conceptual or temperamental difference between them – a tendency to seek similarities and connections, on the part of James, as opposed to a tendency to make distinctions and divisions, on the part of Dilthey. A nice illustration is provided by James's ([1890] 1981) description of a *tension* between the "transitive" and "substantive" elements embedded alike in the stream of consciousness (ibid., vol. 1, 233–240) and Dilthey's subsequent conversion of this tension into a radical *antinomy*, which he then used as a reason to reject introspection in favor of a hermeneutic understanding of human expression (Ermarth 1978, 212f.). This possible difference is further suggested by James's reaction to Windelband's distinction between fact and value. Writing to F.C.S. Schiller in September of 1904, James noted that he did not feel that he had missed much by not attending the Second International Congress on Philosophy in Geneva because the chief event had apparently been "a discourse by Windelband on the *distinction* of Fact & Value. . . .If so it shows how arrieré they are on the Continent. For we at least have got the question of their *connexion* clearly raised" (James 1992–2004, vol. 10, 473).

This is a telling comment. As James ([1912] 1976) said elsewhere, "life is in the transitions as much as in the terms connected" (James [1912] 1976, 42), and he sought out transitions or connections, sometimes even of an ethereal sort, before he accepted apparent divisions and differences. Again, this may reflect his more "organic" or "biological" premises, especially as these supported the pragmatic viewpoint that he shared with Schiller. We can see this viewpoint and related tendency in his erasing of a clear distinction between subjective and objective as well as in his denying any necessary differences between explanation and description – or *Erklären* and *Verstehen*.[12]

So we can understand the difference between James and Dilthey, as regards *Erklären* and *Verstehen*, in two ways: (1) James did not believe there was an adequate reason to suppose a radical distinction between description and explanation. (2) This principled conclusion accorded with his temperamental preference to see relations among even seemingly disparate views. Perhaps this suggests a good way to think about the difference between James and Dilthey themselves: not so radical that we should fail to appreciate the common spirit and overlap in their ways of thinking.

[12] To be fair, James sometimes made hard-and-fast distinctions, and Dilthey sometimes cautioned against over-reliance on distinctions. For instance, Dilthey ([1910] 1985), like James, commended the balancing of subjectivity and objectivity in Goethe's thought and work (Dilthey [1910] 1985, 278), and he is known to have opposed "one-sidedness" (Ermarth 1978, 78), as illustrated by his *Idealrealismus*. Still, I think there is a point to be made about their relative emphases in this regard.

## References

Browning R ([1868–1869] 1971) The Ring and the Book. Altick RD (ed.). Yale University Press, New Haven, CT

Campbell D (1960) Blind variation and selective retention in creative thought as in other knowledge processes. Psychol Rev 67:380–400

Croce P (1995) Science and Religion in the Era of William James, vol 1. Eclipse of Certainty, 1820–1880. University of North Carolina Press, Chapel Hill, NC

Darwin C (1859) On the Origin of Species by Means of Natural Selection. John Murray, London

Dilthey W ([1883] 1989) Introduction to the human sciences. In: Makkreel RA, Rodi F (eds) Selected Works, vol I. Translated by Michael Neville. Princeton University Press, Princeton, NJ

Dilthey W ([1894] 1977) Ideas concerning a descriptive and analytic psychology. In: Zaner RM, Heiges KL (eds) Descriptive Psychology and Historical Understanding. Translated by Richard M. Zaner and Kenneth L. Heiges. Martinus Nijhoff, Den Haag, pp 21–120

Dilthey W ([1910] 1985) Goethe and the poetic imagination. In: Makkreel RA, Rodi F (eds) Selected Works, vol V. Translated by Christopher Rodie. Princeton University Press, Princeton, NJ, pp 235–302

Ebbinghaus H (1895) Über erklärende und beschreibende Psychologie. Zeitschrift für Psychologie und Physiologie der Sinnesorgane 9:161–205

Emerson RW ([1841] 1983) The over-soul. In: Essays and lectures. Porte J (ed). The Library of America, New York, pp 38–400

Emerson RW ([1850] 1983) Representative men. In: Essays and lectures. Porte J (ed). The Library of America, New York, pp 611–761

Ermarth M (1978) Wilhelm Dilthey: The Critique of Historical Reason. University of Chicago Press, Chicago, IL

Fisch MH (1964) Was there a metaphysical club in Cambridge? In: Moore EC, Robin RS (eds) Studies in the Philosophy of Charles Sanders Peirce, second series. University of Massachusetts Press, Amherst, MA, pp 3–32

Guerlac H (1977) Where the statue stood: Divergent loyalties to Newton in the eighteenth century. In: Essays and Papers in the History of Modern Science. Johns Hopkins University Press, Baltimore, MD, pp 131–145

Hacking I (2002) Language, Truth, and Reason. In: Historical ontology. Harvard University Press, Cambridge, MA, pp 159–177

Hughes HS (1961) Consciousness and Society: The Reconstruction of European Social Thought 1890–1930. Random House, New York

James W ([1875] 1987) Chauncey Wright. In: Burkhardt F (ed.) Essays, Comments, and Reviews. Originally published in The Nation (Sept 1875). Harvard University Press, Cambridge, MA, pp 15–17

James W ([1876] 1978) The teaching of philosophy in our colleges. In: Burkhardt F (ed.) Essays in Philosophy. Harvard University Press, Cambridge, MA, pp 3–6

James W ([1878] 1983) Brute and human intellect. In: Burkhardt F (ed.) Essays in Psychology. Harvard University Press, Cambridge, MA, pp 1–37

James W ([1878] 1978) Remarks on Spencer's definition of mind as correspondence. In: Burkhardt F (ed.) Essays in Philosophy. Harvard University Press, Cambridge, MA, pp 7–22

James W ([1878a] 1988) Johns Hopkins lectures on 'The senses and the brain and their relation to thought.' In: Burkhardt F (ed.) Manuscript lectures. Original lectures delivered 1878. Harvard University Press. Cambridge, MA, pp 3–16

James W ([1878b] 1988) Lowell lectures on 'The brain and the mind'. In: Burkhardt F (ed.) Manuscript Lectures. Original lectures delivered 1878. Harvard University Press, Cambridge, MA, pp 16–43

James W ([1881–1882] 1988) Notes for philosophy 4: Contemporary philosophy. In: Burckhardt F (ed.) Manuscript Lectures. Original lectures delivered 1881–82. Harvard University Press, Cambridge, MA, pp 178–182

James W ([1890] 1981) The Principles of Psychology. In: Burkhardt F (ed.) 2 vols. Harvard University Press, Cambridge, MA
James W ([1892] 1984) Psychology: Briefer Course. In: Burkhardt F (ed). Harvard University Press, Cambridge, MA
James W ([1902] 1985) The Varieties of Religious Experience. In: Burkhardt F (ed.). Harvard University Press, Cambridge, MA
James W ([1907] 1975) Pragmatism. In: Burkhardt F (ed.). Harvard University Press, Cambridge, MA
James W ([1909] 1975) The Meaning of Truth. In: Burkhardt F (ed.). Harvard University Press, Cambridge, MA
James W ([1909] 1977) A Pluralistic University. In: Burkhardt F (ed.). Harvard University Press, Cambridge, MA
James W ([1911] 1979) Some Problems of Philosophy. In: Burkhardt F (ed.) Originally published posthumously. Harvard University Press, Cambridge, MA
James W ([1912] 1976) Essays in Radical Empiricism. In: Burkhardt F (ed.) Original work, composed of previously published essays, gathered and re-published posthumously. Harvard University Press, Cambridge, MA
James W (1992–2004) The Correspondence of William James. In: Skrupskelis IK, Berkeley EM (eds) 12 vols. University Press of Virginia, Charlottesville, VA
Ladd GT (1892) Psychology as so-called 'natural science'. Philos Rev 1:24–53
Leary DE (1990a) Psyche's muse: The role of metaphor in psychology. In: Leary DE (ed.) Metaphors in the History of Psychology. Cambridge University Press, New York, pp 1–78
Leary DE (1990b) William James on the self and personality: Clearing the ground for subsequent theorists, researchers, and practitioners. In: Johnson MG, Henley TB (eds) Reflections on the Principles of Psychology: William James After a Century. Erlbaum, Hillsdale, NJ, pp 101–137
Leary DE (1992) William James and the art of human understanding. Am Psychol 47:152–160
Mandelbaum M (1971) History, Man, & Reason: A Study in Nineteenth-Century Thought. Johns Hopkins University Press, Baltimore, MD
Menand L (2001) The Metaphysical Club: A Story of Ideas in America. Farrar, Straus and Giroux, New York
Mill JS ([1872] 1973) A System of Logic: Ratiocinative and Inductive. In: Robson JM (ed.) Reprint of the 8th edn. University of Toronto Press, Toronto
Ostwald FW (1909) Grosse männer: studien zur biologie des genius. Akademische Verlagsgesellschaft, Leipzig
Peirce CS ([1893] 1992) Evolutionary love. In: Houser N, Kloesel C (eds) The Essential Peirce, vol 1. Indiana University Press, Bloomington, IN, pp 352–371
Peirce CS ([1907] 1998) Pragmatism. In: Houser N (series ed) The Essential Peirce, vol 2. Original work written but not published in 1907. Indiana University Press, Bloomington, IN, pp 398–433
Perry RB (1935) The Thought and Character of William James, 2 vols. Little, Brown & Company, Boston, MA
Richards RJ (1977) The natural selection model of conceptual evolution. Philos Sci 44:494–501
Wiener PP (1949) Evolution and the Founders of Pragmatism. Harvard University Press, Cambridge, MA
Wordsworth W ([1814] 1977) The excursion. In: Hayden JO (ed.) The Poems, vol 2. Oxford University Press, Oxford, pp 35–289
Wright C ([1873] 1877) Evolution of self-consciousness. In: Norton CE (ed.) Philosophical Discussions. Henry Holt, New York, pp 199–266

# Chapter 8
# *Erklären, Verstehen*, and Embodied Rationalities: Scientific Praxis as Regional Ontology

**Katherine Arens**

> *The goal of scholarly research is not only the cognition, but also the understanding of phenomena. We have gained cognition of a phenomenon when we have attained a mental image of it. We understand it when we have recognized the reason for its existence and for its characteristic quality (the reason for its being and for its being as it is) (Menger [1923] 1976, 43).*

Today's typical formulation of the *Erklären/Verstehen* differential rests on a stylized conceptual dichotomy preserved, for example, in the idea of the *two cultures*, or in Windelband's classic distinction between nomothetic and ideographic sciences (those based on generalized laws, and those exploring a single case in depth, with knowledge based on empirical evidence treated inductively or deductively). It is commonplace to follow Windelband and to differentiate among the sciences and set them over and against historical or historicized understandings (*Erfahrungswissenschaften*).[1] This practice projects the sciences as if they were complementary rationalities, each a formalized paradigm with its own distinct analytics and field of data.

The present discussion will take a less strictly historicist view to guide its analysis of science, based on my own earlier work on a strand of nineteenth-century Kant reception outside of university NeoKantianisms, preserved in theories of education[2] and in specific critiques of scientific epistemology (especially those of Ernst Mach, as treated briefly below). For historical reasons, that inheritance seems to have resonated most strongly in the sciences in the Austrian and Austro-Hungarian Empires of the latter nineteenth century (even in the nascent social sciences and in the characteristic modernizing impulses for the humanities proper).

K. Arens (✉)
E.P. School 3.102, Department of Germanic Studies, University of Texas at Austin, Austin, Texas 78712
e-mail: k.arens@mail.utexas.edu

[1] This appeal to *Erfahrungswissenschaften* we find in Windelband's "Geschichte und Naturwissenschaft" ([1894] 2006).

[2] See Arens (1989) for an account of that inheritance.

U. Feest (ed.), *Historical Perspectives on Erklären and Verstehen*, Archimedes, Vol. 21,
DOI 10.1007/978-90-481-3540-0_8, © Springer Science+Business Media B.V. 2010

Following this tradition allows a different image of the era's empiricism to emerge, and to recover a different model of hermeneutics than that associated with dominant readings of the *Verstehen* and *Erklären* divide – somewhat different than in the philosophical–philological models most familiar from hermeneutics since Schleiermacher that try to bridge experience and mind. The alternate tradition traced here defines science not as research, but rather specifically as a hermeneutics based on analyses of phenomenal fields of experience. The understandings and/or explanations which science produces, in turn, do not base themselves on patterns of pure logic and theory-formation, but rather are heavily implicated in cultural rationalities which heavily overwrite any baseline experience or logic available. This alternate *science* is more heavily implicated with learning practices and with community norms for designation and communication rather than with problems of knowledge production and retrieval; it acts as a hermeneutics of cultural understandings more than it refers to *a priorist* cognitive acts of mind.

To argue for the existence of such an alternative to the discussions of science in *Erklären/Verstehen* debates, I will take up three brief case studies from different branches of science (natural, social, and human) to show how they take up knowledge as questions of embodied and socialized rationalities formalized as distinct acts of cognition. Each science is a formalization offering only one reading of a pool of empirical evidence shared within a culture; each is based on acts of designation which rely on the culture's inheritance of language; and each is influenced by the individual performing acts of cognition.

In this paradigm for science, to even consider dividing experience and understanding (or *Erklären und Verstehen*) is a methodological error reifying "the empirical" and privileging cognition. Both the bottom-up cognition of induction, moving from specifics to generalities, and the deduction that begins with general principles and ends at specifics are two moments within shared acts of understanding implicated in culture – not in "empirical evidence," but rather in material experience always already implicated in cultural interests and how those interests have shaped the mind–body interface, and in designations as pre-structured fields of cognition. All science is thus more properly to be described as an exercise in a kind of semiotic realism (my term), aimed not at discerning the individual contributions of mind and experience to understanding, but rather at uncovering how the knowledge it produces represents a strategic extension of the interests already present in any act of designation.

Thus acts of understanding enacted as "science" are immediately with community interests, the limits of the individual observers/analysts, and the persistence of larger-scale logics of embodied practice (habits, neurophysical programming, disciplinary practices) and culture. *Values* are hence not seen as separated from *logic*; the *subjective* and *objective* sides of most hermeneutic models cannot be divided from each other; historical, theoretical, and practical knowledge are produced simultaneously. Each act of knowing, in consequence, is modeled as a situated production or retrieval of knowledge operating within a "frame of reference" chosen by the science to enact its truths, yet also implicated with multiple others.

The cases which follow, therefore, are rereadings of three projects of science which, I believe, cannot be adequately understood in terms of the *Erklären/Verstehen* debate, because they operate in an alternate yet wide-spread nineteenth-century paradigm for *situated* and *embodied* practices of science as world understanding, one that has not heretofore been acknowledged as a coherent approach to the philosophy of science.

My discussion will take as its first case the work of the physicist Ernst Mach (1838–1916), professor for the history and theory of the inductive sciences at Vienna. More than the phenomenalist he is too often dismissed as, he redefines science's epistemology by setting as primary the field of phenomenal data which constitutes the corpus that a science evolves to cognize and individual sciences' rationalities, and thus simultaneously historicizing the rationality that emerges as a science as practiced in a particular context. In this sense, Mach grounds science in a community's experience and in the cognitive habits and concepts *inherited* from prior generations as situated narratives of experience – a semiotic problem rather than an empirical one. As such, he posits each science as a relative, contingent logic of local concepts, situated within a communication community and coextensive with various forms of *praxis*, including technology, social habits, and spaces.

After outlining how Mach defines science, I will then turn to two further cases addressing how a particular science – a systematic analysis of a distinct field of data – needs to be theorized against these basic assumptions.

The second case is the economics of Carl Menger (1840–1921), who distinguishes between the theoretical, historical, and practical sciences of economics as related but not identical understandings of the whole phenomenal field which is the domain of economics. His work takes a science paradigm like Mach's into the realm of value-driven social sciences. The third example is the art historian Alois Riegl (1858–1905), whose *Historical Grammar of the Visual Arts* (written 1897–1898, but not published at the time) models a modern scientific art history on the basis of aesthetic experience (a specific category of empirical evidence, codified in a specific semiotics). Riegl defines art history as the science concerned with various cultures' *Kunstwollen*, the collective will (intentionality) to create works of art as artifacts articulating an era's self-understanding. Like Mach and Menger, Riegl outlines his "science" as an analytic logic working within a historicized horizon of knowledge and intentionality reflecting the *physis* and *praxis* of its community.

As noted above, these three discussions operate along distinctions different than the traditional *Erklären* and *Verstehen* differentiation. They factor in ideology and historical situatedness as foundational to understanding rather than privileging either *a priorist*, logical strategies for *explanation* or a naïve realism of *understanding*. Both understandings, in their accounts, exist as commonly grounded in a culture's material semiotics and praxis – an assumption common to many other disciplines in the *fin de siècle* context of science theory and practice.

Before I proceed, a methodological point is in order. The following discussion does not uphold historical conventions by organizing its narrative around origins, schools, or influences. Instead, what I am tracing might best be described as a scientific community sharing a particular epistemological strategy – not a discipline, but a set of procedural assumptions about the status of empirical data (it is not only phenomenal, but cultural material), and about how community praxis, not just mind, grounds all knowledge production. As such, I will most often point to the tools they use to make their arguments, rather than to the ideology of a particular set of terminology (often the "signature" of their "movement" or "school"). My justification for so doing is that, in many cases, historically attested debates between schools and experts are taken to be more serious breaks in strategies for knowledge production than they actually are: Lenin, Bukharin, and Bogdanov may diverge in their chosen implementation of their respective interpretations of Marxism, even while accepting a common source for their thought and much the same etiquette of praxis in pursuing the cause of the proletariat. One basic theory administered differently will create evidence of ideological difference and disciplinary privilege. One theory or set of related theory administered under different ideologies of practice will yield ethical claims on the social and political networks in which it is situated.

My case studies here – Mach, Riegl, and Menger – share an understanding of how science produces knowledge from designated data sets, thus defining science differently than in the debates around Windelband. Their models share the basic assumption that cognitive modeling is incomplete without attention to the social, political, and praxiological contexts in which it is enacted. Just as critically, these three cases approach science as expert hermeneutics designed to unfold systems of material designations, not "empirical evidence" or cognition in general. As such, they straightforwardly assume that phenomena never present themselves neutrally, but always in forms influenced by culture, history, and interest of the group supporting acts of knowing. Science, in turn, takes as it charge the analyses of these data sets as situated cultural knowledge and praxis.

## 8.1 Mach and the Cultural Implication of Phenomena

Ernst Mach is today remembered as both a physicist and philosopher.[3] Public memory preserves the significance of his experimental proof of the Doppler Effect and his influence on Einstein's technique of thought experiments; philosophically, his work on frame of reference and the role of the observer lead him to be classified as a phenomenalist and often as a major vector between Kant's work and the Vienna Circle of the 1920s. Mach's project was an analysis of the cognitive and psychological underpinnings of the sciences as well as their technical data sets, purportedly in a synthesis of empiricism and idealism. Yet Mach distanced himself from epistemology proper, "not claiming [to be] a philosopher," as the preface to his *Analysis of the Sensations* (Mach [1900] 1959) expresses. Instead, he approached the problem of knowledge as a working physicist interested in bringing common understanding and experimentalism together without succumbing to a naïve realism of experience.

Overall, Mach's approach takes on another face, since his *Analysis* offers a kind of critical foucauldian genealogy of scientific concepts *avant la lettre*. In this text, he belies any simple separation between understanding through reference to a formalism (*erklären*) or through direct experience (*verstehen*). Experience and mind are united in the Kantian act of grasping – *begreifen* – in which concepts are

[3] He was Professor of Physics at Graz from 1864–1867; at Prague from 1867–1895; and finally, at Vienna, from 1895–1901.

produced, later to be involved in theories and more elaborate culture-bound knowledge. In this sense, concepts and laws are not logically prior to experience, and experience itself is subject to the intentional logic inherent in the concepts used and in the embodied practices which instantiate understanding. Thus experience emerges as knowledge only in an act of *interested* cognition.

In Mach's accounts, then, all logics are historically bound and conditioned, tied into the notational systems in which the laws of nature and the observed reality are recorded; all science, in turn, produces knowledge that is situated historically and culturally. In consequence, a science like physics cannot be defined as a science of principles. Instead, the science's epistemological claim rests not only on what knowledge it produces within its formalism, but also as it reflects a culture's space of cognition and its strategies for designation. A science can never produce knowledge that can claim to transcend history. In consequence, Mach consistently stresses how even the supposedly "pure" sciences originate in mind as part of history.

Mach's project reveals itself across a series of texts, starting from his *Analysis of the Sensations* (Mach [1900] 1959) and *Knowledge and Error: Sketches Toward a Psychology of Investigation* (Mach [1905] 1975), and then moving on to his great textbooks, *Physics for High School* (Mach 1894), *Mechanics* (Mach 1896), and *Optics* (Mach 1921), and to his *Popular-Scientific Lectures* (several differing editions since 1896) (Mach [1896] 1910).

The textbooks focus on how cherished concepts such as measurement scales and mass have been formulated under historically relative conditions, which must be approached Kantian-critically, since inherited scientific nomenclature and problem-solving techniques represent explanations in specific contexts under specific observational conditions. In consequence, Mach sees little or no distance between the acts of *Verstehen* and *Erklären*, since hermeneutic–philological acts of understanding fall together with those uncovering how formalisms produce rule-bounded knowledge. Scientific nomenclature gives material to give form to acts of understandings formulated with respect to a specified observed set of data designated as reality and annotated in terms satisfying stated historical–cultural purposes, including intentionalities and other more overt ideological skews.

For example: various temperature scales (Centigrade, Fahrenheit and Kelvin) facilitate rule-governed analyses of data, each in their own ways, since each scale is indexed to a different purpose and to a different set of physical properties designated as important. By extension, the operative concepts of a physical science – concepts such as "wave," "mass," and "causality" – implicate not only data sets but also specific rule-governed interests, each reflecting a specific purpose. Mach's foreword to the *Analysis of the Sensations* ties such designations to the understanding of phenomena and their relations. In this prolegomena to any specific science, Mach describes the general acts of mind which generate scientific concepts off the basis of observed facts under the pressure of culture, before these concepts are extended into systematic disciplines. In these explications, *Verstehen* and *Erklären* fall together, because both experience and explanation function in reference to the cultural third, designation, which encompasses history, group experience, and habit as it contributes to knowledge production.

In this move, Mach posits all scientific understanding as an "analysis of the sensations" that integrates history and culture into cognition, which occurs historically-imminently, in a fundamental "antimetaphysical thrust" (Mach [1900] 1959, vii). From this perspective, a single formalism (a single discipline of any sort whatsoever) can never be said to function as a single master framework of understanding (*a priori* or *a posteriori*), because the narrower realm of a single science like physics will never be conducive to the progress of knowledge past a certain point: "in spite of its considerable progress, physics is only a part of the greater total knowledge, and cannot, with its one-sided intellectual means created for one-sided purposes, exhaust the materials presented by the physical world" (Mach [1900] 1959, 1). At the very least, the act of *designating concepts*, thus shaping the data of experience, is a second critical factor modifying scientific understanding by turning experience into a product of culture and into culture's networks of communication. Science seen as a logical formalism, in consequence, must be seen as offering only a single perspective on one already-skewed set of the data of experience.

Mach thus pursues science as imminent to the observer's individual and cultural location, based on phenomenal evidence, on the "elements of sensation" available. Most critically, these elements are implicated not only with seemingly objective properties appearing in consciousness (colors, tones), but also with those properties usually attributed to the individual observer (moods, feelings, and volition). These two frames of understanding, however, need to be seen as already present in the framework of the observer's acculturation, including the patterns of knowledge encoded in designation (Mach [1900] 1959, 2). Each concept – "scientific" ones like color and tone, and "psychological" ones like moods or volition – is simply a series of complexes, habituated into observable patterns. These data sets, in turn, exist simultaneously in individual and cultural memory in no small part because interest has been called to them:

> Relatively stable and persistent [entities] stand out from this network; they become imprinted in memory and expressed in language. As relatively more persistent, *complexes* of colors, tones, and pressures which are linked spatially and temporally are distinguished primarily; they therefore have received special names and are designated *bodies*. Such complexes are by no means absolutely persistent. (Mach [1900] 1959, 2)

Thus by definition, any science must be seen as a formalism built around acculturated designations and producing knowledge within its field of complexes, according to the purposes for which that field was designed, rather than as a more or less invariant logic or stable field of experience: "Thus, then, is constituted the great cleft between physical and psychological investigation for a habituated, stereotyped observational perspective.... Not the data, but the direction of observation, differs in the two fields" (Mach [1900] 1959, 11f.).

Twentieth century philosophical phenomenologists often deride Mach's methodological move as "psychophysics," a naïve conflation of empiricist and idealist perspectives. Yet here I underscore that Mach is rendering material an ideal very much like Husserl's notion of intentionality, where designated phenomena appear multiply as parts of different states of affairs – a considerably less *a priorist* notion of phenomenology than that made familiar in Heidegger's tradition, but a phenomenology nonetheless, albeit one with claims of factoring in embodiment, not just ideas. A science is a frame of reference centered around an observer, but not solely determined by an individual perceiving consciousness. Any conclusions reached about the constitution of an "ego" or mind beyond an analysis of how these concepts guarantee a kind of stable analysis of a field of elements are equally as false as any ontological argument in science: "[t]he world does not consist of puzzling entities for us" (Mach [1900] 1959, 21). Overall, science is best seen as a set of management strategies for this field of data, shared by a community at determinate sites:

> Science always originates in a process of the adaptation of thoughts to a specific region of experience. The results of the process are the elements of thought which are capable of representing the entire region. The results will naturally differ according to the type and magnitude of the region. If the region of experience expands, or if a number of regions unite which had been separate to this point, then the inherited, habituated elements of thought no longer suffice for the expanded region. In the struggle of acquired habits with the attempt to adapt, *problems* originate, which disappear with the completed adaptation, in order to make room for whatever others will arise. (Mach [1900] 1959, 22)

"Habituated elements of thought" are the basis of science; any science progresses by a constant comparison of the known to data, in a pattern of analysis that has, according to group experience in history, proved itself suited to its chosen region of data or elements and a particular set of goals to be fulfilled or discarded. Each science is thus actually an act of interested analysis, locally and culturally contingent: "No point of view has an absolutely permanent applicability; each is only important for a designated purpose" (Mach [1900] 1959, 27). The material semiotics of signs used to designate concepts plays a significant role in this local situation. A technical lecture delivered to a lay audience, for example, must be shifted into a different frame of material designation, its concepts adapted to the interests of the new audience and to perspectives often dormant or peripheral to those in the original formalism.

Mach's science, then, is anything but the representation of world data ordered by mind. Instead, even the concepts of "ego" and "self" only serve as reference points within a given region of elements of the sensation. An individual mind does not create the world; instead, perceptions instantiate already present meanings, their limits, and logics, within the field of sensation managed by the group. "The world" is an idealized representation of such interested concepts that might even reify itself and hinder the free progress of knowledge. An individual "consciousness" is the designated center of a set of perceptions, neither individual nor group nor natural. Instead, it is simply a center of data regions grasped as concepts – made, as the root for the word fact, *factum*, suggests – and interrelated to serve interests. In this reading, "laws of nature" are inductions specific to the perceptions of a particular time and place that do not *prescribe* any necessity to nature, but only *describe* a possible contingency ("Facts are not required to order themselves according to our thoughts" (Mach [1905] 1975, 447f.).

Thus as Mach describes it, natural science is neither nomothetic nor ideographic, nor even opposed to the *Geisteswissenschaften* in Dilthey's sense. Instead, natural science is the first of the sciences of culture. Science is little more than a systematized attention paid to a region of data identified as significant within a culture;

any particular science serves one set of interests within that region. "Theories" of the natural sciences straightforwardly summarize a particular culture's point of view toward and means of orientation within experience. They are neither permanent, nor ontic; instead, they are complex abstractions and functional models made to serve specific community interests and to articulate a specific ideal of knowledge within a realm of sense data, and as such, are passed down as frames that reify understandings of that data and a group's orientation within the frame as a communicable view of reality.

The link Mach makes therefore is that between cognition and the material existences of bodies, habits, and designations. He thus limits the distance between *Verstehen* and *Erklären* to the point of its disappearance. All cognition is tied to the purposes and intellectual inheritance of a group, not to the actions of absolute mind or to a principle of rule-governedness in and of itself. Mach immediately ties science to social and ethical purposes, as well as to mediality – to the designations which mark a group's cognitions and which render them communicable. His is a semiotic realism anchored in bodies and designation, not a transcendental or materialist one.

Mach's contribution to the era's *Wissenschaftsdebatte*, the debate about the sciences, is thus more than a mere historicization of sciences theories and forms, as is commonly assumed. More fundamentally, he associates science with culture, as a formalization and clarification of a group's experience and purposes, and as a redefinition of objectivity as existing semiotically. His view of science is that it is nomothetic only insofar as general rules and principles derive not from mind alone, but principally from patterns inherent in a group's understanding (acculturated, learned, and habituated in bodies and signs as well as minds). Yet neither is it classically ideographic, because it takes experience as a prestructured field of potential experience, not a one-time constellation of experience or state of affairs. Each understanding is indeed relative to a moment in history, but it is absolutely linked in terms of intentionality to a specific frame of reference, and hence objectively present for each of the users and in each of the users.

For the overall case I am making here, it is critical to note that there is no two-culture divide in Mach's account. He has instead defined natural science as the *first* of the sciences of culture – the most important of the realms of science because of the concepts it, in any of its iterations, frames for the observer as a world activity. To make this point in another way, it is worthwhile to take on another kind of science – in this case, an example drawn from the social sciences cast using a similar model of systematic knowledge within a group: Carl Menger's economic theory.

## 8.2 Menger's Economics as Phenomenology of Needs

Carl Menger (1840–1921) is widely remembered today as the founder of the Austrian School of economics, as well as for a writing career culminating in his *Principles of Economics* (*Grundsätze der Volkswirtschaftslehre* (written 1871)

(Menger [1923] 1976) and his great methodological treatise, his *Investigations into the Method of the Social Sciences with Special Reference to Economics* (*Untersuchungen über die Methode der Sozialwissenschaften und der politischen Oekonomie insbesondere*, written 1883) (Menger [1883] 1985).

The early phases of his career found him working primarily in the civil service, writing first for the semi-official *Wiener Zeitung* and then as professor of economics at the University of Vienna. The respect given his work was evidenced in his commission for a series of lectures to be delivered to Crown Prince Rudolf. Later, after serving 4 years in the Austro-Hungarian army in World War I (1 year as a staff officer in economics), he worked out a theory of the economics of war. But the Menger school moved out of Austria in the early 1930s, displacing its influence to the United States, particularly to the University of Chicago. Among his most famous followers were Ludwig von Mises and Friedrich von Hayek, the latter of whom received the 1974 Nobel for economics as an opponent of Keynesian economics.

In the popular account, Menger's economics is a science of social action, based on a modified deductive logic at least partially compatible with socialist theories of economics such that that of Otto Bauer and other theorists of Red Vienna. Menger's most memorable contribution is what is termed his subjective approach to economics, particularly his "theory of marginal utility," which ties the value of a unit of a commodity to an individual's psychological perception of its value. In its simplest form, that theory outlines how the number of commodity units owned by an individual affects the goods' value. The more units owned, the less the owner will value each individual unit, thus making commodity value an unstable number.

Working out from such general principles, Menger defines money and the economy as correlated with needs and desires, showing in each case how price and money is tied to human value, choice, needs, and their satisfaction, not to commodities in themselves or to abstract systems of capital appreciation. Menger thus consciously combined aspects of ethnography and philosophy into his economics (Menger [1883] 1985, 36), yet still described it as a science, "the solution of a problem that is directly related to human welfare, to serve a public interest of the highest importance, and to enter a path where even error is not entirely without merit" (Principles, Menger [1923] 1976, 46). His contemporaneous adversaries are identified as the German Historical School, which was trying to provide the state with objective data about price and value (Menger [1883] 1985, 1).

That attempt to make economics a particular kind of engaged applied science led Menger to intense work on theoretical models. His preface to *Principles of Economics*, for example, thus begins to explain his "empirical method" as lying apart from a classical natural science (Menger [1923] 1976, 47). Yet the science he describes is drawn much in the spirit of Mach: a science of principles originating in imminent experience, aimed at a systematic investigation of phenomena, using a field of data not taken up scientifically to that point. He recognizes that this attempt implicates his work in something like the era's monist project, the idea that there is only one core set of cognitive acts at the core of a science: "the task of our age is to establish the interconnections between all fields of science and to unify their most important principles" ([Menger [1923] 1976, 47).

As he works out his project, Menger reveals himself as working very much in Mach's vein, treating phenomenal data a conscientious phenomenologist of a very specific sort:

> In what follows I have endeavored to reduce the complex phenomena of human economic activity to the simplest elements that can still be subjected to accurate observation, to apply to these elements the measure corresponding to their nature, and constantly adhering to this measure, to investigate the manner in which the/more complex economic phenomena evolve from their elements according to definite principles. (Menger [1923] 1976, 46f.)

His economics is not "natural-scientific" because it deals with phenomena of human behavior rather than objects of nature, but it still hopes to provide laws describing the economic behavior of individual humans with free will, on the basis of empirical evidence.

Menger's first methodological problem parallels Mach's: describing the elements of his economics, with the field of data properly circumscribed. His preliminary description of his science of economics already has a clear methodological focus:

> Whether and under what conditions a thing is *useful* to me, whether and under what conditions it is a *good* [commodity], whether and under what conditions it is an economic good, whether and under what conditions it possess *value* for me and how large the *measure* of this value is for me, whether and under what conditions an *economic exchange* of goods will take place between two economomizing individuals, and the limits within which a *price* can be established if an exchange does occur . . . economic theory is concerned, not with practical rules for economic *activity*, but with the *conditions* under which men engage in provident activity directed to the satisfaction of their needs. Economic theory is related to the practical activities of economizing men in much the same way that chemistry is related to the operations of the practical chemist. (Menger [1923] 1976, 48f.)

He is describing an economics that respects human choice and needs ("freedom"), even while casting them as laws expressing regularities in behavior: "Things that can be placed in a causal connection with the satisfaction of human needs we term *useful things*" (Principles, Menger [1923] 1976). Such terminological distinctions lead to the evolution of scientific logics modeling regularities in behavior correlated with economic concepts.

Menger's introduction to the later *Investigations into the Method of the Social Sciences* situates his project with respect to the *Methodenstreit* between German and Austrian Economics. There has been, he notes, a tradition of defining political economy as part of national economy. From a methodological point of view, this is inadequate theorizing, because it conflates two different fields of data (and hence two different logics): "The methods of theoretical economics and of the practical sciences of national economy cannot be the same" (Investigations, Menger [1883] 1985, 23f.). The projects that have been lumped together under the common rubric of "economics" thus actually are two different disciplines, as Menger will ultimately describe. Each discipline is a different logical projection, a different understanding about human need and goods at the basis of systems of economic exchange. Menger's own theory project "is primarily concerned with determining the nature of political economy, of its subdivisions, of its truths, in brief, with the goals of research in the field of our science" (Menger [1923] 1976, 26); it is a social science, parallel to "linguistic research, of political science, and of jurisprudence" (Menger [1923] 1976, 29).

Menger's text is a long discourse on practical method, based on the kind of experiential phenomenology we have already seen in Mach – a science based on observation and physical experience, within a cultural context. Menger first notes that his project is *empirical* – that is, aimed at the general, not just at the individual and concrete forms of appearance. That distinction is critical: *empirical* is the word he uses to talk about *patterns of phenomena*, the "general," or something closer to the *intentionality* of patterns that reveals itself through particular concrete appearances in a time and space. As he summarizes:

> The world of phenomena can be considered from two essentially different points of view. Either there are concrete phenomena in their position in space and time and in their concrete relationships to one another, or else there are the empirical forms recurring in the variation of these, the knowledge of which forms the object of our scientific interest. The one orientation of research is aimed at cognition of the concrete, or more correctly, of the *individual* aspect of phenomena; the other is aimed at cognition of their *general* aspect. (Menger [1923] 1976, 35)

Concrete phenomena are individual, as opposed to the general knowledge of empirical types; both poles are mutually implicated in the science he constructs.

How Menger describes science in general is critical to seeing how he bridges (circumvents) a possible distinction between ideographic and nomothetic sciences. He outlines ideas and their phenomenological-empirical forms as mutually predicating, not one prior to the other. And as science emerges from such mutually predicating acts, phenomena are reveled as interconnected:

> not every single phenomenon exhibits a particular empirical form differing from that of all others. Experience teaches us, rather, that definite phenomena are repeated, now with greater exactitude, now with lesser, and recur in the variations of things. We call these empirical forms *types*. The same holds true of the relationships among concrete phenomena. These also do not exhibit a thorough individuality in every single case. We are able, rather, to observe without much difficulty certain relationships among them recurring now with greater, now with lesser regularity (e.g., regularities in their succession, in their redevelopment, in their coexistence), relationships which we call *typical*. (Menger [1923] 1976, 36)

These "typical" relationships might also be characterized by reference to gestalt, a state of affairs or potential relation to be realized once tied to a historical time and place.

The phenomena to be investigated by economics in their typological connections are social in nature; they exist in typical relationships that emerge as the general laws of these phenomena – their science. In contrast, individual phenomena in typological relationships are just concrete, individual moments within the greater whole. Note, in addition, that "general" does not mean collective, just as "individual" does not mean belonging to an individual subject of knowledge. Scientific comprehension of these typical empirical forms is directed at a more comprehensive understanding of the "real" world, defined more precisely as the lived world of the knowers, including their ability to plan (Menger [1923] 1976, 36f.).

Menger's differentiation between the *concrete* and the *empirical* is critical. In economics, the empirical forms around which knowledge is generated are concepts such as "price" or "ground rent." Through them in the aggregate of their historically attested forms, we gain cognition of phenomena, but then also an understanding of it, "the reason for its existence and for its characteristic quality" (Menger [1923] 1976, 43). Ultimately, "we become aware of the basis of the existence and the peculiarity of the nature of a concrete phenomenon by learning to recognize in it merely the exemplification of a conformity-to-law of phenomena in general" (Menger [1923] 1976 45). This he terms a "theoretical understanding" of the matter in question. Yet that theoretical understanding is not yet a theory: the former is simply a different form of historicized understanding, while the latter is a logic which may or may not have relevance to any particular context of practice.

In moving toward a theory of economics, Menger also clearly uses these general relationships to break it into different subdisciplines. In so doing, he distinguishes first between historical and pure theoretical sciences (Menger [1923] 1976, 38), with the former being individual (one time, situated), and the latter, general (focused on types, recurring across historical appearance). A historical science deals with concrete phenomena (patterns of particular appearance); the theoretical discipline is *empirical*, the study of types. Beyond these, there is a third kind of science, "practical sciences or technologies" (Menger [1923] 1976, 38), which include practical applications like economic policy, national economy, and finance. Each of these "sciences" (systematics for knowledge production) also reflects a different interest inherent in an act of understanding, as well. History explores "the individual nature and the individual connection of economic phenomena"; theory investigates "their general nature and general connection (their laws)"; practical science (like national economy), "investigating and describing the basic principles for suitable action (adapted to the variety of conditions)" (Menger [1923] 1976, 39). In consequence, Menger will identify *three* different economic sciences, each arising from different interests in and experiences of economic phenomena.

In any such science, *laws* – relationships – are *not* to be considered permanent or *a priori* valid. Instead, a law is imminent to a state of affairs, to the organizing pattern inherent in a set of phenomena, "regularities in the coexistence and in the succession of phenomena" (Menger [1923] 1976, 50). When they seem to be without exception, they are called laws of nature; if they admit of exceptions, they are "empirical laws" (Menger [1923] 1976, 50). Exceptions, as Mach also noted in his analyses of frames of reference, are simply different classes of phenomena, in different relationships, seen from different points of view (Menger [1923] 1976, 51).

Those phenomena designated as part of "nature" seem to yield more exact understandings because they are based on simple, relatively stable patterns of data. In contrast, when one deals with phenomena of culture, then the science will also be dealing with a gap between "truths" of nature and those of culture – with typology issues that work differently in these different frames (again, as Mach showed). These truths are nonetheless regularities of cognition (Menger [1923] 1976, 60), a situation that is posited as stable, in which the same conditions of understanding should produce the same output. That is, "truth" is a condition where typologies can be understood as patterns emerging regularly from a region of data as it is being cognized.

The difficulty faced by each science in producing understandings of this sort is: how does one establish what regularities actually implicate the "same conditions" that produce the "same output"? Menger finds an answer in a recourse to procedure, in a systematic analysis of phenomena that precedes any reconstruction or synthesis of a pattern grounding understanding. Like Mach, Menger also starts with the "simplest elements of everything real" and then analyzes them to determine "empirical forms which *qualitatively* are strictly typical" (Menger [1923] 1976, 60): "these results correspond to the specific task of the exact orientation of theoretical research and are the necessary basis and presupposition for obtaining *exact laws*" (Menger [1923] 1976, 61). Theoretical research, then, is most properly defined as the cognition of *typical relationships*, the *laws* of phenomena (Menger [1923] 1976, 61), or patterns of *empirical-realistic analysis*, which underlie the specific individual experience of the concrete. Even research into ethics must take this path, if it is to be scientific. Yet this analysis can never be taken as primarily cognitive.

The patterns or regularities sought by Menger's science, to be sure, derive in no small part from cognition; but they also are coincident with habits of experience, as that experience is grasped and designated by a group habituated to them. Interrelationships among the resulting concepts are then explored for their patterns. In economics, phenomena are investigated for patterns of *Wirtschaftlichkeit* – for how they manifest the conditions of the possibility for something appearing and being understandable within the field of interest for economics. Yet as this happens, an investigator's physical and cultural habits are revealed as well – science is not only investigating the real, but rather elements of cognition, influenced "by human art" (Menger [1923] 1976, 61). Judgment and habit – physical and mental patterns, concepts inherited from different frameworks, and other patterning elements – all interact in cognition, as patterns evolve and certain elements are designated as relevant, and others not. "Thus we arrive at a series of sciences which teach us strict types and typical relationships (exact laws) of phenomena, and indeed not only in respect to their nature, but also to their measure" (Menger [1923] 1976, 61f.).

Here, Menger sets his science of economics off from natural sciences. "Exact" cognition of the relationships in a field of data will typically only derive from a field designated as natural phenomena, the simpler, more stable ones; "realistic-empirical" cognitions are the ground for the more complicated ones in this theoretical mode (Menger [1923] 1976, 56). He addresses this gap explicitly in Book 3, "The Analogy between Social Phenomena and Natural Organisms" (Menger [1923] 1976, 129). In Menger's usage, the designation "natural" again points to the mechanistic patterns characterizing objects, while elements of a very different kind comprise social patterns, especially those dealing with human will (Menger [1923] 1976, 133). Thus while both nature and social patterns can be the data addressed in sciences, the analogy between them is pragmatic, never complete (Menger [1923] 1976, 135), in no small part because of historical accidents (Menger [1923] 1976, 139). In consequence, Menger cautions his readers that the *nature* part of a "natural science" is a function of human will and choice of data, not an ontic reality.

History is indeed critical for Menger's idea of science, because repetitions over time of identified phenomena provide new data that help to trace patterns and show their cycles of development and death (Menger [1923] 1976, 102). Here, then, another difference emerges between the purportedly exact sciences – the natural ones – and his *empirical-realist* science in historical context: each must develop a different notion of change, "modif[ying] the aims of research" (Menger [1923] 1976, 113). Additionally, each science will have its own history, leading Menger to discuss various parts of his newly discriminated sciences of economics, from their base elements up through the various patterns of interest on the basis of which will be built the canon of the various scientific disciplines associated with economic interests.

Out of the *Methodenstreit* with German national economists, Menger has crafted a *Methodenlehre* for economics – a tool box out of which sciences of economics can emerge. Menger's strategies for defining science parallel Mach's analyses of individual natural sciences, starting with phenomenal evidence and group interest (intent and intentionality), moving through history (and cultural determination), and locating the science's production of knowledge within the space of culture. He outlines a moment of origin of a science, where Mach addressed the knowledge production of existing sciences, but both stress the contribution of experience (not just phenomena) to knowledge production – how data comes to be isolated and processed with specific interests in mind.

Menger offers an intensified description of how a single field of phenomenal evidence – in this case, economic interests – can actually be interpreted within multiple scientific frameworks. Classic hermeneutic definitions of understanding stress how two frames of reference, a past to be understood and a present in which the interpreter stands, are to be brought into a stable relationship. Menger's sciences, however, suggest that what an interpreter accesses in trying to understand knowledge produced in the past is not a stable frame, but rather a set of data that can be construed multiply, and even possibly disjunctively. Concomitantly, "culture" cannot be considered a stable frame of reference guaranteeing understanding, but rather a set of human purposes united by the existence of a community through its designations– an embodied rationality rather than a more abstract one centered in mind. As such, culture comprises the memory (in body and mind) of those engaged in any systematic understanding.

Comparing Menger's work with Mach's begins to show a common definition of science. Both Mach and Menger stress how designation influences the production of knowledge. Mach, however, takes as his example a set of experimental common-places, not the elements interest that Menger did. A third period treatise, Alois Reigl's work on art history, opens out the picture of the cultural horizon in which understanding takes place.

## 8.3 Riegl and the Hermeneutics of Art History

Alois Riegl's *Historical Grammar of the Visual Arts* (Riegl [1966] 2004) is seminal for the modern discipline of art history as practiced by the so-called Vienna School. Never completed, but present in two extensive drafts, Riegl's *Historical Grammar* is redefines the artwork and recaptures art as a historically determined practice of knowledge, iconic or legible as a product within a particular ideological framework. Riegl here defines semiotics as the method for a cultural science, specifying how culture's signs are to be read as interested, as Menger defined that concept.

Riegl's art history addresses cultural artifacts whose meaning is more than functional. In his account, such artifacts have a material dimension that reveals and reproduces a culture's interests and knowledge. Their forms speak in terms of a kind of autopoesis, pre-structuring knowledge by appealing to their audience's acculturated style preference for form, their physical habits, and their will to innovate; they must be understood as instantiating and propagating a culture's world understanding.

Riegl posits the central interest behind art as a *Kunstwollen*, a collective will to create art(ifacts). As he explains it, a work of art, like any made artifact, instantiates its makers' individual and collective horizon of understanding, as an intentional object reifying in concrete form the grid (re)producing human consciousness (a historicized horizon). The *Kunstwollen* inhering in such a horizon structures artifacts in ways that reveal an era's (self-) understanding and interests. First among those is the conceptual distinction between "nature" and "culture."

As Riegl how such artifacts are to be understood as a science, he specifies history as the artwork's primary form (mode) of appearance. In consequence, he divides Europe's art into three longer periods, *world views* of longer duration underlying any specific sites where artworks emerge. The historian is charged with identifying the artworks' basic elements of appearance within these historical frames. Among those elements are their era's characteristic *Purpose* (we might say the conscious ideology they fill), and the *Motifs*, *Form* and *Surface Elements* appearing in them. His "Introductory Remarks" thus define art as both a *techné*, a practice of production and a specific content area:

> The human hand fashions works from lifeless matter according to the same formal principles as nature does. All human art production (*Kunstschaffen*) is therefore at heart nothing other than a contest (*Wettschaffen*) with nature. The sense of delight with which a work of art fills us is commensurate with its human maker's capacity to bring to clear and convincing expression the respective formal laws of natural creation. (Riegl [1966] 2004, 51)

This passage accommodates both traditional philosophical aesthetics and a more modern science methodology, echoing the traditional dictum that a work of art delights and educates, as well as Kant's (and Schiller's) insistence that an artwork is a made and bounded object, set apart from nature. Yet Riegl does not relate the work directly to an idea, but rather to construction patterns, "the basic principle of crystallization" which form it within a specific historical material grid (Riegl [1966] 2004, 51). Riegl's "principle of crystallization" is an image from science: crystalline structures form around a fleck within the substance which centers the crystal's growth – an image expressing how a moment of history is bounded—while positing a *more* to its material form.

More significant is Riegl's statement that art's fundamental interest, its *Kunstwollen*, is creating objects in competition with nature (Riegl [1966] 2004, 52). Such acts of form-giving intensify nature, "demonstrat[ing] man's ability to conjure up a particular visual effect of nature" (Riegl [1966] 2004, 52). Thus art is neither a representation nor a deception or counterfeit (*Täuschung*, *Fälschung*); instead, it is defined as an intention to make artifacts out of a particular material (a craft), displaying the era's characteristic intentions (the motifs, forms, and surfaces it prefers). Riegl's science pursues such historical structures: "the following questions must be considered when evaluating a work of art: (1) For what purpose was it made? (2) From what? (3) Through what? (4) How?" (Riegl [1966] 2004, 53). These art elements parallel Menger's economic ones, designations framed within a particular group's history.

A culture's art objects instantiate phenomenally a *Kunstwollen*, their epistemological ground and historical particularity, as implicating a worldview, "the understanding of man's relation to matter" (Riegl [1966] 2004, 55). Since worldviews shift over time, Riegl divides the art data of Europe into three major historical divisions: the Roman Empire, the Medieval and Renaissance era, and a Third Period extending into his present. Each period, each worldview, develops through a predictable cycle of a development, a peak, and a collapse (Riegl [1966] 2004, 55). Each period manifests a particular understanding (not an explanation) of how the human relates to nature, "explaining" that relationship as a mediated understanding of distinctions between art and nature.

Riegl's science follows the *Kunstwollen*'s three great forms of appearance as different historical forms of human interest, three great ideological forms of general community. Chapter One's "First Period: Art as Improvement of Nature Through Physical Beauty" "requires that man consider himself superior to nature" (Riegl [1966] 2004, 57), and aligns art and nature to privilege sensation and perfection. Chapter 2 introduces the "Second Period: Art as Improvement of Nature Through Spiritual Beauty," where the sensual form of art is construed as referring to an underlying ideal more important than its physical presence, when it is seen as "spiritual" and referencing abstract ideals. In the "Third Period: Art as Reproduction of Transitory Nature," human art activity evolves a "profound interest in transitory matter for its own sake" (Riegl [1966] 2004, 95), when the artifact is assumed to function as a counter to the ephemeral timeframe of the natural world.

Riegl models these three historical worldviews as three versions of the one *Kunstwollen*, the one general principle constitutive of art as a recognizable class of objects. As such, each era's art manifests what Deleuze might term the ideological striations of an era's aesthetic discussions, schools, movements, and stylistic conventions (all terms which Riegl avoids because they are too-simple parsings of art objects). Art history as a systematic discipline must thus archive art's historical appearances as a specific category of cultural production, investigating art's relation to the materiality of the human and its interests:

> All works of visual art are created to fulfill some function. The purpose of any work of art must therefore be sharply distinguished from its artistic *habitus*. The relation between these aspects is not fixed and immutable and is thus subject to historical investigation.... On the one side stands a complete preponderance of functionality, next to which any aesthetic effects seem to arise spontaneously and without conscious consideration.... On the other side is the total supremacy of the intent of art (*Kunstabsicht*), next to which a work's functional intent merely provides a pretext for its existence. (Riegl [1966] 2004, 109)

His subsequent discussion of the "Elements" identifies the purposes at play in each worldview, summarizing how works of art need to be understood according to a particular era's purposes. Thus the artwork's intentional form must be accounted for in terms of its "Motifs" taken from nature (Riegl [1966] 2004, Chapter 5): the principles embodied in them, such as symmetry or crystallinity. Decorative or practical motifs are specifically inorganic in origin; organic motifs are organizing principles like motion. Such "motifs" organize the artwork as an entity opposed to nature, as preferred in particular eras or disciplines of art. For instance, "the law of crystallization [is said by Jakob Burckhardt and Gottfried Semper] to be a constant in architecture" (Riegl [1966] 2004, 125).

Similarly, "Form and Surface" address not the content but the dimensionality of the artwork, how it is projected into space, as a visual projection in two dimensions that evokes in the viewer the sensation of three dimensions and its appearance as real. *Every* artwork presents itself historically in terms of surface and form; each era has a different ideology about what is a "natural" surface:

> For any given thing, we have thus to distinguish three elements: (1) *form*, which is fundamental to the object; (2) *objective surface*, which, being an intrinsic component of form, is equally fundamental to the whole; and (3) *subjective surface*, which is a mere illusion of the visual faculties. (Riegl [1966] 2004, 189)

Each historical field implicates different relationships between surface and form, and thus a different ideology. Antique anthropomorphic polytheism, Christian polytheism, and the natural scientific worldviews characteristic of Riegl's three longer periods each help redefine what the era's art looks like, cementing alternate visions of what nature "is" and thus what is at stake in the contest between nature and art. These ideologies also render the artwork historically specific as an act of designation.

Thus the "historical grammar" at the basis of art history is the history of the elements of art's "language" (Riegl [1966] 2004, 292): "We will be dealing with (1) elements; (2) the developmental history thereof; (3) the factors that determined that development" (Riegl [1966] 2004, 293). As he summarizes the elements:

Thus we have five elements:

> (1) Purpose: *why?* Usually this is too narrowly defined as the practical, utilitarian function meant to satisfy the needs of our five senses.
> (2) Raw materials: *from what?*
> (3) Technique: *in what way?*
> (4) Motif: *what?*
> (5) Form and surface: *how?* (Riegl [1966] 2004, 295).

Purposes can include utility, decoration, and abstract concepts.

Overall, Riegl's scientific art history works in the models traced here, an investigation of a particular class of phenomena in terms of the rules of its knowledge production, within a particular cultural field. However, he also defines art history as a theoretical science, not just a historical science, noting that there is probably more than one science addressing art objects, beyond just art history. Still, this art history is new:

> One now poses the following questions: Does the Parthenon obey the developmental laws of architecture exclusively? Do the sculptures of Phidias follow only the developmental laws of sculpture, and Attic vases solely those of painting? Is there really no transverse connection between these laws at all? Or do there exist, above them all, certain universal principles to which all works of art conform equally and without exception – underlying principles from which the laws of architecture, sculpture, and painting represent mere secondary deviations? (Riegl [1966] 2004, 291)

Riegl emphasizes the ideology of an era's representations to specify "culture" as a definable context for the work's interpretation. Thus like Mach, Riegl posits interests as the central elements in a culture's horizon of understanding and communication. In his view, phenomena are effectively real only in interested, ideological forms, never neutrally.

While following Menger in distinguishing more than one type of science addressing art phenomena (one historical, one theoretical), and Mach in the insistence that the phenomenal field is interested, Riegl provides a more extensive methodological meditation about how phenomenal elements express historical–cultural differences. To understand such a category of artifacts, one must see how they are formed as both cognitive phenomena and rule-governed behavior: *Erklären* and *Verstehen* are conjoined acts of scientific understanding.

## 8.4 Conclusion: Modeling a Unified Epistemology of Science

These three cases all originate in Vienna and its science, including a factor not often discussed as relevant to the formation of a scientific community: the question of cognitive style as reinforced by education. Austria-Hungary's school system espoused a pedagogy derived from the work of Johann Friedrich Herbart (1776–1841), a dissenting student of Kant whose work anticipates the strategies associated with Montessori education in the United States, and with Friedrich Froebel (1746–1827) and his kindergarten learning toys. Both models stress that knowledge must be acquired through all the senses, not just cognitively, and that individuals will each have different learning styles, depending on their personal dispositions of senses.

The Viennese context of science around these three case studies straightforwardly can be seen as an extension of such learning paradigms into epistemology, especially as they refuse to distinguish experience and understanding. All three models insist on both psychological and physical habits as part of knowledge production, and on historically attested patterns of designation as structuring the knowledge encoded within it. Thus knowledge is tied to more than cognition at the moment of its production.

Just as critically, every act of knowing has an automatic link to a community and its history, to the limits (physical, psychological, and social) of the mind enacting it or acting with it, and the values (interests), cultural, and disciplinary traditions inherent in that frame of action. Culture, not mind, constitutes the main frame of reference in which all knowing emerges; bodies, not minds alone, sponsor its work, its communication, and the patterns within which it is generated.

In this sense, sciences can never be seen simply as either nomothetic or ideographic, inductive or deductive, because they are implicated by more than logical rules: they are embodied and historical, formed in one piece with the interests of their user groups. As Mach, Menger, and Riegl define their sciences, each reflects a group interest in a field of data, under the influence of the history of experience designated and passed to the individual knower. The epistemological paradigm of any science is therefore also a field of values – an ethics of experience – and a representation of values as well as logic, as well as of the physical existence of that group.

Overall, then, the strategy for defining science shared across all three cases comes much closer to the late twentieth century's post-structural project of uncovering the interests inherent in all acts of knowledge, rather than to the classic hermeneutic traditions. This is an alternate construal of the Kantian legacy, and one that needs to be pursued as conditioning the emergence of many new forms of science, especially social sciences, after the turn into the twentieth century.

## References


Arens K (1989) Structures of Knowing: Psychologies of the Nineteenth Century. Kluwer, Dordrecht

Mach E (1894) Lehrbuch der physik für das gymnasium. G. Freytag, Leipzig

Mach E (1896) Die principien der wärmelehre: historisch-kritisch entwickelt. J.A. Barth, Leipzig

Mach E ([1896] 1910) Populär-wissenschaftliche vorlesungen, 4th edn. J.A. Barth, Leipzig

Mach E ([1900] 1959) The analysis of sensations, and the relation of the physical to the psychical. Translated by C.M. Williams; revised by Sydney Waterlow. New introduction by Thomas S. Szasz. Originally published as Die analyse der empfindungen, und das verhältnis des physischen zum psychischen. 2., vermehrte Aufl. Gustav Fischer, Jena

Mach E ([1905] 1975) Knowledge and error: Sketches on the psychology of enquiry. Translated by Paul Foulkes. Originally published as Erkenntnis und irrtum: skizzen zu einer psychologie der forschung. J.A. Barth, Leipzig

Mach E (1921) Die prinzipien der physikalischen optik: historisch und erkenntnispsychologisch entwickelt. J.A. Barth, Leipzig

Menger C ([1883] 1985) Investigations into the method of the social sciences with special reference to economics. Schneider L (ed) Translated by Francis J. Nock. Originally published as Untersuchungen über die methode der socialwissenschaften, und der politischen oekonomie insbesondere. Duncker & Humblot, Leibzig

Menger C ([1923] 1976) Principles of economics. Translated by James Dingwall and Bert F. Hoselitz. Originally published as Grundsätze der volkswirthschaftslehre. Holder-Pickler-Tempsky, Wien

Riegl A ([1966] 2004) Historical grammar of the visual arts. Translated by Jacqueline E. Jung. Originally published as Historische grammatik der bildenden künste. Swoboda KM, Pächt O (eds). Böhlau, Graz

Windelband W ([1894] 2006) Geschichte und naturwissenschaft: rede zum antritt des Rektorats der Kaiser-Wilhelms-Universität Straßburg, gehalten am 1. Mai 1894. Reprint. Originally published by Straßburg:heitz. <http://www.fh-augsburg.de/~harsch/germanica/Chronologie/19Jh/Windelband/win_rede.html>, accessed Oct 30, 2006

# Chapter 9
# British Thought on the Relations Between the Natural Sciences and the Humanities, c. 1870–1910

**Roger Smith**

The fact is that debate about explanation and understanding in science, and the related debate about relations between *Naturwissenschaften* and *Geisteswissenschaften*, was a German-language debate conducted in German philosophical terms. There was no late nineteenth-century English-language argument about the same matters. There is no direct comparison to make, unless, that is, we were to take an essentialist view of philosophical questions and, as a result, look for what English-language writers missed, transformed into local idiom or substituted for the issues of substance rightly formulated by German-language philosophers and wrongly ignored by British ones. By contrast, if we take our cue from what Uljana Feest, in her introduction, refers to as 'the thicket' of issues and debates in the late nineteenth century, rather than from imagining that there was one debate, possibilities for comparison open up.

I make a generalisation about intellectual culture in Britain between about 1870 and the First World War, and I then use this as the framework for a description of the relations between the different sciences and areas of learning, or, to use the modern classification in the English-speaking world, between the natural sciences, the social sciences and the humanities. (The use of the word 'science' is something that the paper will have to address.) The main issue, I suggest, for English-language writers in the philosophy of scientific knowledge was *naturalism*. This was a late Victorian term: 'we may know "phenomena" and the laws by which they are connected, but nothing more ... the World with which ... alone we can have any cognisance, is that World which is revealed to us through perception, and which is the subject-matter of the Natural Sciences' (Balfour 1895, Vol. 1, xiii).[1] This definition contained a much debated ambiguity. 'Naturalism' was at times a metaphysical

R. Smith (✉)
Lancaster University, UK, and independent scholar, Russian Academy of Science, Moscow
e-mail: rsmith@mail.ru

[1]The *Oxford English Dictionary* (completed 1928) described naturalism as 'a view of the world, and of man's relation to it, in which only the operation of natural (as opposed to supernatural of spiritual) laws and forces is admitted or assumed' (*The Compact Edition of the Oxford English Dictionary* (1979, 1,899)). The variety of meanings of 'naturalism' were recognised in the late nineteenth century: Andrew Seth, 'Note A. The use of the word "naturalism"' (Seth 1897, 289–303).

U. Feest (ed.), *Historical Perspectives on Erklären and Verstehen*, Archimedes, Vol. 21, 
DOI 10.1007/978-90-481-3540-0_9, © Springer Science+Business Media B.V. 2010

claim that the world (at least, the one that we can know) is exclusively as the natural sciences describe it, and it was also an epistemological claim about the limits to knowledge. The latter aspect connected with the then much discussed question of 'agnosticism', a term which rapidly caught on after its introduction by T. H. Huxley.[2] It would be another paper adequately to describe the cluster of debates around naturalism. But we may at least note that it was also Huxley who, in passing, introduced the term 'scientific naturalism' as a synonym for naturalism, which historians now often use, since it points to the role the natural sciences played in the spread of naturalistic viewpoints.[3] When Huxley used the word, he opposed it to 'supernaturalism', the explanation of the world, which for naturalists emphatically includes the human world, by some kind of non-natural, or spiritual, agency. He portrayed naturalism as the inescapable consequence of the right use of reason, by which he meant reaching conclusions about the world based on experience, which he thought exemplified by the natural sciences, and not on a priori reasoning. He rejected both the deductively argued naturalism of Herbert Spencer and the idealism which became prominent in British philosophy and political thought in the last decades of the nineteenth century. For Huxley, as for other naturalists, the stance was not only epistemological but ethical, a commitment to seeking the truth and not relying on past authority, speculating or making claims about knowledge which we simply cannot have.

'Naturalism' denoted the belief that since everything is 'nature', there is only one kind of knowledge, the kind that the natural sciences in fact exemplify. The idealist philosopher Bernard Bosanquet therefore negatively characterised 'the ideas of Naturalism, which means, those of uncritical metaphysic founding itself on conceptions current within the natural sciences' (Bosanquet 1912, 74).[4] Actually, a large number of different positions tended to be conflated in polemics, and in particular there were significant differences between the public stance of writers like Huxley and the physicist John Tyndall in 1870, rhetorically striking out against their religious opponents, and the more thoughtful, philosophical writings of scientists (including Huxley and Tyndall). Nevertheless, there was a substantial difference between the idealists, like Bosanquet, who gave logical reasoning priority over natural science, and the natural scientists, like Huxley, who thought the empiricist methods of science the starting point for knowledge and quite incapable of leading to conclusions about metaphysics. In response to the empiricists, authors writing in general periodicals quoted Hamlet's line – to an extent which turned it into a cliché – that 'there are more things in heaven and earth, Horatio/Than are dreamt of in

[2] In the course of discussions at the Metaphysical Society: Huxley, 'Agnosticism' (Huxley [1889] 1892, 352–357; Brown 1947; Lightman 1987).

[3] Huxley, 'Prologue' (Huxley 1892b, 35), portrayed 'the principle of the scientific Naturalism of the latter half of the nineteenth century' as the culmination of the Scientific Revolution. In the title of his 1892 book, Huxley (1892) referred to 'controverted questions'; he had in mind 'naturalism' versus 'supernaturalism'. Also, Paradis (1978, 178–180) and, more widely, Turner 1974).

[4] I use 'idealist' consistently to name metaphysical positions, in contrast to the 'idealism' which denotes commitment to ideals.

our philosophy'. (Writers sometimes cited this against opponents as 'dreamt in *your* philosophy'.) The allusion was not necessarily to the possibility of miracles but to the possibility of knowing something beyond phenomenal relations and the material, mechanistic causes supposedly underlying them.

For present purposes, to make possible comparison with German-language debates, the point is to stress two things. Firstly, the claims of naturalism directed discussion to the basis of knowledge *in general* rather than to *distinctions* between the foundations of the different areas of learning, in the way the *erklären/verstehen* distinction did. There was an interest in what we can say with authority exists in the world rather than with the epistemology of the different disciplines. Clearly, though, this was a matter of emphasis rather than absolute contrast – all the more so since much of the discussion took place in a non-specialist setting and had the form of debate about world view rather than formal metaphysics. Nevertheless, we can say that naturalism was a metaphysical belief, with the uniformity of nature (including humankind) its principal claim.

Secondly, debates about naturalism *expressed moral and religious concerns*; they were not dry, putatively neutral exercises in abstract argument but of living concern to the protagonists. Ethical and spiritual values, even the possibility of such values, were felt to be at stake. For many people, this was what really mattered. A. J. Balfour, a future prime minister, for example, noted that naturalism 'is nothing more than the result of rationalising methods applied with pitiless consistency to the whole circuit of belief', thus excluding, in his view, moral and spiritual knowledge and a way of life based on it (Balfour 1895, 172f.). Historians, especially Stefan Collini, have noted a peculiarly British preoccupation, in the arena of politics and not only in philosophy, with the moral content of belief (Collini 1991). Noel Annan commented on 'the passion of the English for discussing everything in terms of [individual] morality' (Annan 1959, 15). In what we might pick out as philosophy of science, ethical rather than epistemological questions frequently had priority.

There is, I should note in passing, an alternative view about the contrast between British and German philosophical culture, which a number of English authors have appealed to, turning an absence into a positive trait. This alternative is the conviction that philosophy is simply 'outside the reach of the native genius' and that 'the English mind has baulked and stumbled at the fence of philosophy' (the historian of political thought, Ernest Barker, in 1949, quoted in Collini 1991, 358).[5] However, whatever the ambivalent response of British empiricist academic culture in the twentieth century to so-called 'theory', that is, non-empiricist discourse of any kind, it is simply untrue to say that the late Victorians 'baulked' at philosophy; though, as we shall see, certain groups, notably historians, kept their distance.

Gadamer commented that 'all the arduous work of decades that Dilthey devoted to laying the foundations of the human sciences was a constant debate with the logical

[5] For the argument that just this kind of characterisation of Englishness created a stereotype at the end of the nineteenth century, see Colls and Dodd (1986). Ignorance of philosophy was Perry Anderson's accusation in his much cited Left critique: (Anderson [1968] 1992). For the questionable standing of 'the intellectual' in Britain, see Collini (2006).

demand that Mill's famous last chapter made on the human sciences' (Gadamer [1960] 1998, 6f.). Whatever the accuracy of the statement, Mill's 'famous last chapter', 'On the logic of the moral sciences' (actually, Book VI, from *A System of Logic* (Mill [1843] 1900)), remained in the nineteenth century the single most influential statement of the view that the same inductive method of arguing from phenomenal experience to law-like generalisations holds in the sciences of man (or 'moral sciences') as in the sciences of nature, and hence that it is possible to advance the latter in the manner of the former. He argued in the context of reform-minded politics. Since, he assumed, 'the laws of the phenomena of society are, and can be, nothing but the laws of the actions and passions of human beings united together in the social state', knowledge of the laws of individual character is the basis for a science of society. 'The science of Human Nature may be said to exist in proportion as the approximate truths which compose a practical knowledge of mankind can be exhibited as corollaries from the universal laws of human nature on which they rest, whereby the proper limits of these approximate truths would be shown, and we should be enabled to deduce others for any new state of circumstances, in anticipation of specific experience' (Mill [1843] 1900, 573, 575). Originating in the British controversy of the 1820s and 1830s between idealist and utilitarian approaches to the sciences of man, by 1870 Mill's book was a standard university text; at Cambridge, for example, Mill's *Logic* was 'the orthodoxy of reading men' (Harvie 1976, 33).[6] This did not mean, however, that there was consensus about his theory of knowledge; indeed, there was a persistent division between empiricist and idealist epistemologies. In 1843, however, Mill had good reason to think that his conception of the moral sciences was a minority position, in comparison with the Christian-idealist alternative that started with the presumption that there are revealed or intuitive moral principles.[7] This changed: by the 1860s and 1870s, Mill's position, aligned with the naturalism for which it appeared to provide the formal epistemology, had started to set the agenda; those who disagreed felt compelled to say why.

The 1860s appear a pivotal decade; in later years, the onus of proof lay with those who would put forward either supernaturalist explanations or a priori arguments about human affairs. There was no logically necessary connection between inductive reasoning, the formal counterpart of the empirical methods of practicing scientists, and naturalism, but much of the literature, notably including Huxley's essays, linked them. And again, though not a matter of logical entailment, writers such as Huxley closely tied induction and naturalism to agnosticism, stating limits to knowledge, whether of a metaphysical or a religious character. The central argument of the idealist critique was, therefore, that naturalists mistook the nature of experience

[6] For Mill's involvement with argument about cultural values, which he saw as the argument between 'experience' and 'intuitionism', see Rothblatt (1968, Chapter 3).

[7] For shifting positions on what we would call ethics, see Schneewind (1977, Part I). Mill himself actually excluded ethics from the sciences on the grounds that it is a practical discipline.

[8] Naturalism, as a set of claims about the world, and empiricism, as an epistemology, were often but not necessarily associated. It is appropriate for the philosopher F. H. Bradley to be discussed in a separate paper, by Christopher Pincock, in this volume. Bradley's position combined idealism

and proceeded inductively from abstractions rather than from 'the real'; true attention to 'the real', experience, revealed the spiritual.[8]

Mill's philosophy increased the importance of history, as it required the study of the local particularities of individual and social development as the basis for the sciences of human nature and society. Reacting against Bentham and his father, James Mill, he concluded that social analysis must draw on particular historical knowledge before any attempt is made to apply it; nevertheless, he still understood explanation to be the search for general laws (Collini et al. 1983, 131–148). Parallel to and supporting Mill, a historical and comparative method in social thought became very widespread in the 1860s, much influenced by Henry Maine's *Ancient Law* (1861). Spencer, who began issuing his 'Synthetic Philosophy' in installments in 1860 and ended only thirty years later, made the understanding of human affairs subsidiary to the understanding of the single pattern of development running through cosmic, biological and historical processes alike (Burrow 1966; Peel 1971). His call for a unified science on common principles became a standard reference point in philosophical literature, even if many readers criticised his deductive manner of reasoning from first principles, his relegation of spiritual values to 'the unknowable', or both. Darwin, it scarcely needs saying, published *On the Origin of Species* in 1859, and the book was at once the focus of a debate about the method of right reasoning in science as well as about its substantive claims (Hull 1973). The argument for the origin of species, especially humans, by descent with modification from previously existent kinds became *the* exemplar of the naturalist world view.

The sciences of development, biological and historical alike, focused attention on the underlying importance of the principle of uniformity – the principle of the uniform operation of natural law – and its implication: miracles and human action contrary to natural law are inconceivable. As H. T. Buckle observed, 'the marked tendency of advancing civilization is to strengthen our belief in the universality of order, of method, and of law' (extract from Buckle [1857–1861] 1973, 125).[9] Buckle's attempt to survey the facts of history and arrive at general laws of physical and moral development, however, earned him the scorn of academic historians and a reputation for determinism.

Mill, who had to be taken much more seriously, had meanwhile, in the 1840s, materially and morally supported two younger writers who contributed to the argument for uniformity, Alexander Bain, who brought the psychology of James Mill into relation with the new physiology, and Spencer (Young 1970). Spencer built continuity between the natural and human sciences on first principles.

[8](continued) (supported by philosophical argument) with his own interpretation of naturalism (preserving the autonomy of scientific knowledge, including history and psychology, freed of metaphysical intentions). Bradley's severe, intensely philosophical and original writings make him difficult to fit into any generalisation – except as disproof of the supposed ineptness of the English for philosophy. His contributions were a sign of the increasingly academic context of philosophical work, and it is notable that the writers for a wider audience, with whom this paper is largely concerned, rarely cited him. In later life, Bradley actually became a recluse.

[9]For an assessment of Buckle's lack of influence in Britain, see Stephen (1880).

His thought reached a large audience, especially through *The Study of Sociology* (1873, published in the North-American and British 'International Scientific Series', a major venture in the public understanding of science). As Spencer wrote, 'if there is natural causation ... it behoves us to use all intelligence in ascertaining what the forces are, what are their laws, and what are the ways in which they co-operate' (Spencer [1873] 1892, 47). He therefore lambasted the continuing practice of politics on the basis of tradition, prejudice and wishful-thinking, and he called for historians and social scientists to combine and establish the empirical laws of social structure and function in an evolutionary framework, thus laying the objective basis for political action. Mill, Buckle, Bain and Spencer perceived themselves as advancing the rights of rationalism against the irrationalism of belief in supernatural intervention in the world of natural law. Political and social thought therefore ran parallel to an intense critique of belief in miracles and contributed to highly charged debates around science and religion.[10] In this connection, it is important to remember both that Oxford and Cambridge academics, until the reforms of the 1870s, were required to agree to the Articles of the established Anglican Church, and that rationalism was, and remained, closely linked to working-class radical culture.

Debate about naturalism as a theory of knowledge was a direct expression of the expanding claims and authority of the natural sciences. Tyndall put the matter bluntly, stating that any systems of thought which infringe 'upon the domain of science' must submit to its control, declaring that scientists 'claim, and we shall wrest from theology, the entire domain of cosmological theory' (Tyndall [1874] 1892, 197).[11] The outcome decidedly supported Tyndall's stance: the onus of proof shifted, and in the last decades of the nineteenth century it was up to scholars who wished to maintain that there is part of the world where natural physical law does not operate to show the grounds on which this could possibly be the case. Natural scientists gained for themselves the authority to pronounce on matters of natural science unfettered by religious critics.

Here a note on the word 'science' is necessary. The growth of authority of natural science during the Victorian period accompanied the spread, in English, of the now customary usage, equating science with natural science or learning claiming to follow the pattern of natural science methods and forms of understanding.[12] Earlier writers referred to natural philosophy and moral philosophy (or, in the nineteenth century, to moral science, or, where the stress was on the mind's operations, to

[10] There is a large literature on the Victorian relations of science and religion; for an introduction, see Brooke (1991, Chapters 7 and 8). An expansion of the present paper would have to go into the place of theology in classifications of science and knowledge and not just only into the place of Christian belief.

[11] Also quoted in Lightman (2004, 201). More widely, see Turner (1978).

[12] The peculiarity of conflating 'natural science' and 'science' has not been adequately explored, yet it lies at the heart of the subsequent debates about the relations among different forms of knowledge. See Ross (1962). For exemplification of the argument that historians should replace studies of relations between forms of knowledge by studies of construction of the identity of individuals and communities, see White (2003).

mental science), the latter indicating the sphere of learning about human actions. Usage was neither settled nor consistent. But it was more common, towards the end of the nineteenth century, to discuss the relations between the sciences and moral philosophy (or moral science or mental science) rather than the relations among the sciences, in the way German-language literature discussed relations among the different branches of *Wissenschaft*. Yet, as we shall see, it was also common enough to refer to history or political economy, for example, as science. 'Man of science', however, firmly meant 'natural scientist' or someone promoting the advance of human science as natural science. ('Scientist' spread only at the turn of the century, associated with practical, specialist pursuits rather than gentlemanly learning.)

Some remarks by the Oxford historian E. A. Freeman well illustrate the changing meaning of the word 'science' and the problems its gradual identification with 'natural science' caused some scholars with a traditional classical education. 'I use the word *science* at all simply to assert our right [i.e., the right of historians] to use it, if we please; for our own use I far better like such plain English words as *knowledge* and *learning*.... When I was young, science in this place [the University of Oxford] meant chiefly the knowledge of man's moral faculties, the lore which we learned from Aristotle and [Bishop] Butler. It has now taken on itself other meanings ... [and] some special merit and dignity is often claimed for those branches on the strength of this unfair monopoly of a name' (Freeman 1886, 117).[13] Indeed, equation of 'science' and 'learning', alongside usage opposing 'science' and 'the arts' or, later, 'humanities', persisted, clouding discussion of what was at stake. In 1899, the Royal Society of London, participating in a meeting to found an international association of academies, was embarrassed to expose the fact that there was no British body which could speak for learning beyond the natural sciences, which it represented. There was then a debate among British scholars about whether the Royal Society should create a section to include 'literature, antiquities and philosophy' or whether there should be an entirely new academy. The outcome was the new British Academy for the Promotion of Historical, Philosophical and Philological Studies, cementing the institutional separation of the natural sciences and other areas of scholarship (Anonymous 1903–1904; also Collini 1991, 21–27). At the same time the first President, Lord Reay, perpetuated the older usage of the word 'science', committing the Academy to the study of the moral sciences and arguing that 'the unbiased attitude of the mind towards ethical and metaphysical problems is one of the conditions of our existence as a scientific body' (Reay 1903–1904, 14f.).[14]

In 1870 there were hopes for a unified intellectual culture; by 1900, there were many specialists in different disciplines who got on with research without giving the issue much thought. Thus, in 1869, the intellectual elite based in London took the trouble to form the Metaphysical Society, a dining and discussion club, in order to bring together religious, philosophical and scientific leaders of opinion specifically

[13] The critic John Ruskin, on similar grounds, was a vehement opponent of the restricted meaning of the word 'science'; see Ross (1962, 70f.).

[14] Lord Reay was also President of the Royal Asiatic Society.

to seek common ground (Brown 1947; Young 1985). A number of contributors dealt with naturalism, analysing causation and seeking to show, while fully accepting the principle of uniformity, that natural laws describe the activity of a world in which purpose, both Divine and human, does have real existence. The result was a variety of different philosophies of nature, no one of which commanded widespread assent. This was a sign of the persistent failure of attempts to reconcile naturalists and their critics with an agreed, coherent alternative metaphysics.

These debates in natural philosophy provide evidence that a concern, shared in common by naturalists and their critics, with the moral content of human action directed attention to the foundations of knowledge in general rather than to philosophies separating human action from other processes in the world (though such arguments were also present). Educated opinion accepted the principle of continuity of explanation of human and natural occurrences. Idealists in philosophy, equally with empiricist scientists, opposed dualisms separating thought and external reality, mind and nature. But scholars divided over the question about how far reason could proceed beyond empirical knowledge and, if it could, about what metaphysical knowledge there is. Crucially, this left the question of the source and vindication of moral values open for dispute, and this appeared to many writers, who were worried about the national culture, to be the really urgent matter.

By the 1870s, a substantial number of British academics professionally occupied in philosophy had turned towards idealism, some towards Hegelian idealism. The opposition of these scholars to naturalism was entirely general, founded in the logic of reasoning and not primarily concerned with the failings, or falsely extended arguments, of any one science. Moreover, even non-idealist philosophers interested in the natural sciences, like C. Lloyd Morgan, Samuel Alexander and, later, A. N. Whitehead, turned to general metaphysics rather than the particular character of different branches of knowledge. Alexander is an interesting case. Having arrived at his own approach to philosophy, partly through the psychology of the German Hugo Münsterberg, he sharply supported naturalism in opposition to idealism. Yet he tried to find ways to treat 'minds related to bodies as a particular instance of a more *general* kind of fact to be found in nature, and as an instance in which we ourselves "enjoy" this fact from within', and this led to the metaphysical theory of emergence (Emmet [1920] 1966, vol. 1, xiii; emphasis added).

The conservative, elite response to naturalism, involving general criticism of it as a theory of knowledge, is well illustrated by Balfour's writings. In *A Defence of Philosophic Doubt* (1879), his argument, implicitly against Mill, was destructive: 'Has Science any claim to be thus set up as the standard of belief? ... If there is, it cannot be drawn from the nature of the scientific system itself' (Balfour 1879, 302; more generally, see Jacyna 1980). Reason, he implied, must rest on some principles transcending reason. Later, he elaborated in a more constructive spirit, arguing that our notions of what has value require belief in something beyond what naturalists would allow us to believe. 'If, then, naturalism, is to hold the field, the feelings and opinions inconsistent with naturalism must be foredoomed to suffer change; and how, when that change shall come about, it can do otherwise than eat all nobility out of our conception of conduct and all worth out of our conception of life, I am wholly unable to understand' (Balfour 1895, 81). Characteristically for his time, Balfour engaged the theory of knowledge as the intellectual means with which to underpin ethical authority and, by implication, social order.

For participants in the German-language debate about relations between the sciences, the very terms of the questions asked about the different forms of human understanding were 'Kantian' questions. It is therefore relevant to ask about the reception of Kant's philosophy in Britain, and perhaps even to ask whether knowledge of Kant was a precondition for the differentiation of different forms of knowledge.[15] It is not immediately obvious, however, that Kant was an important direct influence on thought about the relations among fields of knowledge, though philosophers frequently cited his work in the second half of the century. Thus, when the Oxford examinations in 'Greats', that is *Literae Humaniores*, began in the 1860s to expand to include questions on modern as well as ancient philosophers, Kant was the first modern philosopher to receive serious attention. But it is not clear how far this went. At Oxford, T. H. Green was the major influence in bringing philosophy out from the dominance of Classics, and he did so in order to show that reason leads not to naturalism but to idealism. His turn to idealism was a project to show that Christian ethical beliefs and obligations are grounded in reason itself, not vulnerable to historical questioning of revealed truths. 'Human knowledge and moral experience can be understood only on the supposition that a spiritual principle, an eternal consciousness, is manifested in men' (quoted in Richter 1964, 177). It was this belief in 'an eternal consciousness ... manifested in men' which led young men and women to the practical social concern expressed in charity work and, in the 1890s, to the politics of 'the new liberalism', which favoured state intervention in social problems on moral grounds (society has an obligation to create the appropriate conditions for each person's self-development) (Collini 1979; Freeden 1978). But these interests and social ethics were a long way from Kant.

Green's influence once again throws into relief the British concern with the relations of knowledge in general and ethical action rather than the possible existence of different forms of understanding. Both empiricists, following Mill, and idealists, like Andrew Seth (later Andrew Seth Pringle-Pattison), Bosanquet and D. G. Ritchie, following Green, looked to knowledge of the formation of individual character, in historically specific settings, as the basis for social policy. In philosophy, utilitarianism and idealism opposed each other – Henry Sidgwick, for example, regarded Green's idealism as the position against which utilitarianism had to establish itself – but neither school of thought had an interest in differentiating different forms of knowledge, separating knowledge of human beings from knowledge of nature. It was therefore appropriate that G. E. Moore, representing a new generation of severely critical, analytic philosophers, should have attacked both idealism and naturalistic ethics in one year (1903). 'All the most important philosophic doctrines have as little claim to assent as the most superstitious beliefs of the lowest savages', he remarked (Moore 1903b, 436; also Moore 1903a; Regan 1986).

---

[15] I have not found a considered historical study of these questions.

Though notions of 'the good' took contrasting idealist and utilitarian forms, shared aspirations for a moral polity motivated social thought in Britain. Intellectuals across the political spectrum united around belief that the goal of society is the creation of the conditions for the individual's self-development. As L. T. Hobhouse (appointed to the Martin White Chair of Sociology at the London School of Economics in 1907) wrote, 'true liberty ... is found when each man has the greatest possible opportunity for making the best of himself' (quoted in Collini 1979, 69f.). Social theorists held different views, however, about how far moral development required the growth of the state and about the manner in which they thought the state should act. Spencer's brand of individualism, fiercely opposing state action, for example, in education, reached a high-point of influence in the 1880s; thereafter, many of those who shared his political liberalism and his commitment to sociology, like Hobhouse himself, moved towards accommodation with the idealists, and they favoured the growth of the state as the collective framework for individual development. Neither post-Spencer liberals nor idealists felt it necessary sharply to distinguish knowledge of nature and knowledge of society; they presumed that a common line of progress united these spheres. Hobhouse fused biological and social thought, seemingly as the natural way to think – evolution and human action achieve the same kind of progress. The reforming impulse 'is not in conflict with immovable laws of evolution but is continuous with the line of advance which reduced the higher from the lower animal forms, which evolved the human out of the animal species and civilized from barbaric society' (Hobhouse, quoted in Collini 1979, 146). Such belief in the continuity of natural law, understood teleologically to include a law of progress, was extremely widespread in the social thought of the period. Spencer described the law in naturalistic terms, Bosanquet in idealist terms, but progress of a moral character they took to be fundamental in the world.[16] There was no tension in this respect between natural science and social science.

What was vulnerable about this belief in progress – besides its obvious vulnerability to political events provoking scepticism – was the grounding of values. On the one hand, the idealists made values dependent on a metaphysics of the Absolute which many people found highly implausible. On the other hand, the naturalists appeared to derive values from particular factual claims about evolutionary nature, leaving themselves open to the empirical possibility that their facts might be wrong and the logical possibility that factual and value statements belong to different registers. It was no wonder that a committed rationalist like Sidgwick, who attempted to rethink utilitarianism in opposition to idealism, and at the same time clearly distinguished facts and values, notoriously never reached a settled position. The idealists were in a way more consistent, since they derived the evolutionary process itself from the ethical or ideal sphere (see Boucher 1997, Part I). Moreover, with some reason, they 'claimed to articulate and make defensible the central

[16] Historians once assumed that precisely this consensus did not survive the Great War. But it can be argued that a belief in common biological and moral progress survived much longer in Britain. See, e.g., Smith (2003).

elements of that consciousness as it already existed' in people's experience, meaning that they claimed to ground rationally what ordinary people believed and felt to be the principles of good conduct and the good society (Vincent and Plant 1984, 7). None of this, however, made different forms of understanding the primary focus of concern.

These intellectual arguments took place in a context of institutional change which decisively turned philosophy and other disciplines into professional occupations. Balfour, for example, to use later parlance, was an 'amateur', writing in the interstices of political life. Around him, however, the social landscape of the intellectual world was changing, bringing in the system of academic professions which lasted at least until the 1980s. This was of fundamental importance to the way relations between different areas of knowledge were discussed and the way, which I am stressing is so characteristic of Britain, self-appointed keepers of the national culture assessed the contribution of different fields in moral terms.

There was reform of Oxford and Cambridge between the 1850s and 1870s (latterly under the influence of the German university model), University College and King's College in London expanded as London University (joined in 1902, after many years of planning, by Imperial College), and city colleges like those in Manchester, Liverpool and Bristol transformed into universities. Something like the pattern of research and teaching which lasted until the 1980s emerged in England and Wales. The Scottish universities had for some time taught a much broader curriculum, including philosophy and the natural sciences. A career structure, linking teaching and research, developed for academics in the natural sciences and what are now called the humanities, the latter including philosophy, history, the study of English language and literature, as well as Classics, and to some extent also political thought, anthropology and economics. This was the social basis for a more self-consciously 'disciplined' view of academic practice, encouraging attention to methods and regulation of standards within particular fields, and encouraging specialisation rather than general questions, such as the question of relations between areas of knowledge. For example, the practice came into existence among philosophers, which was to be taken for granted in the twentieth century, of writing philosophy for an audience of peers not for a general educated audience. One sign of this was that philosophers called their field 'philosophy', separating off what had before been a branch of the moral sciences, and they also increasingly corralled questions about moral principles in a specialist branch of philosophy called 'ethics'. In addition, local struggles, such as those within Oxford and Cambridge to create the conditions for liberal scholarship unfettered by religious tests, sometimes preoccupied reformers to such an extent that they mistook the local issues for the substance of philosophical debate (Harvie 1976, Chapters 3 and 4; Rothblatt 1968).

All the same, though the reforming generation of the 1860s and 1870s wanted to transform scholarship in the universities, it never repudiated, even when it secularised, the moral task which Edward Pusey had so uncompromisingly stated in the 1850s: 'the special work' of the university is 'not how to advance science, not how to make discoveries, not how to form new schools of mental philosophy, nor to invent new modes of analysis, not to produce works in Medicine, Jurisprudence, or even Theology; but to form minds religiously, morally, intellectually, which shall discharge aright whatever duties God, in his providence, shall appoint to them' (quoted in Harvie 1976, 30). That is, while the new professional academics committed their energies to reform of academic practice, they did not really question the principle that the purpose of knowledge is the cultivation of the moral life. For example, the politician and administrator Richard Burdon Haldane, who took a significant lead in expanding the university system between the 1890s and 1920s, as a service to industry and empire, still emphasised 'the double function of our educational institutions, the imparting of culture for culture's sake on the one hand, and the application of science to the training of our captains of industry on the other' (Haldane 1902, x; see Vincent and Plant 1984, Chapter 8). Expansion was a moral as well as imperial programme, the creation of a system of education that would enable people to contribute politically and to fulfil themselves in so doing.

A further dimension of the educated elite's focus on moral action was that social thought developed first and foremost as a *field of practical action*. There is a literature in the history of British sociology directed to explaining why, though there was much interest in scientific methods, usually understood as a basic, even common-sense, form of empiricism, theoretical work and an academic science barely developed. According to this view, the rationalism and individualism of Victorian writers on social affairs created an intellectual environment in which debates about *social* knowledge, of the kind which took place in Germany, could hardly have occurred (Annan 1959). Philip Abrams argued that the permeability of government to information from the social survey led to the soaking up of energies, which elsewhere might have gone into an academic discipline of sociology, into national administration – with clear implications for the resources committed to high theory (Abrams 1968, 1985).[17] Other historians, however, have criticised such theses, since they seem to assume that social thought was, or should have been, one kind of thing, and they moreover downplay the considerable interest which both social evolutionists and idealists had in general theories of social development. Moreover, in a number of provincial universities, social philosophy expanded as a subject intended to re-invigorate classical studies, and the root of this was not in practical action (Collini 1978; Harris 1996). What was generally absent, however, was a focus on the epistemology of social knowledge, on the sort of issue that became the staple diet of the philosophy of the social sciences in the 1960s.

If we turn from social thought to history, we find a field in change, the outcome of which was the specialisation of methods, content and readership characteristic of the modern academic discipline in Britain. Right through the nineteenth century, and in history written for a public audience well beyond this, history teaching and writing had moral and political goals. History, Lord Acton told the Cambridge audience at his inaugural lecture on his appointment to the Regius Professorship of Modern History in 1895, 'is a narrative of ourselves, the record of a life which is

[17] Compare Hawthorne ([1976] 1987, 170): 'Sociology was virtually absent in England as an intellectually and academically distinctive pursuit because it was virtually everywhere present as part of the general liberal and liberal-socialist consensus.'

our own, of efforts not yet abandoned to repose, of problems that still entangle the feet and vex the hearts of men' (Acton [1906] 1960, 23). Historians, well into the 1870s, explicitly structured narrative around the themes of individual moral character and collective progress towards current ideals. History, like Classics (where many of its roots lay), was a source of maxims and political wisdom. William Stubbs, Regius Professor of Modern History at Oxford (and later Bishop of Oxford), in *The Constitutional History of England in Its Origins and Development* (1873–1878), told the institutional history of the country with 'a Burkean endorsement of the historically given' (Collini et al. 1983, 193; also Burrow 1981, Part II). If there was little philosophy of history as a self-reflective practice, conventional narrative forms nevertheless implicated Providence and presupposed a philosophy. Stubbs, however, also placed a premium on facts, and his patient and methodical use of primary sources became a model of good scholarly practice. He exemplified the research activity that became established as the academic occupation of history expanded, with ever increasing specialisation and attention to empirical evidence.

In the 1860s, influenced by comparative philology and anthropology, there was a swing towards embedding historical writing in the more 'scientific' setting provided by theories of development. John Seeley, at this point (1863) inaugurating a London chair in Classics stated that history writing, firstly, provides the student with models of human action for emulation, but secondly (here he followed Mill), creates 'a collection of observations intended to serve as data for general propositions concerning the nature of man' (Seeley 1864, 19).[18] Seeley went on, in 1869, to the chair of modern history at Cambridge, preceding Acton, from which position he declared history, by virtue of this commitment to inductive generalisation, to be a 'science' – the basis, indeed, of moral science. In the 1870s and 1880s, this focus on facts, equated with making history a science by rigorous attention to the primary sources, became central to the professionalisation of the academic discipline. The editorial to the *English Historical Review*, founded in 1886 as a forum for professional historians in danger of becoming isolated from each other because of specialisation, stated simply that 'the object of history is to discover and set forth facts'. The context of the remark was not an interest in the philosophy of knowledge nor any discussion about whether history is a science, but, rather, concern to secure history's independence from religious or political views. The author of the remark wanted the journal to keep to statements of facts, because such statements 'can usually escape the risk of giving offence' (*English Historical Review* [1886] 1973, 176). Gentlemanly conduct, not philosophy, mattered.

When Seeley used the phrase 'the philosophy of history', he used it as a name for the moral sciences, not to refer to either a grand view of historical development or the theory of historical knowledge. There was very little of the latter kinds of writing about history in the way which was so characteristic of German-language scholarship. All the same, whatever the stress on facts, a number of historians

[18] See Collini et al. (1983, 214). Mill ([1843] 1900, 598) had written: 'History accordingly does, when judiciously examined, afford Empirical Laws of Society.'

maintained belief that there is a historical process and that it expresses a *telos*. Acton (a liberal Catholic), instituted the multi-volume *Cambridge Modern History* (1902–1910) in order to reconcile 'universal history' with specialist scholarship for a general audience. And 'by a universal Modern History, we mean a narrative which is not a mere string of episodes, but displays a continuous narrative ... [The stories of nations] will accordingly be told here, not for their own sakes, but in reference and subordination to a higher process.' (Ward et al. 1902, v).[19] All the same, it is far from clear that the large volumes which resulted at all achieved this goal. Moreover, there was no claim that it was knowledge of the 'higher process' which specifically constituted the 'science' of history. When J. B. Bury, Acton's successor in the Cambridge chair, stated that 'it has not yet become superfluous to insist that history is a science, no less and no more', he said nothing about what constitutes a field as a 'science'. What he appears to have had in mind was the contrast between the new historical scholarship, based on the assembling of tested facts, and the older view of history as literature: 'history is not a branch of literature' (Bury 1903, 7, 16). Historical science, he implied, constituted itself through its rigorous methods of data collection. Yet, like Acton, though in a much more secular way, he also presupposed that there is a larger story to tell, the story of human progress, which he equated with the right use of reason. 'If the history of civilization has any lesson to teach it is this: there is one supreme condition of mental and moral progress which is completely within the power of man himself to secure, and this is perfect liberty of thought and discussion' (Bury 1913, 239).

The work of Frederick Maitland, who had a great reputation among fellow historians in the period and later, is instructive. He was impressively active in providing an informed reading of medieval legal documents, setting new standards for using such sources as the basis for knowledge of social and political developments. But he repeatedly diverted his attention from writing a promised synthesis to arguments about the accuracy and interpretation of details which the lay person could scarcely understand let alone feel to be important. He opposed generalisation without better knowledge of the sources, making extended narrative a virtual impossibility, at least for the present. He was, in the English phrase, a historian's historian. The judgement of a later historian, G. R. Elton, holds: 'In Maitland's day historians, especially English historians, virtually never reflected on their activities. ... They did not feel called upon to philosophize about it; at most they would stake claims for the role their calling played in the formation of public men' (Elton 1985, 19). S.R. Gardiner, an Oxford historian, stated in 1902: 'History is not a matter of beautiful expression but of absolute science, whose results are obtained by careful

---

[19] The quotation is from the editors' introduction to *The Cambridge Modern History*, using words taken from Acton's 'Letter to the contributors to the *Cambridge Modern History*' (of which an extract is reprinted in Stern ([1956] 1973, 247–249). This study of 'modern' history began with the Renaissance; the chairs in the field, however, at both Cambridge and Oxford, dated modern history from the fall of the Roman Empire, reflecting the academic roots of history in Classics.

observation, correct reasoning, and proper methods of investigation' (quoted in Kenyon 1983, 178).[20]

Mill had thought that to be a science, and not just an assemblage of empirical facts, history had, firstly, to reach general laws of social development, and secondly, show that these laws could be deduced from the more fundamental laws of human nature. Seeley, like others in his generation of liberal scholars interested in the spread of an enlightened democracy, trusted that history would find its purpose and justification, like social thought, in a comparative science of development.[21] But the fact is that academic history and social thought, which sought to use historically-based generalisations, went separate ways. Moreover, a number of historians, including G. M. Trevelyan, doubted whether historians ever arrived at empirical laws, let alone succeeded in deducing such laws from laws of human nature. 'An historical event cannot be isolated from its circumstances, any more than the onion from its skins, and because an event is itself nothing but a set of circumstances, none of which will ever recur.' And if historians did not arrive at laws, this appeared to leave historical knowledge without status as a science, unless one were to equate science simply with the statement of authenticated facts. 'The alleged science does not exist, and cannot ever exist in any degree of accuracy remotely deserving to be described by the word "science"' (Trevelyan 1913, 7, 6).

Trevelyan's declarations were a direct, and angry, response to Bury's inaugural lecture, and the contrast between the two historians draws attention to the fact that for them, and for most British historians, the question of the status of historical knowledge was not a matter of comparing knowledge in the natural and human sciences, let alone a matter of epistemology, but of *comparing history and literature*. What Seeley emphasised was that there is a difference between academic history, based on facts and the separation of what can be stated as fact and what not, and 'literature'. This was not a dispute about epistemology, as Trevelyan's riposte made clear; it was a dispute about the primary purpose and intended audience of historical writing. Trevelyan appears never to have forgiven his teacher, Seeley, for describing the great exponents, in the early Victorian period, of history as literature, Macaulay (Trevelyan's great-uncle) and Carlyle as charlatans for using literary skill, with risk to the facts, to create a public audience for their subject. Hearing the same message in Bury's lecture, Trevelyan turned on academics who want 'to drill us into so many Potsdam Guards of learning' and outlaw the empathetic, and literary, interpretation of historical actors. Both Bury and Trevelyan had a moral conception of their discipline and promoted it for its contribution to the public good. But Bury's conception of this good was a rationalist one, while Trevelyan's point was that 'historical writing was not merely the mutual conversation of scholars with one another, but was the means of spreading far and wide throughout all the reading classes a love and knowledge of history, an elevated and critical patriotism and

[20] Kenyon (1983, 175) himself simply dismissed historians' claims to 'scientific pretensions'.

[21] Collini et al. (1983, Chapter 7; Harvie 1976; Rothblatt 1968, 155–180). For the historians' moral mission and claim to educate a liberal elite fit for government, see Soffer (1994).

certain qualities of mind and heart'. He concluded: 'the analogy of physical science has misled many historians during the last thirty years right away from the truth of their profession.... In short, the value of history is not scientific. Its true value is educational' (Trevelyan 1913, 4, 12).[22]

This did indeed align history with literature and not science in the British context. During the period under review, academics and reformers of education made large claims for the study of English language and literature, and they did so with precisely the same sentiments and cultural values which Trevelyan articulated for history. To a significant extent, the academic discipline of English began to replace Classics as the mainstay of higher education for citizenship. In this context, it was common enough to show hostility to science, equated with natural science or social science modeled on natural science, for being incapable of addressing the deeper questions facing each person (Baldick 1983, 199–203). By contrast, it was argued, literature – and Trevelyan thought history also – engages the moral and spiritual values underlying civilised life. Criticism of science was especially strong when academics struggled to reassert civilised values in the aftermath of the barbarism of the Great War (Mayer 1997).

Trevelyan, and historians who thought like him, like J. A. Froude earlier, and like novelists, wrote to bring out the place of individual human feeling, judgement and character in events. In this respect, their commitments can be compared to the German philosophers who stressed the idiographic form of knowledge, the significance of biography and the hermeneutic relationship between the scholar-historian and his subject. But for the British writers, the matter was a moral one, a question of the historian's obligation to the public audience, not a subject for philosophical reflection. In the idiom of English expression, the issue, which divided historians, was whether history is a science or not, not whether there are differences between the science of history and the natural or social sciences. On the one side, there were those like Bury who thought history a science which exemplifies the purposes of higher education in general, 'the training of the mind to look at experience objectively, without immediate relation to one's own time and place'; on the other side, there were historians like Froude, who thought the word 'science' out of place in history, since 'if it is free to a man to choose what he will do or not do, there is no adequate science of him.... Mankind is but an aggregate of individuals – History is but the record of individual action' (Bury 1903, 30; Froude 1867, 11).[23] All the same, all historians conceded that 'in collecting and weighing evidence as to facts, something of the scientific spirit is required for a historian' (Trevelyan 1913, 30). What had 'dug so deep a trench between history as known to our grandfathers and as it appears to us, is ... [the modern historian's] dogma of impartiality' (Acton [1906] 1960, 31).

The social process which produced modern disciplinary structures established philosophy, history, Classics and English language and literature as specialised

---

[22] Also Annan (1955); Collini (1999). For further English opposition to German scholarship, see Freeman's (1886, 288) remark about 'the fashionable idolatry of the latest German book'.

[23] For Froude, see Burrow (1981, Part IV).

branches of the academic profession.[24] One characteristic of professional practice was mutual respect for different types of expertise. This grew on the back of a gentlemanly culture, to which nearly all academics, in principle, subscribed, and it encouraged scholars not to venture statements in fields not their own. This was, I suggest, not an academic environment conducive to comparison of the forms of knowledge in different disciplines. Philosophers, who at times in the German academic context made the claim that their goal was a comprehensive overview, or even synthesis, of knowledge, did not claim this in the British setting. (That Spencer had indeed done so was part of what brought his work into disrepute.) Rather, British philosophers created for themselves *one specialised discipline* among others, distinguished by its analytic as opposed to empirical methods, a discipline far removed from history, and exemplified in the writing of F. H. Bradley or Moore.

In addition, the analytic distinction between factual and evaluative statements played a larger and more prominent part in intellectual culture.[25] For many academics, in all fields, this legitimated the distinction between modern professional scholarship, grounded in facts, and the older, 'amateur' practice (which some historians thought Trevelyan perpetuated) that shaped knowledge according to the evaluative context of facts (such as their place in national self-realisation). Where academics drew a distinction between forms of knowledge, it was between factual, scientific knowledge and knowledge falsely so called, because overly influenced by evaluation. Most scholars decided to leave philosophy to philosophers. Indeed, the more empirical the customary discourse of a field – which perhaps reached an extreme with historians – the more non-empirical discourse, such as philosophy, let alone speculation, appeared something which ought to be avoided.

There was, therefore, no systematic, and very little formal, debate about the philosophy of historical knowledge in Britain. Whether or not writers held history to be a 'science' depended in some contexts on how people opposed 'science' to 'art', objective fact to fiction or speculation, and in other contexts on the way historians argued that historical narrative would illuminate general development, laws of human nature or moral purposes as opposed to establishing a mere record of events. In the latter context, it was possible to relate history and natural science, as Buckle had done, but most British historians thought it none of their business to say anything about natural science at all, even if only for purposes of comparison with their own activity. The situation was interestingly different in the case of psychology: various conceptions of psychology as a subject area placed it right on the problematic boundary between the natural sciences and the moral sciences. Psychology was an ill-defined area with little institutional presence, in contrast to philosophy and history, which had significant disciplinary identity by 1900.

---

[24] I have attempted to develop an extended account of the relation between 'the moral sciences' and the more recent conception of 'the human sciences', and to relate both to the natural sciences, in Smith (2007).

[25] Henry Sidgwick's studies of ethics made the formal statement of the distinction familiar; see Schneewind (1977). Sidgwick was Knightsbridge Professor of Moral Philosophy in the University of Cambridge.

The situation also contrasted with that in Germany and the United States. By the 1890s, in Göttingen, Leipzig, Berlin, Würzburg and elsewhere, there were distinct research programmes, though in the institutional setting of philosophy departments, to develop experimental psychological research. In Britain at this time there was very little comparable activity – only a small amount of work in sensory psycho-physiology at Cambridge and London from the late 1890s. Attempts to introduce universities to experimental work on mental processes met a double resistance: from conservatives, like the Cambridge dons who, in 1891, reluctantly voted £50 for psycho-physical apparatus, but for the physiological laboratory not moral sciences faculty, or the Oxford philosophers who thought experimental work trivial, to the physical scientists who opposed the presence of psychology in the British Association for the Advancement of Science on the grounds that subjective states could not be made the subject matter of a quantified science.[26] Such parochial, knee-jerk academic politics was hardly fertile ground for a collective debate about epistemology; but the sense in which psychology is a 'science' was debated.

The journal *Mind* was founded in 1876 by the philosopher-psychologist-pedagogue Bain and the journal's first editor, George Croom Robertson, to provide a forum for 'mental science', and it did not clearly differentiate psychology and philosophy (Neary 2001). The journal reflected Bain's own wide range of interests, and this included new experimental work as well as traditional analytic approaches. After 1892, however, with a new editor, the journal gradually became more obviously identified with philosophy, establishing the topics that later blossomed into 'the philosophy of mind'. Yet the new editor, G. F. Stout, was also the author of one of the most used textbooks for psychology students until the 1930s (Stout [1899] 1938). The fact was that the 'analytic psychology', which Stout articulated with his teacher at Cambridge, James Ward, persisted as an approach in Britain, at least until the Second World War, making it very questionable where, if at all, to draw a line between philosophy and 'science'. There continued to exist a significant body of opinion, led by Ward, which did not equate psychological knowledge with natural science knowledge, and which regarded the principal methods of the field as analytic and not empirical, but which nevertheless claimed the status of knowledge, or 'science'. Ward, indeed, who supported the establishment of psycho-physiology, indicted the new experimental psychology insofar as it claimed to be a 'science' and not just a new method (Ward 1918, vii).[27] What, for Ward, separated psychology as a science from the natural sciences was that the former was concrete, 'since it deals in some sort with the whole of experience', while the latter were 'abstract' (Ward 1886, 38).

---

[26] For Cambridge, see Robertson (1891), and more generally Hearnshaw (1964, 171–172, 174–175, 181). (The anecdote, which has Cambridge dons claiming that psycho-physical apparatus would 'insult religion by putting the human soul in a pair of scales', appears apocryphal.) For the BAAS, Myers (1932).

[27] As Ward noted, this book was an extended version of the already long article he had written on 'Psychology' for the *Encyclopaedia Britannica* in the 1880s (Ward 1886).

Ward's and Stout's starting point was 'the unity' of experience. 'The point from which the psychologist must start is the unity of experience of itself. It is that which primarily constitutes the mind ... It is in virtue of forming part of this peculiar unity that its components are called mental ... But within this unity we find something which is central and invariably present whatever other things may come and go. This we may call the self in the narrower sense of the term' (Stout [1899] 1938, 19).[28] Stout's point was a philosophical one, but it lent authority to the assumption, so prevalent at the time, that knowledge of being human must begin with the moral self. Thus, a number of writers on psychological topics took as their point of departure what, in biological terms, was the end-point of the evolutionary process, the moral mind, rather than the supposed elementary, or evolutionarily primitive, elements of psychological processes.[29] When Hobhouse, for example, wrote on psychology (in 1902), he suggested a 'top-down' as opposed to 'bottom-up' approach: 'The study of mind ... takes us at once to the highest thing that evolution has produced, and when we compare the different phases of mental growth, we get into the way of judging the lower by the higher, and viewing the process in relation to the result' (quoted in Collini 1979, 181). In other worlds, the definition of psychology as the science of *mental life* imposed on psychology from the outset a form of understanding appropriate for a moral subject matter. This differentiated psychological science from natural science.

Ward's interest in psychology was part of a larger philosophical concern with naturalism and its limitations. His career began with training for the ministry in the nonconformist Congregational Church, and though he became a philosopher not a preacher, he remained troubled by the implications of naturalism for spiritual aspirations. His basic criticism was perfectly general: experience, logically, requires a prior unitary subject, and 'it is certain – not as a matter of theory but as a matter of fact – that the characteristics of the side of life and mind are *prima facie* essentially teleological'. Psychology, he argued, is the science of experience from the special point of view of the individual as experiencing agent. An idealist metaphysics lay behind this defence of analytic psychology's autonomy from natural science: 'Mind is not the impotent shadow of Nature as thus shaped forth [in natural science], but this shaping is itself the work of mind' (Ward 1899, 209, 247).[30]

William James, though based in Boston, was very much a contributor to the British debates on psychology, contributing regularly, for example, to *Mind*, and he gave the lectures published as *The Varieties of Religious Experience* (1902) in Edinburgh.

---

[28] Stout also maintained that the ideal form of psychological knowledge would take a 'genetic', that is, a developmental, viewpoint, but he eschewed this as inappropriate for his textbook.

[29] I do not mean to deny the contribution of those like Darwin, followed by G. J. Romanes, and also of Spencer and the physiologists who extended the sensory-motor analysis of the nervous system to the functions of the brain, all of whom suggested an evolutionary approach to mind based on the study of elementary functions. See Richards (1987); Young (1970).

[30] For Ward's troubled reactions to naturalism, see Turner (1974, Chapter 8). Ward and Stout attempted to focus psychological analysis on 'activity', and in this context Bradley's work was important. Herbart, Lotze and Brentano were significant German influences.

No one was more thoughtful about the implications of the argument from evolutionary continuity for knowledge in psychology. He contributed to the debate in the 1870s, given prominence by the rhetoric of scientific naturalists like Huxley and Tyndall, about whether it is correct to describe human actions as automatistic. In this context, James argued that consciousness, as an outcome of evolution, must have an adaptive function, as indeed experience suggests it does, and cannot be a mere epiphenomenon (James [1879] 1983; MacKenzie 1980; Neary 1999). James thereby made an important contribution to naturalism, suggesting a way in which it might be possible to conceptualise human agency as a natural process. Yet James's psychology, as he himself acknowledged, remained dualistic, recognising both mental activity and physiological determinism, and his own discourse articulated different forms of knowledge.[31] Indeed, his interest in spiritualist and religious experience, to which he devoted much study, contributed to a kind of informal phenomenology as a human science, far from the physiological psychology with which much of his *Principles of Psychology* (1890) was concerned. James therefore differentiated scientific knowledge, with its 'impersonality' and 'certain magnanimity of temper', from the kind of knowledge which literature, for example, exhibits, and which appears in daily life to be so important as knowledge. He even called the former, scientific knowledge, 'shallow' in contrast to the latter. The 'reason is that, so long as we deal with the cosmic and the general, we deal only with the symbols of reality, but *as soon as we deal with private and personal phenomena as such, we deal with realities in the completest sense of the term*' (James [1902] 1982, 498).[32] Once again, I suggest, we have evidence of the way arguments for naturalism set the agenda for writers insofar as they considered the relations among the different kinds of learned knowledge.

British psychology after 1900 took a number of different paths, and for the main ones both the idealists and James were substantially irrelevant. The identification of psychology as a science, that is a natural science, increasingly depended on the status of its objective methods, firstly, the experimental testing of hypotheses, and secondly, statistical analysis of test results. The two methods were associated with two influential teaching schools, the first at Cambridge under Frederic Bartlett, and the second in London under Charles Spearman. Neither school was noted for an interest in epistemology; specialists in philosophy and in psychology went separate ways.

This is a necessarily brief overview of certain aspects of British social thought, philosophy, history and psychology in the closing decades of the nineteenth century. I have suggested that the general, or metaphysical and moral, question of naturalism preoccupied reflective writers in these fields, with the result that there was comparatively little formal attention to possible differences in epistemology in different fields of knowledge. At the same time, specialisation and professional concern with the social and institutional infrastructure of disciplines, including philosophy

---

[31] This James attempted to resolve through his philosophy of pure experience; see Myers (1986).

[32] For James's highly personal involvement with the status of scientific knowledge, see David Leary's contribution to this volume.

itself, made philosophical questions appear a matter for specialist judgement. The moral sciences broke up into their constituent parts. These generalisations go only so far – psychologists, for example, unlike historians, did write philosophically, and Ward explicitly wrote on epistemology. I have also emphasised the significance of the taken-for-granted belief in *the moral subject of the human sciences* for the way writers treated the problem of knowledge.

A large number of questions, of course, have not been addressed. Social thinkers of the period characteristically combined resources from anthropology, political economy, political thought, social psychology, the study of religion and comparative history, all areas about which I have said little or nothing. At the same time, a number of new disciplines, such as economics, economic history and field-based social anthropology, acquired a degree of institutional identity. What were the consequences for the philosophy of knowledge of the tensions between the penchant for eclectic theorising and the social process of discipline formation? What, if any, was the British reception of the contemporary German-language literature about forms of understanding? My suggestion – which requires empirical research – is that this would have been seen as a specialist philosophical topic. There was considerable British interest in German-language scholarship, with many reviews in journals both general and specialist; and the idealists took up Hegel, Acton (who studied with and became close to Ignaz von Döllinger in Munich) admired historicism, Ward studied with Lotze. All this requires much more analysis.

The general thesis of this paper, that English-language debate about knowledge in the human sciences had a different character from German-language debate, receives confirmation if we jump forward seventy years and look at the later philosophy of the social sciences. Beginning in the 1950s, there began to be a flood of English-language criticism of the theory of knowledge of the logical positivists. Then, in the 1960s, a specialty called 'philosophy of social science' took institutional form for the first time (the journal, *Philosophy of the Social Sciences*, for example, was founded in 1971). So-called ordinary language philosophy, in the writings of Stuart Hampshire, R. S. Peters and others, argued that understanding human actions requires reference to intentions; and the later Wittgenstein deeply influenced 'the idea of social science', in Peter Winch's phrase, as a form of knowledge of rule-following within a linguistic community.[33] This work created interest in and receptivity for the earlier German-language debates. The subject matter of the workshop in which this paper was first presented became part of the standard intellectual repertoire of philosophers of the human sciences. Translations of German scholars like Dilthey, Rickert and Simmel appeared, and there was a major, still continuing, re-assessment of Weber. Habermas's *Knowledge and Human Interests*, in translation (Habermas [1968] 1972), introduced a large audience to the Kantian language of forms of knowledge.[34] Philosophers of history responded with

[33] E.g., Outhwaite (1975). A helpful survey is Bernstein ([1976] 1979).

[34] Earlier, translations, as well as the late writings in English, of Cassirer had had importance in this way.

incredulity to Carl G. Hempel's inclusion of historical knowledge under the general law model of explanation, and they turned elsewhere for insight.[35] My point is obvious: when English-language authors found the need critically to re-assess relations between the natural sciences, the human sciences and the humanities, and to consider the possible existence of different forms of knowledge, part of their response was to turn to the earlier German-language debates for the necessary intellectual resources. They did not turn to an English-language literature from the decades around 1900. There was no such literature which appeared to have the same depth and relevance to *these* topics.

## References

Abrams P (1968) The origins of british sociology 1834–1914: An Essay with Selected Papers. University of Chicago Press, Chicago, IL

Abrams P (1985) The uses of British sociology, 1831–1981. In: Bulmer M (ed.) Essays on the History of British Sociological Research. Cambridge University Press, Cambridge, pp 181–205

Acton, Lord [J.E.E.D. Acton, 1st Baron Acton] ([1906] 1960) Inaugural lecture on the study of history [Delivered 1895]. In: Lectures on Modern History. Reprinted with an introduction by H. Trevor-Roper. Collins Fontana, London, pp 17–41

Anderson P ([1968] 1992) Components of the national culture. In: English Questions. Verso, London, pp 48–104

Annan N (1955) The intellectual aristocracy. In: Plumb JH (ed.) Studies in Social History: Atribute to G. M. Trevelyan. Longmans, London, pp 243–287

Annan N (1959) The curious strength of positivism in English political thought. L. T. Hobhouse Memorial Trust lecture, no. 28. Oxford University Press, London

Anonymous (1903–1904) A brief account of the foundation of the academy. Proc Br Acad 1:vii–ix

Baldick C (1983) The Social Mission of English Criticism 1848–1932. Clarendon Press, Oxford

Balfour AJ (1879) A Defence of Philosophic Doubt: Being an Essay on the Foundations of Belief. Macmillan, London

Balfour AJ (1895) The Foundations of Belief: Being Notes Introductory to the Study of Theology. Longmans, Green, London

Bernstein RJ ([1976] 1979) The restructuring of Social and Political Theory. Methuen, London

Bosanquet B (1912) The principle of individuality and value: the Gifford lectures for 1911 delivered in Edinburgh University. Macmillan, London

Boucher D (ed) (1997) The British Idealists. Cambridge University Press, Cambridge

Brooke JH (1991) Science and Religion: Some Historical Perspectives. Cambridge University Press, Cambridge

Brown AW (1947) The Metaphysical Society: Victorian Minds in Crisis, 1869–1880. Columbia University Press, New York

Burrow JW (1966) Evolution and Society: A Study in Victorian Social Theory. Cambridge University Press, Cambridge

Burrow JW (1981) A Liberal Descent: Victorian Historians and the English Past. Cambridge University Press, Cambridge

Bury JB (1903) An inaugural lecture delivered in the Divinity School Cambridge on January 26. Cambridge University Press, Cambridge

Bury JB (1913) A History of Freedom of Thought. Williams & Norgate, London

[35]Thus stimulating an interest in narrative: (Danto 1965; Gallie 1964).

Collini S (1978) Sociology and idealism in Britain: 1880–1920. Archives européenes de sociologie 19:3–50
Collini S (1979) Liberalism and Sociology: L. T. Hobhouse and Political Argument in England 1880–1914. Cambridge University Press, Cambridge
Collini S (1991) Public Moralists: Political Thought and Intellectual Life in Britain 1850–1930. Clarendon Press, Oxford
Collini S (1999) Writing 'the national history': Trevelyan and after. In: English Pasts: Essays in History and Culture. Oxford University Press, Oxford, pp 9–37
Collini S (2006) Absent Minds: Intellectuals in Britain. Oxford University Press, Oxford
Collini S, Winch D, Burrow J (1983) That Noble Science of Politics: A Study in Nineteenth-Century Intellectual History. Cambridge University Press, Cambridge
Colls R, Dodd P (eds) (1986) Englishness: Politics and Culture 1880–1920. Croom Helm, London
Danto AC (1965) Analytic Philosophy of History. Cambridge University Press, Cambridge
Elton GR (1985) F. W. Maitland. Weidenfeld, London
Emmet D ([1920] 1966) Foreword. In: Alexander, Samuel, Space, Time, and Deity: The Gifford Lectures at Glasgow, 1916–1918. 2 vols. Macmillan, London
English Historical Review ([1886] 1973) Prefatory note to the first issue of the English Historical Review. In: Stern F (ed.) The Varieties of History: From Voltaire to the Present, 2nd edn. Vintage Books, New York, pp 174–177
Freeden M (1978) The New Liberalism: An Ideology of Social Reform. Clarendon Press, Oxford
Freeman EA (1886) The methods of historical study: Eight lectures read in the University of Oxford in Michaelmas Term, 1884, with the inaugural lecture on the office of the historical professor. Macmillan, London
Froude JA (1867) The science of history: A lecture delivered at the Royal Institution. February 5, 1864. In: Short studies on great subjects, vol 1. Longmans, Green, London, pp 1–36
Gadamer H-G ([1960] 1998) Truth and method. Translation of the fifth German edition, revised by Joel Weinsheimer and Donald G. Marshall. Continuum, New York Originally published as Wahrheit und methode. Grundzüge einer philosophischen hermeneutik. J.C.B. Mohr, Tübingen, 5. edition published1986 as Hermeneutik. Wahrheit und Methode J.C.B. Mohr, Tübingen
Gallie WB (1964) Philosophy and the Historical Understanding. Chatto & Windus, London
Habermas J ([1968] 1972) Knowledge and human interests. Translated by Jeremy J. Shapiro. Originally published as Erkenntnis und Interesse, Suhrkamp, Frankfurt am. Main Heinemann, London
Haldane RB (1902) Education and Empire: Addresses on Certain Topics of the Day. John Murray, London
Harris J (1996) Platonism, positivism and progressivism: Aspects of British sociological thought in the early twentieth century. In: Biagini EF (ed) Citizenship and Community: Liberals, Radicals and Collective Identities in the British Isles, 1865–1931. Cambridge University Press, Cambridge, pp 343–360
Harvie C (1976) The Lights of Liberalism: University Liberals and the Challenge of Democracy 1860–1886. Allen Lane/Penguin, London
Hawthorne G ([1976] 1987) Enlightenment and Despair: A History of Social Theory, 2nd edn. Cambridge University Press, Cambridge
Hearnshaw LS (1964) A Short History of British Psychology 1840–1940. Methuen, London
Hull DL (1973) Darwin and His Critics: The Reception of Darwin's Theory of Evolution by the Scientific Community. Harvard University Press, Cambridge, MA
Huxley TH (1892a) Essays upon Some Controverted Questions. Macmillan, London
Huxley TH (1892b) Prologue. In: Essays upon Some Controverted Questions. Macmillan, London, pp 1–53
Huxley TH ([1889] 1892) Agnosticism. In: Essays upon Some Controverted Questions. Macmillan, London, pp 329–377
Jacyna LS (1980) Science and social order in the thought of A. J. Balfour. Isis 71:11–34

James W ([1879] 1983) Are we automata? In: Essays in Psychology. Reprinted with an introduction by William R. Woodward. Harvard University Press, Cambridge, MA, pp 38–61
James W ([1902] 1982) The Varieties of Religious Experience: A Study in Human Nature. Penguin, New York
Kenyon J (1983) The History Men: The Historical Profession in England Since the Renaissance. Weidenfeld & Nicolson, London
Lightman B (1987) The Origins of Agnosticism: Victorian Unbelief and the Limits of Knowledge. Johns Hopkins University Press, Baltimore, MD
Lightman B (2004) Scientists as materialists in the periodical press: Tyndall's Belfast address. In: Cantor G, Shuttleworth S (eds) Science Serialized: Representations of Science in Nineteenth-Century Periodicals. MIT Press, Cambridge, MA, pp 199–237
MacKenzie L (1980) William James and the problem of interests. J Hist Behav Sci 16:175–185
Mayer A-K (1997) Moralizing science: The uses of science's past in national education in the 1920s. Br J Hist Sci 30:51–70
Mill JS ([1843] 1900) A system of logic ratiocinative and inductive: Being a connected view of the principles of evidence and the methods of scientific investigation, 8th edn. Longmans, Green, London
Moore GE (1903a) Principia Ethica. Cambridge University Press, Cambridge
Moore GE (1903b) The refutation of idealism. Mind (new series) 12:433–453
Myers CS (1932) On the nature of mind [Presidential Address, Section J]. British Association for the Advancement of Science: Report of the centenary meeting, London – 1931. John W. Parker, London, pp 181–195
Myers GE (1986) William James: His Life and Thought. Yale University Press, New Haven, CT
Neary F (1999) Consciousness, evolution and morals: Some critical perspectives on a history of the human automatism debate 1870–1910, Ph.D. dissertation. Lancaster University, England
Neary F (2001) A question of 'peculiar importance': George Croom Robertson, *Mind* and the changing relationship between British psychology and philosophy. In: Bunn GC, Lovie AD, Richards GD (eds) Psychology in Britain: Historical Essays and Personal Reflections. BPS Books, the British Psychological Society, Leicester, pp 54–71
Outhwaite W (1975) Understanding Social Life: The Method Called 'Verstehen'. George Allen & Unwin, London
The Compact Edition of the Oxford English Dictionary (1979) 2 vols. Book Club Associates, London
Paradis J (1978) T. H. Huxley: Man's Place in Nature. University of Nebraska Press, Lincoln
Peel JDY (1971) Herbert Spencer: The Evolution of a Sociologist. Heinemann, London
Reay, Lord [DJ Mackay, 11th Lord Reay] (1903–1904) First annual general meeting. Address by the President, Lord Reay. Proc Br Acad 1:1–16
Regan T (1986) Bloomsbury's Prophet: G.E. Moore and the Development of His Moral Philosophy. Temple University Press, Philadelphia, PA
Richards RJ (1987) Darwin and the Emergence of Evolutionary Theories of Mind and Behavior. University of Chicago Press, Chicago, IL
Richter M (1964) The Politics of Conscience: T. H. Green and His Age. Harvard University Press, Cambridge, MA
Robertson GC (1891) Valedictory. Mind 16:557–560
Ross S (1962) Scientist: The story of a word. Ann Sci 18:65–85
Rothblatt S (1968) The Revolution of the Dons: Cambridge and Society in Victorian England. Faber & Faber, London
Schneewind JB (1977) Sidgwick's Ethics and Victorian Moral Philosophy. Clarendon Press, Oxford
Seeley JR (1864) Classical studies as an introduction to the moral sciences: An introductory lecture delivered on October 13, 1863, at University College, London. Bell & Dalady, London
Seth A (1897) Note A. The use of the word 'naturalism'. In: Man's Place in the Cosmos and Other Essays. Blackwood, Edinburgh, pp 289–303
Smith R (2003) Biology and values in interwar Britain: C. S. Sherrington, Julian Huxley and the vision of progress. Past and Present 178:210–242

Smith R (2007) Being Human: Historical Knowledge and the Creation of Human Nature. Manchester University Press, Manchester
Soffer RN (1994) Discipline and Power: The University, History, and the Making of an English Elite, 1870–1930. Stanford University Press, Stanford
Spencer H ([1873] 1892) The Study of Sociology, 16th edn. Kegan Paul, London
Stephen L (1880) An attempted philosophy of history. Fortnightly Rev (new series) 27:672–695
Stern F (ed.) ([1956] 1973) The Varieties of History: From Voltaire to the Present, 2nd edn. Vintage Books, New York
Stout GF ([1899] 1938) A Manual of Psychology, 5th edn., revised by C. A. Mace. University Tutorial Press, London
Trevelyan GM (1913) Clio, a Muse. In: Clio, a Muse and other essays literary and pedestrian. Longmans, Green, London, pp 1–55
Turner FM (1974) Between Science and Religion: The Reaction to Scientific Naturalism in Late Victorian England. Yale University Press, New Haven, CT
Turner FM (1978) The Victorian conflict between science and religion: A professional dimension. Isis 69:356–376
Tyndall J ([1874] 1892) The Belfast address. In: Fragments of Science, vol 2, 8th edn. Longmans, Green, London, pp 135–201
Vincent A, Plant R (1984) Philosophy, Politics and Citizenship: The Life and Thought of the British Idealists. Basil Blackwell, Oxford
Ward J (1886) Psychology. In: Encyclopaedia Britannica, vol 20, 9th edn. Adam & Charles Black, Edinburgh, pp 37–85
Ward J (1899) Naturalism and agnosticism. The Gifford Lectures delivered before the University of Aberdeen in the years 1896–1898, vol 2. Adam & Charles Black, London
Ward J (1918) Psychological Principles. Cambridge University Press, Cambridge
Ward AW, Prothero GW, Leathes S (eds) (1902) [planned by the late Lord Acton], The Cambridge Modern History, vol 1. The Renaissance. Cambridge University Press, Cambridge
White P (2003) Thomas Huxley: Making the 'Man of Science'. Cambridge University Press, Cambridge
Young RM (1970) Mind, Brain, and Adaptation: Cerebral Localization and Its Biological Context from Gall to Ferrier. Clarendon Press, Oxford
Young RM (1985) Natural theology, Victorian periodicals, and the fragmentation of a common context. In: Darwin's Metaphor: Nature's Place in Victorian Culture. Cambridge University Press, Cambridge, pp 126–163

# Chapter 10
# Accounting for the Unity of Experience in Dilthey, Rickert, Bradley and Ward

**Christopher Pincock**

I. The history of philosophy differs from other kinds of history mainly in its attempts to understand historical change exclusively through the examination and evaluation of philosophical arguments. Of course, nobody writing the history of philosophy is likely to deny that philosophical arguments are given by people and that the context and aspirations of the philosopher will shape her arguments. Despite this concession it remains common practice to ignore such contributions in the reconstruction of a historical figure's philosophical views. Reflecting an extreme version of this approach, Scott Soames has recently written in response to criticism of his *Philosophical Analysis in the Twentieth Century*,

> if progress [in philosophy] is to be made, there must at some point emerge a clear demarcation between genuine accomplishments that need to be assimilated by later practitioners, and other work that can be forgotten, disregarded, or left to those whose interest is not in the subject itself, but in history for its own sake. The aim of my volumes was to contribute to making that demarcation (Soames 2006, 655).

The limitations of this conception of history for resolving questions about long-standing philosophical distinctions like Erklären/Verstehen are easy to see. As Peter Simons has argued concerning the equally vexing distinction between analytic and continental philosophy, "No individual or movement, no dramatic event is solely or simply responsible for the opening up of the analytic-continental rift. A constellation of circumstances, compounded by political and historical events, accidents of human destiny, are responsible" (Simons 2001, 307). Thus opposing what he calls the "more or less deterministic internal explanations" typically prized in the history of philosophy, Simons uses the evidence that "the rift is an accident" (Simons 2001, 296) to conclude that "what we make of the divide is our business" (Simons 2001, 308), i.e., the distinction has no lasting philosophical significance. More generally, it appears that we are unlikely to find arguments of the sort that Soames is after that will clearly vindicate the course that the history of philosophy happened to take.

While Simons' position is certainly more interesting and historically informed, we can see him as simply accepting the other side of the coin offered by Soames.

C. Pincock (✉)
Department of Philosophy, Purdue University, West Lafayette, IN 47907, USA
e-mail: pincock@purdue.edu

U. Feest (ed.), *Historical Perspectives on Erklären and Verstehen*, Archimedes, Vol. 21,
DOI 10.1007/978-90-481-3540-0_10, © Springer Science+Business Media B.V. 2010

That is, unless a distinction can be supported by conclusive philosophical argumentation or is, as Soames says of Kripke's distinction between necessity and analyticity, "almost irresistible" (Soames 2003, 354), that distinction becomes void of philosophical interest.[1] Both alternatives are unavailable for a philosopher who wishes to understand the opposition of Erklären and Verstehen. On the one hand, nobody is likely to be able to come up even with a univocal characterization of this distinction, let alone a conclusive philosophical argument for its presence. On the other hand, we cannot embrace an explanation of the distinction and its significance that rests content with an enumeration of the various historical contingencies that led to its inception. For if such an account exhausts the point of Erklären/Verstehen, then it really does seem to lack continuing philosophical significance.

The main goal of this essay, then, is to begin to offer a way of tackling the history of such philosophical distinctions that avoids these two extremes. My working assumption is that the Erklären/Verstehen distinction is motivated by a philosophical problem and that we can better understand the distinction by seeing both how various ways of drawing it can contribute to a solution to that problem and how the problem was solved by those who neglected to draw the distinction in question. This turns our focus away from simply considering the reasons that some particular philosopher offers for drawing the distinction in their own special way. But, the hope is, it restricts the resulting account to properly intellectual issues, most especially the issue of whether or not the distinction really solves the problem it set out to solve.

The problem that I focus on in connection with the Erklären/Verstehen distinction is what I will call the unity of experience, and in the next section I will explain why I take Mill's empiricism to have raised this problem in an especially forceful way. In section three I then turn to a consideration of how Dilthey and Rickert, in very different ways, proposed to account for the unity of experience by invoking the opposition between Erklären and Verstehen. This sets the stage for a consideration, in section four, of two British philosophers, Bradley and Ward, who sought to account for the unity of experience without making any appeal to Erklären versus Verstehen. I conclude by arguing that it is unlikely that the distinction in question is necessary to solve the unity of experience problem and that accepting it can blind the historian to the interesting similarities and differences between philosophers like Dilthey, Rickert, Ward and Bradley.

II. Historians continue to quarrel over the causal significance of the publication of Mill's *System of Logic* (Mill [1843] 1973) in 1843 for the Geisteswissenschaften debates later in the nineteenth century, often emphasizing independent developments within the sciences themselves as well as properly German philosophical traditions (Anderson 2003). In line with the orientation described in the last section, I do not propose to enter into these questions. Instead, it will be sufficient for our purposes to establish a point that is much weaker than the claim that the Geisteswissenschaften debates would not have occurred if Mill had not published his *Logic*. I argue only that Mill's *Logic* posed the problem of the unity of experience in an especially clear and urgent way.

[1] See also Leiter (2004).

What is this problem? Simply put, the problem of the unity of experience is to explain how experience can contribute to scientific knowledge. As we will see, Mill offered a surprisingly crude account of the connection between experience and scientific knowledge. Many feel, upon encountering this proposal, that there must be some aspects of experience that Mill is ignoring which may be loosely characterized as the unity of experience, and it is these features of our experience that are in fact crucial to the integration of experience and knowledge in the sciences. Different solutions to the problem of the unity of experience will draw attention to different features of our experience that Mill supposedly missed, and use these features to explain how experience fits with scientific knowledge. One might suspect that there is, in the end, no single problem of how to relate our experience to scientific knowledge given that there is no agreement on what experience or scientific knowledge amount to. Still, I hope to show that this sort of problem can serve as a useful framework for our historical investigation.

Early on in his *Logic* Mill offers an argument for a sweeping conclusion about not only the extent of our scientific knowledge, but also the range of circumstances that we can even describe. His premise is an epistemic one: "of the outward world, we know and can know absolutely nothing, except the sensations which we experience from it" (Mill [1843] 1973, vol. 7, 62; I, iii, §8). This argument quickly follows:

> For if we know not, and cannot know, anything of bodies but the sensations which they excite in us or in others, those sensations must be all that we can, at bottom, mean by their attributes; and the distinction which we verbally make between the properties of things and the sensations we receive from them, must originate in the convenience of discourse rather than in the nature of what is signified by the terms (Mill [1843] 1973, vol. 7, 65; I, iii, §9).

Let us call this combined epistemological and semantic claim "phenomenalism". Phenomenalism is maintained not only as all that science requires, but as all the metaphysics that is possible for beings like us. The facts of experience, thought of as discrete sensations standing in various relations of association, stand as the ultimate epistemological and metaphysical foundation of all the sciences. Laws concerning their succession and coexistence are all that science should hope for.

If we take Mill's phenomenalism seriously, then we have a nearly trivial argument for an extreme form of the unity of science. For if all we can even talk about, and so therefore all that we can know about, are laws concerning the coexistence and succession of a unified kind of thing, our sensations, then all sciences will take on essentially the same form. They will give laws that purport to govern these sensations. Much of the *Logic* can be seen as further articulating this picture by arguing that a small group of methods suffices to establish all the laws that we can ever hope to know. But it should not be surprising that the same methods are employed across the sciences if we have already accepted that sensations are the subject matter of the sciences quite generally. To be sure, Mill allows himself some wiggle room by contrasting "the convenience of discourse" which presumably may invoke entities besides sensations with "the nature of what is signified", but the impression remains that all genuine science is a science of sensations.

The unsatisfactory features of Mill's phenomenalism were to occupy Mill's critics for decades, and it seems that these criticisms have been so effective that a contemporary philosopher finds it hard to believe that Mill really meant to defend it. To return to the problem of the unity of experience, there is a glaring gap between any plausible interpretation of the range of our sensations, or more broadly our experience, and the sorts of things that the mature and successful natural and human sciences give us information about. For one who feels this gap, it is hard to even know where to begin when confronted by an advocate of phenomenalism. For surely, unless the phenomenalist proposes to reconstruct all of science so that it will meet the strictures imposed by phenomenalism, the phenomenalist owes us some account of exactly how the discrete sensations making up our experience can hope to contribute to the knowledge of all the different sorts of things that the sciences invoke. The suspicion shared by Mill's critics is that this was simply not possible. As we will see, Dilthey, Rickert, Bradley and Ward offered four different accounts of what was missing and how these missing aspects could integrate experience and scientific knowledge.

Interestingly enough, Mill himself seems to have overstepped the bounds of his phenomenalism, both in his discussion of chemistry and in the concluding book of the *Logic*, "On the Logic of the Moral Sciences". For in both chemistry and the moral sciences like psychology and history, Mill argued that our sensations could combine to form entities of a new kind whose causal laws were not reducible to the laws that govern the original sensations. As our concern for much of this essay will be psychology, I will focus on what Mill has to say here. After beginning with the confident assertion that "The subject, then, of Psychology is the uniformities of succession, the laws, whether ultimate or derivative, according to which one mental state succeeds another – is caused by, or at least is caused to follow, another" (Mill [1843] 1973, vol. 8, 852; VI, iv, §3), Mill goes on to clarify how a complex mental state might relate to the simple sorts of sensations invoked by empiricists. Comparing the process of chemical composition discussed earlier in the *Logic* to the case "when the seven prismatic colours are presented to the eye in rapid succession [and] the sensation produced is that of white", Mill notes that

> so it appears to me that the Complex Idea, formed by the blending together of several simpler ones, should, when it really appears simple, (that is, when the separate elements are not consciously distinguishable in it,) be said to *result from*, or *be generated by*, the simple ideas, not to *consist* of them (Mill [1843] 1973, vol. 8, 854; VI, iv, §3).

Psychology, then, and the other moral sciences, can make progress only by granting this sort of "mental chemistry" and tailoring its methods to help us understand how it works.

The possibility of this sort of chemical composition greatly weakens the kind of unity that Mill can ascribe to the sciences, and in the end it seems in tension with his phenomenalism. For if new sorts of entities can result from the combinations of simpler entities, then there is no hope for a single category of entities whose laws can be used to explain all natural phenomena. Instead we will have a series of increasingly complex entities where each level obeys its own collection of causal laws, and there is no prospect of reducing these higher-level laws to the laws governing the simpler entities. The tension with phenomenalism becomes clearer when we see Mill speaking of entities arising through this sort of chemical composition that have no clear connection to our sensations. For example, Mill includes beliefs and desires in his psychology, but unless these are included in the category of sensations, phenomenalism has been violated even in the case of psychology.

If phenomenalism fails even here, what chance is there that our talk of bodies or social phenomena could be reduced to talk of sensations?

Even allowing chemical composition, and thereby weakening somewhat his conception of the unity of science, Mill's empiricism raised in a particularly urgent form what I have been calling the problem of the unity of experience. If we accept Mill's account of the relation between experience and scientific knowledge, then we have great difficulty understanding how scientific knowledge is even possible. For example, how can we know about the origins of the solar system if all we experience are sensations? The four writers that I will turn to next proposed to bridge the gulf between our experience and scientific knowledge by emphasizing the unity of our experience that they claimed Mill had missed. But, as we will see, no consensus emerged on what exactly the unity of experience amounted to or how it contributed to scientific knowledge. Our four philosophers chose instead to argue that experience or scientific knowledge or both had unanticipated features which would facilitate their integration.

III. The most direct way to resolve our problem of how experience relates to scientific knowledge is to insist that we are always directly in contact with the subject matter of science, i.e., a unified, law-governed natural world. This sort of naïve realism bridges our gap by fiat and is not pursued by any of the writers we will consider. But we can see Dilthey as opting for the next best option in writings like *Introduction to the Human Sciences* (Dilthey [1886] 1989) and "Ideas Concerning a Descriptive and Analytic Psychology" (Dilthey [1894] 1977).[2] For Dilthey insists that our experience always begins with an original awareness of a unified structure which combines not only our representations, but also our feelings and volitions. This "original nexus, a unity which is not put together out of separate elements and functions" (Dilthey [1894] 1977, 104/224),[3] is what a genuinely scientific psychology must begin with. As our experience starts with such a unity, psychology employs a different method than a science like physics. For while the subject matter of psychology is given as a whole system, the subject matter of physics lies outside of our immediate experience and so initially lacks the systematic character required by science. To resolve this, Dilthey argues, physics must employ a series of hypotheses in an effort to explain those regularities that we do have access to in our otherwise disjointed experience of the physical world:

> hypotheses do not all play the same role in psychology as in the study of nature. In the latter, all connectedness [Zusammenhang] is obtained by means of the formation of hypotheses; in psychology it is precisely the connectedness which is originally and continually given in lived experience [Erleben]: life exists everywhere only as a nexus or coherent whole. Psychology therefore has no need of basing itself on the concepts yielded from inferences in order to establish a coherent whole among the main groups of mental affairs (Dilthey [1894] 1977, 28/144).

---

[2] I am indebted to Uljana Feest for providing me with her (Feest, 2007). This paper was instrumental in helping me to see a connection between Dilthey and what I am calling the problem of the unity of experience.

[3] References to Dilthey ([1894] 1977) include the page number of the German original in Dilthey (1959, vol. 5).

Psychology should become descriptive or analytic rather than imitating physics through cycling through hypotheses in an attempt to explain mental phenomena. Analysis, beginning with the given unity of our experience, can productively distinguish the different aspects of our mental life that are initially combined.

Dilthey has a complicated position concerning when hypotheses will enter into his analytic psychology and what function they will serve. Beginning with hypotheses is pointless because no system can ever be confirmed or refuted (Dilthey [1894] 1977, 29/145, 104/224). One suggestion for why this happens in psychology, and not in physics, is that the original unity of our experience makes any system of hypotheses superfluous, and so it has no function to serve in conferring unity on our experience, unlike in physics. The situation is apparently quite different, however, once an initial analysis is made of our experience and its different aspects. For at that point analytic psychology can take on the hypotheses considered in the explanatory psychology that Dilthey attacks: "Descriptive and analytic psychology ends with hypotheses, whereas explanatory psychology begins with them" (Dilthey [1894] 1977, 57/175). By ending with hypotheses Dilthey may mean that these hypotheses are formulated only after analysis delivers the general and shared structures of human experience. This fits with his practice later in the "Ideas" essay, where after distinguishing different aspects of our mental life, he propounds what can only be called a hypothesis: that the qualitative differences in character that we find across individuals are due entirely to the quantitative combinations of a fixed list of these different aspects (Dilthey [1894] 1977, 110/230). Presumably if we had not followed Dilthey in his careful analysis of the given nexus of our experience, this hypothesis would not be possible to formulate, let alone evaluate.

Unsurprisingly, Mill's approach to psychology is singled out as a particularly useless example of explanatory psychology that begins with hypothetical elements and tries to reconstruct the original unity of lived experience through hypothetical mechanisms of association and combination. At least in this case, Dilthey has a fairly plausible diagnosis of why an explanatory psychology can never lead to a well confirmed system of hypotheses. His criticism focuses on the possibility of the chemical composition discussed in the last section: sensations can combine so that the result appears simple and where the result obeys new causal laws. But if we allow without restriction the possibility of this sort of combination, our hypotheses become so flexible that we can handle whatever patterns within our experience we happen to encounter. Such an appeal "allows us to depend on certain regular antecedents and to fill the gap which separates them from the subsequent state by the psychical chemistry. But the latter cannot fail as well to reduce to zero the degree, already quite low, of persuasive force which this construction and its results have" (Dilthey [1894] 1977, 44/160).

Once Dilthey's analytic psychology is in place, it is a short step to the full blown distinction between the natural and the human sciences, and the associated thesis that explanation is central to the natural sciences while understanding is the proper goal of the human sciences. For as Dilthey had argued in the earlier *Introduction to the Human Sciences*, psychology lies at the basis of all the human sciences like history, and in a somewhat different fashion, also sets limits on what the natural sciences can achieve. One helpful summary of the distinction is that

> We explain [erklären] by purely intellectual processes, but we understand [verstehen] through the concurrence of all the powers of the psyche in the apprehension. In understanding we proceed from the coherent whole which is livingly given to us in order to make the particular intelligible to us. Precisely the fact that we live with the consciousness of the coherent whole, makes it possible for us to understand a particular sentence, gesture or action. All psychological thought preserves this fundamental feature, that the apprehension of the whole makes possible and determines the interpretation of particulars (Dilthey [1894] 1977, 55/172).

That is, the distinctive cognitive attitude of understanding extends not only throughout psychology, but also to the interpretation of the actions and utterances of others that are central to history, legal theory and aesthetics. In each case, we exploit the assumed similarities between the given lived nexus of our own experience and the structure of the mental life of the people we are interpreting. By contrast, in the natural sciences, we are forced to begin with the hypothetical properties of elements like atoms whose features we are not intimately aware of. These "heuristic constructions" (Dilthey [1886] 1989, 199/365)[4] remain, in the end, constrained by the structure of inner experience. In particular, Dilthey argues that natural science results from abstracting out the intellectual aspects of our experience and projecting them onto the world, as when we assume that the natural world is governed by a system of universally valid causal laws (Dilthey [1886] 1989, 202/369).[5] Natural scientific explanation is, then, not only distinct from the understanding of the human sciences, but the former is radically dependent on the latter. For nothing that we can put into our hypotheses concerning genuine reality can transcend the materials that we can abstract out of our original lived experience.

Dilthey's dissatisfaction with any broadly neo-Kantian solution to the problem of the unity of experience is revealed in his remark that even the theory of knowledge smuggles in psychological facts, and so is not actually independent of psychology: "it is evidently impossible to connect the spiritual data which form the matter of epistemology without relying on some idea or other of the psychic nexus ... Thus it happened that the fundamental concepts of Kant's critique of reason belong throughout to a definite psychological school" (Dilthey [1894] 1977, 32/148), in particular the defective explanatory approach of faculty psychology. The most sophisticated neo-Kantian reply to this charge was offered by Rickert in his *Limits of Concept Formation in Natural Science* (Rickert [1902] 1986).[6] While conceding that Dilthey was right to deny the unity of the sciences, Rickert claims that "Dilthey ... unfortunately did not succeed in providing a *clear conceptual* analysis of the essential feature of his principle of demarcation" (Rickert [1902] 1986, 146). While the differences between Dilthey and Rickert are not our main concern, the contrast is useful in showing how Rickert's way of drawing the Erklären/Verstehen distinction figures into his own solution to the problem of the unity of experience.

---

[4]References to Dilthey ([1886]1989) include the page number of the German original in Dilthey (1959, vol. 1).

[5]The same strategy is employed in Dilthey's criticism of metaphysics.

[6]All quotations are from the fifth edition of 1929, abridged and translated by Guy Oakes. I draw liberally on Oakes' helpful introduction.

Much of Rickert's book concerns what he calls methodology, by which he appears to mean the systematic and largely logical investigation of the presuppositions of the various sciences (Rickert [1902] 1986, 65). Chapter 5 is occupied with epistemology proper, where Rickert argues that the presuppositions of the various sciences are in fact met. For Rickert, all scientific representations depend on concepts which are derived from experience. Here there is an apparent agreement with Dilthey, but Rickert will argue that a central feature of experience that Dilthey missed points to the need for significant non-experiential assumptions in order to account for scientific knowledge. For Rickert claims that experience does not begin with the basic unity that Dilthey invoked, but that instead experience presents us with an infinitely complex manifold of individuals. This complexity has two aspects. On the one side, each individual that we experience is related in numerous ways to other individuals. But, on the other side, even if we consider one individual by itself, we realize that it has infinite internal complexity. The richness of our experience precludes any attempt to develop a science that would reproduce or copy the natural world in all of its fine detail, and Rickert argues more generally that any approach to scientific knowledge in terms of correspondence with reality is bound to fail. Instead, the natural scientist sets herself the goal of producing laws which are universally valid in the sense that they hold for all events of a certain kind. This goal requires that the scientist abstract from the character of the individuals that she experiences by developing concepts that group them together in appropriate ways. The natural scientist, then, must de-individualize her subject matter and consider the natural world as composed of entities of different types, but which have no further specific features. This is the price that the natural scientist must pay if she is to generate any effective scientific representations consistent with her goals.

In taking up this generalizing attitude, Rickert argues, the natural scientist runs into certain limits on what she can scientifically comprehend. For the natural scientist is precluded from taking a scientific interest in the particular individuals that she began with. To take an example that Rickert sometimes appeals to, the natural scientist cannot consider Luther as an individual, but only as an entity of a particular type. For example, a natural scientific psychologist could include Luther in her research only as a more or less arbitrary subject for the verification of her psychological laws. But what of our interest in Luther as that particular person who revolutionized the practice of religion? If we restrict ourselves to the natural sciences, then we must reject such questions as unscientific.

Rickert is not willing to accept this consequence, and argues instead that there is an alternative sort of goal that the scientist can adopt when she works in what Rickert calls the historical sciences.[7] Here the goal is basically the opposite of the natural sciences. Instead of thinking of the individuals as instances of a kind, and seeking general regularities, we investigate the individual as such and try to

[7] Rickert prefers the terms "historical science [Geschichtswissenschaft]" or "cultural science [Kulturwissenschaften]" to Dilthey's "human science [Geisteswissenschaft]". See Makkreel (1969), cited by Feest (2007).

understand its origins and effects. Immediately, though, Rickert confronts a problem based on the infinite complexity of our experience that he took as an initial premise. Wouldn't this make a historical science of individuals impossible? For example, in Luther's case, there are so many features of Luther, his origins and his effects, that it seems like any genuine scientific understanding of Luther is ruled out.

It is here that Rickert appeals to values. Values for Rickert are not individuals that we experience, and so in this sense they are "not real" or, as he prefers to put it, values are "nonreal [*irreal*]." It is only when individuals are related to nonreal values that they can be considered as objects of scientific historical investigation.[8] The connection between the individual and the value gives the historian a non-arbitrary basis to select only certain features of the individual. These are the features that are responsible for the individual being related to the value that the historian is interested in and they confer a unity on that way of considering the thing. It is helpful to return to our example of Luther at this point. The value in question here is religious value, and quite clearly not all of Luther's features are significant to the history of religion. The historian of religion, then, would try to discern the features that were responsible for Luther's importance in the history of religion, and to this extent ignore some features of Luther the individual. Still, this sort of abstraction is completely different from the generalizing tendency of the natural scientific psychologist. The historian is still interested in that particular person for his own sake, we might say, and would never slip into considering Luther merely as a representative of a general type.

It is only after distinguishing the natural and historical sciences in this way that Rickert turns to the connection between the historical sciences and minds along with the distinction between explanation and understanding (Rickert [1902] 1986, 157–174). A central point of his discussion is that Dilthey misunderstood the basis of this distinction because he did not appreciate the logical presuppositions of the genuinely historical or cultural sciences. The root of the unity presupposed in the historical sciences is not any basic unity of our lived experience, but only the relations to values that confer the requisite unity on historical individuals. Among other things, Rickert concludes that psychology should be classified as a natural science whose special task is to uncover the laws which regulate our experiences.

Still, there is a connection between how individuals are related to values and our minds. Rickert argues that these value relations invariably involve minds grasping nonreal values, or as he often puts it, nonreal meanings. It is not that history exclusively concerns itself with minds, as non-mental things like buildings or oceans can stand in relations to values. But these relations all turn, in the end, on acts of valuation by minds and it is these acts which the historian must presuppose when determining what is of historical significance. Historical understanding is just the attempt to grasp these nonreal meanings, and should be opposed to the sort of explanation sought by the natural sciences:

[8] Rickert opposes ordinary individuals to individuals which result when ordinary individuals are related to values. I will not follow this terminology in my exposition.

> We have pointed out the distinctive "unity" of the nexus or the "totality" characteristic of nonreal meaning in contrast to merely real existence. The distinctiveness of the understanding [Verstehen] of an object, in comparison with the explanation [Erklären] of that object, is also determined on this basis. This is the only way the expression "understanding" acquires a thoroughly precise meaning. Meaningful wholes can be understood only *as* unities or totalities.... The object of understanding as a nonreal meaning configuration always remains a whole or a unity. Realities, on the other had, are decomposed into their parts for the purpose of explanation. Or, in explanation, the path leads from the parts to the whole; in understanding, it proceeds in the opposite direction, from the whole to the parts (Rickert [1902] 1986, 159).

So, an explanation of an event like Luther's nailing of his 95 theses to the door of the Wittenberg Church might focus on the physics of the hammer impacting the nail, and how the resulting physical system was stable, i.e., why the paper stayed on the door. This explanation would begin with a series of events of various physical kinds and culminate in a feature of the whole. This is of course probably not what someone asking a question about this event is likely to be interested in. Instead, the historical significance of the event can only be understood when we grasp its relation to religious value, and this relation depends on treating that event as a unified whole. In virtue of this particular unified whole standing in its relation to religious value we can begin to understand how it came about and what its historical effects were.

Just like Dilthey, then, Rickert makes a connection between wholes or unities and the distinction between explanation and understanding. Both agree that understanding must begin with a unity of some sort that is then decomposed, while explanation starts with distinct parts that are connected together in the process of explanation. On the explanation side, Dilthey and Rickert agree furthermore that this alignment of the parts into a whole in natural science proceeds via laws. But when it comes to understanding, Rickert breaks decisively with Dilthey. For the unity involved in understanding for Rickert is not given in lived experience, but continually made and remade by the acts of valuation of minds. This forces Rickert to give some account of what his nonreal values are and how they can enter into the real world of human minds. Here is a distinctively philosophical project which Rickert sees as the subject matter of epistemology or the theory of knowledge, rather than Dilthey's scientific analytic psychology.

IV. I turn now to two other responses to the problem of the unity of experience. Bradley and one of his critics, James Ward, sought to move beyond Mill's conception of the relation of experience to scientific knowledge, but neither invoked anything like the distinction between Erklären and Verstehen that were central to the proposals of Dilthey and Rickert. As we will see, though, there are important similarities between the views of Bradley and Rickert, on the one hand, and Dilthey and Ward, on the other. The former pair agree that the unity required for scientific knowledge is not given, but made, while the latter pair insist that the unity cannot be made, but must instead be given.

Bradley agrees with Mill, Dilthey and Rickert that all thought begins with experience and that all the materials assembled in judgments must be extracted from experience. But he supplements this picture with two additional commitments which not only undermine Millian empiricism but anything approaching what Dilthey or Rickert would accept. The first commitment seems somewhat benign: Bradley claims that our experience is initially non-cognitive in that it involves only a feeling for a unified mass of experience. For Bradley, the description of experience of Mill and Rickert that both accept of a plurality of entities, be they sensations or individuals, applies only to a second stage of experience. Bradley then raises a second point. What, in the end, licenses us to take the distinctions that *we* introduce into our undifferentiated mass of experience as genuine? That is, if we separate out a general feature, or what Bradley would call an idea, then we need some additional reason to think that this feature could exist independently of the mass of experience with which we began. Here we can see Bradley departing from Dilthey's assumption that an analysis of the nexus of lived experience will reveal the genuine structures of our mind. On the contrary, Bradley argues that all such abstraction or analysis of experience involves distortion and mutilation. Using the term "analytic judgment of sense" to refer to judgments that are composed of ideas that are directly abstracted from an experience, Bradley notes

> But *a–b* by itself has never been given, and is not what appears. It was *in* the fact and we have taken it out. It was *of* the fact and we have given it independence. We have separated, divided, abridged, dissected, we have mutilated the given.[9] And we have done this arbitrarily: we have selected what we chose. But, if this is so, and if every analytic judgment of sense must inevitably so alter the fact, how can it any longer lay claim to truth? (Bradley [1883] 1922, 94).

It is hard to know why Bradley feels he is entitled to adopt the opposite assumption to the empiricist here when he goes from simply questioning the Humean claim that what is distinguishable is separable to asserting that what is distinguishable is never separable. On his behalf we could offer the following argument: an honest examination of our experience reveals that the empiricist assumption is never vindicated, but that instead any abstractions that we introduce into our cognition change the character of what is abstracted. For example, if I start with the perception of a red apple and then isolate an idea of red, the latter is a different thing than the red apple that I began with. Even the redness has been changed to the extent that before it was a feature of a fairly coherent and stable entity, but now it is a free-floating idea that clearly could not exist on its own as a genuine entity.[10]

As this argument suggests, the individuality and unity of a thing plays a central role in Bradley's philosophy. The abstractness and generality of ideas and judgments are sufficient signs that they cannot capture the genuine features of reality. Instead, each judgment *p* must take on the conditional form that if certain unspecified conditions obtain, then *p*. Thought, then, is the attempt to uncover these conditions and incorporate them into the judgment, and thereby to make the

---

[9] Here there is a footnote to Lotze (1888, ii, viii).

[10] Compare "The whole that is given us is a continuous mass of perception and feeling; and to say of this whole, that any one element would be what it is there, when apart from the rest, is a very grave assertion. We might have supposed it not quite self-evident, and that it was possible to deny it without open absurdity" (Bradley [1883] 1922, 95f.).

judgment truly unconditional and categorical. This quest can never be successful because at each stage we will still have a judgment made up of abstract ideas, and we are never able to apply these abstract ideas to reality unconditionally due to the distortions that abstraction introduces.

The preoccupation with the unity and individuality of the real further informs Bradley's metaphysics of the Absolute. His famous regress argument against relational judgments deploys this conception of the real to devastating effect. Suppose, Bradley says, that we judge that two things *a* and *b* stand in some relation *R*: *aRb*. The elements of this judgment are general, abstract ideas. So, as ideas, there must be something more concrete in virtue of which the alleged relational fact obtains. If we think of *R* as an external relation, or as a relation that relates *a* to *b* irrespective of the internal features of *a* and *b*, then we must assume that the more concrete basis of the relational fact includes a new relation *R'* which relates *a* to *R*. This generates a regress. At the same time, if we think of *R* as an internal relation, then the required more concrete basis of the relational fact will involve not just *a* and *b*, but also the internal features of *a* and *b* that are responsible for the tie to *R*. Our alleged relational fact *aRb* then becomes $a(a')R(b')b$. But now a different sort of regress arises in which $a$, $a'$, $b'$ and $b$ are dissolved into further relational facts, e.g., the relation between $a$ and $a'$.

The only resolution of these problems is to subsume any alleged relational fact into a whole that includes all the original constituents, but also whatever else is responsible for the relational appearance. If we apply this way out quite generally, we end up with Bradley's Absolute: a single thing that includes all appearances but only as aspects which cannot exist separately. What drives the argument is clearly a conception of reality as a concrete individual along with the assumption that thought must restrict itself to the abstract and general:

> The end, which would satisfy mere truth-seeking, would do so just because it had the features possessed by reality. It would have to be an immediate, self-dependent, all-inclusive individual. But, in reaching this perfection, and in the act of reaching it, thought would lose its own character. Thought does desire such individuality, that is precisely what it aims at. But individuality, on the other hand, cannot be gained while we are confined to relations (Bradley [1893] 1969, 158).

Bradley is optimistic that we can discern some features of the Absolute, in addition to its individuality. Above all, it must be experiential as this is the only way that Bradley sees how it can genuinely include all aspects of the appearances that we are aware of.[11]

Put in these stark terms, Bradley's contrast between appearance and reality seems to force one to reject science as wholly illusory. This is far from Bradley's intention, though (Mander 1991). What he wants to do is clearly to defend the autonomy of both science and metaphysics. The scientist has her tasks and she should carry them out without either aspiring to displace the metaphysician or being the victim of metaphysical interference. Mill, then, was on the right track

[11] Bradley took this to be a crucial difference with Hegel. See Wollheim (1959, 186f.).

when he insisted that metaphysical constructions are unnecessary for science and that they can do harm if they distract the scientist from her phenomena. Where Bradley would fault Mill is when Mill saddles the scientist with a distorted and impoverished conception of the experience that science must begin with. Bradley seeks to purge metaphysics from science, including the sensationalist metaphysics tacitly offered by Mill.

It remains unclear how the different sciences relate to one another and the Absolute. Bradley is attuned to the differences between a science like history and a science like physics, but at the same time he would also reject any bifurcation of the sciences of the kind we saw in Dilthey and Rickert. In *Appearance and Reality* Bradley argues that we can rank the different special sciences in terms of the degree of truth of their judgments. No completed ranking or metaphysical "system" is presented by Bradley in this book or anywhere else, but he says enough to explain how the ranking would proceed and roughly where he would place the natural sciences and social sciences. Under the heading of a philosophy of nature, Bradley describes how such a ranking would proceed:

> All appearances for metaphysics have degrees of reality. We have an idea of perfection or individuality; and, as we find that any form of existence more completely realizes this idea, we assign to it its position in the scale of being. And in this scale (as we have seen) the lower, as its defects are made good, passes beyond itself into the higher. The end, or absolute individuality, is also the principle. Present from the first it supplies the test of its inferior stages, and as these are included in fuller wholes, the principle grows in reality (Bradley [1893] 1969, 440).

Exactly why the sciences can proceed independently of this ranking is made clearer on the next page:

> How the various stages of progress come to happen in time, in what order or orders they follow, and in each case from what causes, these inquiries would, as such, be of no concern of philosophy. Its idea of evolution and progress in a word should not be temporal. And hence a conflict with the sciences upon any question of development or of order could not properly arise ... Progress for philosophy would never have any temporal sense, and it could matter nothing if the word elsewhere seemed to bear little or no other (Bradley [1893] 1969, 441).

With this sharp separation of progress in metaphysics and change in time, Bradley opens up the possibility for a science of history that would stick just as closely to its phenomena as we will see he urged in the case of psychology. History would become scientific, then, only by freeing itself from metaphysical pretensions.[12]

So where does history sit in Bradley's ranking of the sciences and what distinguishes it from these other sciences? There is good reason to think that Bradley would place history above physics and psychology, which occupy the first and second stages, respectively, but below the supra-scientific domain of religion. Physics occupies the lowest level because it deals with the world in the most abstract way. It attempts to describe phenomena merely in terms of their primary

---

[12] Other discussions of the connection between science and metaphysics are Bradley ([1893] 1969, 251) quoted in Wollheim (1959) and Bradley ([1893] 1969, 109).

qualities, which for Bradley involves considering phenomena as disjointed and unconnected abstractions. The test of reality is always the extent to which the subject matter of the discipline approximates the unity and harmonious character of the Absolute. So, despite rejecting the metaphysical status of the soul, the soul conceived of as a relatively stable center of experience does better than the abstractions of physics: "compared with the physical world, the soul is, by far, less unreal. It shows to a larger extent that self-dependence in which Reality consists" (Bradley [1893] 1969, 316). Continuing upwards, Bradley tells us that a community of persons of an appropriate kind is in turn more real than an isolated soul:

> a social system, conscious in its personal members of a will carried out, submits itself naturally to our test [of Reality]. We must notice here the higher development of a concrete internal unity. For we find an individuality, subordinating to itself outward fact, though not, as such, properly visible within it. This superiority to mere appearance in the temporal series is carried to a higher degree as we advance into the worlds of religion, speculation, and art (Bradley [1893] 1969, 333).

I take this social system to be the primary phenomenon of history: how did this system come about, in virtue of what is it sustained and how does the system develop over time are the main questions for the historian. Higher degrees of truth are no doubt possible if we transcend the domain of science and enter the superior realm of truth where Bradley situates religion and art. Even here ultimate truth is not to be found as an entity like God still remains an appearance that must be subsumed in the Absolute.[13]

The last philosopher that we will consider is James Ward, one of the fiercest critics of Bradley's psychology and metaphysics. Despite his opposition to Bradley, Ward was clearly no friend of Millian empiricism. But in his criticism of Mill Ward emphasized very different alleged aspects of experience than Bradley did, and these differences in turn informed Ward's claims about the shortcomings of Bradley's philosophy. Ward claimed that the core of our experience must involve an active self which attends to the objects that it is presented with.[14] By contrast, Bradley takes the self to be a provisional construction from immediate experience. Despite the centrality of this construction to the science of psychology, and the higher degree of truth of psychology in relation to physics, Bradley strenuously rejects the view that experience presupposes an active, non-experiential self. The self considered in experience is merely a selection and reorganization of what we get in our original felt experience. This becomes clear, for example, in his article "A Defence of Phenomenalism in Psychology," where phenomenalism is defined as "the confinement of one's attention to events with their laws of coexistence and sequence" (Bradley [1900] 1935, 364). In particular, the metaphysical notion of the soul has no place in psychology. For Bradley, psychology considers the experience of one individual after it has been divided into what is experienced and what is doing the

---

[13] Such a system of the sciences played a crucial role in Russell's early idealist Tiergarten program. For extensive discussion, see Griffin (1991).

[14] See Wollheim (1959, 132–138), who cites Brett ([1921] 1965, 675–686).

experiencing. This opposition between the psychological subject and its objects is just as provisional as any distinction that is made within our undifferentiated experience, and so the psychological subject and its features have no connection to the metaphysical soul that would preoccupy some metaphysicians. When we take up a properly metaphysical attitude, Bradley thinks it is easy to argue that the world is not composed of distinct souls by applying his general argument against relations to the relations that would be required between distinct souls (Bradley [1883] 1922, 97; Bradley [1893] 1969, Chapter X).

Ward's argument against this derivative conception of the active self is brief and to the point: "Experience does not divide itself, but is so divided because of the interest of the subject in certain presentations and in certain relations of presentations" (Ward 1887, 569). That is, if we accept the original state of undifferentiated experience posited by Bradley, then we are hard pressed to say in virtue of what any of the complex structures that we actually encounter in the later stages of experience arise. Bradley's whole discussion of abstraction and the distortions that it introduces is very hard to square with the view that the self is nothing but one of these abstractions. What is responsible for the different experiences we have if not the activity of the self? Ward thought that only the primitive activity of the self could help here, and that the function of the self in experience was one of the things that psychology should concern itself with:

> In keeping with this rejection of the subject that acts and feels is Mr. Bradley's further doctrine that activity, as well as feeling, is a mere presentation. Wundt's theory of apperception he holds beneath contempt, and the present use of the term "activity" is, he insists, "little better than a scandal and a main obstacle in the path of English psychology".[15] I should have thought the chief obstacle in the way of English psychology had been its neglect of psychical activity, and the chief merit of psychology in Germany had been essentially that very doctrine of apperception which, as developed by Wundt, Mr. Bradley is loath to criticize. He demands a definition of activity, and offers one of his own. For my part, I doubt if activity can be defined in terms that do not already imply it (Ward 1887, 569).[16]

Activity and other basic concepts of psychology must be taken as undefined or given, rather than built up out of supposedly passive experiences.

For Ward, then, the activity of the self is a basic presupposition of our experience, and without it the unity of experience would disintegrate, as would the ability of experience to underwrite scientific knowledge. Bradley's response to this objection reveals the extent to which he has not really escaped Mill's phenomenalism. In *Appearance and Reality* Bradley mocks an appeal to activity as basic:

> In raising the question if activity is real or is only appearance, I may be met by the assertion that it is original, ultimate, and simple. I am satisfied myself that this assertion is incorrect, and is even quite groundless; but I prefer to treat it here as merely irrelevant. If the meaning of activity will not bear examination, and if it fails to exhibit itself intelligibly, then the meaning cannot, as such, be true of reality (Bradley [1893] 1969, 53).

---

[15]The reference here is to Bradley ([1886] 1935, 196).

[16]Or, more colorfully, Bradley's "procedure is much like bleeding yourself to death to guard against blood-poisoning" (Ward 1887, 575).

As Wollheim has noted, with these sorts of arguments Bradley appears to be assuming that all that we can mean by "activity" must be spelled out in terms of experience (Wollheim 1959, 138). That is, all that I can talk about must have a basis in experience, or what Mill called his sensations. In this case, both Ward and Bradley agree that experience as it is usually understood does not include any constituent clearly answering to the activity of the self. For Bradley this implies that it is a meaningless concept, and to say that such a concept is simple is completely unhelpful. For Ward, as we have seen, psychology can and must appeal to basic concepts like activity that are responsible for our experience, even if they are not properly constituents of it.

The similarities between Ward and Dilthey on the unity of experience are now hopefully clear. Both find a unity to our experience that Mill denied. For Dilthey this unity is the entire nexus of what he called lived experience, while Ward appealed to the more traditional position that there must be an active self at the basis of experience. But when it came to scientific psychology, Dilthey and Ward seem to be in substantial agreement. Both would argue for a psychology that was at least initially occupied with analyzing the basic features of our experience. This sort of analysis was necessary because any constructive or synthetic approach would fail to do justice to the basic features of our experience. In fact, G. F. Stout, one of Ward's students, called his first book *Analytic Psychology*, and largely followed Ward's analytic methodology.[17] Unlike Dilthey, however, Ward went on to attempt to integrate this scientific psychology into an overarching metaphysics, and here again we find an apparent departure from Bradley. It cannot be said that Ward's attempts at a systematic metaphysics were that successful. One recent commentator has noted that it is not clear if Ward meant to defend a form of Absolute Idealism, or whether instead he preferred what is called Personal Idealism (Allard 2003, 53).[18] According to the latter, distinct selves, perhaps even including a personal God, standing together in some kind of harmonious relationship, provide the ultimate basis for reality. The debate between Absolute and Personal Idealism continued to divide the British neo-Hegelians and one of the central issues remained how our experience could be reconciled with a metaphysics emphasizing the individuality of reality.

V. We have reviewed, at least in part, five accounts of the connection between our experience and scientific knowledge. Mill's phenomenalism, even supplemented by his "mental chemistry", sets up a gap between the experience which is supposed to be the basis of our scientific knowledge and the objects that the sciences seem to be invoking in their theories. The gap can be bridged in a number of ways by either putting more into experience or by altering our conception of the objects of science, or by some combination of these two strategies. Dilthey, to begin with, invoked an over-

[17] See Stout (1896, vol. 1, xi) for the claim that "Whatever there may be of value in my work is ultimately due to his [Ward's] teaching." Neither Ward nor Stout show any awareness of Dilthey's conception of psychology.

[18] See Ward (1925) for an unhelpful summary of Ward's mature metaphysics.

arching unity in our lived experience which forces psychology to take on an analytic or descriptive form. With this analytic psychology at its basis, Dilthey's human sciences have as their goal a kind of understanding that is opposed to hypothetical natural scientific explanation. Rickert responds by isolating a different kind of unity in our experience, namely the unity of experienced individuals that results from their relation to a nonreal value. Such historical individuals can be studied as individuals, and again a sort of understanding is possible here that is absent in natural science. When we turn to Bradley, a central theme remains the unity and individuality required for genuine knowledge. But Bradley is willing to give up the assumption so central to both Dilthey and Rickert that genuine knowledge is possible in the sciences. Instead, when compared to the perfect harmony and unity of the only individual there really is, the Absolute, the special sciences get by through a provisional construction of relatively unified and individual things. These constructions and the associated special sciences can be ranked in terms of their degree of truth, with physics at the bottom, followed by psychology in the middle and history towards the top. Ward responds by claiming that Bradley's conception of an experience that lacks an active self is incoherent and incapable of grounding any acceptable form of scientific knowledge, whether in psychology or elsewhere. Ward's own influential version of psychology puts the description of such basic elements of our experience in a central position.

How are we to react to these various positions, thought of as proposed solutions to our problem of the unity of experience, and the associated attitudes towards the Erklären/Verstehen distinction? The first, almost trivial, thing to say is that there is clearly no need to invoke the distinction if we are to offer a solution the problem of the unity of experience. Bradley and Ward made proposals, but these accounts do not lead inexorably to two different cognitive attitudes or any special division between the natural and human sciences. That said, it remains possible that all plausible or reasonable solutions to our problem might require something like the Erklären/Verstehen distinction. To start to support this claim we would need to show that the moves made by Bradley and Ward were unpromising or unreasonable when compared to what we found in Dilthey and Rickert.

An obstacle to such a project is the fact that making the distinction introduces just as many philosophical complications as it promises to resolve. To focus first on Dilthey, it may be helpful to invoke the unity of the nexus of lived experience in order to help ground the human sciences, beginning with Dilthey's analytic psychology, and extending to history, legal theory and aesthetics. But the nexus of lived experience only complicates the story of how our experience relates to the objects studied in the natural sciences like physics. Here Dilthey offers nothing informative and it seems like he has backed himself into a corner by emphasizing the alleged gap between our lived experience and the objects of natural science. What results is an implausible picture of what sort of knowledge we can obtain in the natural sciences.[19]

[19]Indeed, one might suspect that this shortcoming infects the phenomenological tradition more generally. For a more optimistic assessment, see Ryckman (2005, esp. Chapters 5 & 6).

Rickert, on the other hand, faces significant obstacles to vindicating his account of nonreal values and how the acts of valuation of minds are even possible. Some kind of transcendental argument in their favor may be possible, but the prospects do not seem good (Rickert [1902] 1986, 221–225). Rickert, then, solves the problem of unity of experience only by trading it for an even less tractable problem. In both cases, we see that it is not easy to argue for the conclusion that the Erklären/Verstehen distinction is part of the best route to take.

I conclude by making an observation about the danger to the historian of philosophy in taking distinctions like Erklären/Verstehen too seriously when looking back at the history of philosophy and relating this observation back to the contrast between Soames and Simons that we saw in section I. While there is no point in denying the historical efficacy of such distinctions in philosophical debates, granting Dilthey's point that the distinction is necessary or inevitable can lead the historian astray. If we think that something like Dilthey's distinction is inevitable, and we find a philosopher pointing to some differences between, say, history and the natural sciences, then we may ascribe to that philosopher the full apparatus of Dilthey's philosophy of the human sciences. This is what appears to have happened in Rubinoff's introduction to his edition of Bradley's 1874 essay "The Presuppositions of Critical History" (Bradley [1874] 1968). Bradley's focus is the evaluation of historical sources, especially the treatment of biblical texts by theologians like Baur and Strauss. They had argued that biblical texts should be treated as historical documents, which understandably resulted in somewhat skeptical conclusions about the veracity of the testimony they contained. Bradley essentially agrees, arguing that the historian is constrained by his own experience of the world to evaluate reports of extraordinary events in a critical fashion. Typically, this limits me to accepting reports of events that are analogous to what I myself have experienced. But Bradley also allows me to accept reports of new kinds of events by others "by means of an identification of consciousness" (Bradley [1874] 1968, 104). Rubinoff claims that "Bradley's answer is reminiscent of the one given by Droysen and Dilthey. It is that human experience is capable of being expanded by means of an 'imaginative' appropriation of new experience" (Bradley [1874] 1968, 42) and goes on to ascribe much of the apparatus we saw in Dilthey to Bradley. Against this interpretation, I note first that, as Rubinoff admits in a footnote (Bradley [1874] 1968, 42, fn. 74), Bradley does not talk of "imaginative appropriation". More importantly, when we turn to Bradley's discussion of identification of consciousness, we find nothing like Dilthey's account of the understanding of others or the nexus of lived experience that is its basis (Bradley [1874] 1968, 104–108).[20] Rubinoff has been led astray by thinking that an Erklären/Verstehen distinction is inevitable. So, it is no surprise to find him write, in another context, that "No serious attempt to define an autonomous science of human nature can succeed unless it proceeds according to a basic ontological distinction between historical and natural existence" (Rubinoff 1964, 83).

[20] Bradley does cite Droysen's *Grundriss der Historik*, but also adds that it "came too late to be of much use" (Bradley [1874]1968, 78).

Genuine understanding of the history of philosophy is not likely unless we are attuned to the dangers of these sorts of mistakes.

Should we follow Soames and Simons and conclude that because the distinction is not inevitable, and arose as a result of a variety of historical contingencies, that it is devoid of contemporary philosophical interest? Of course not. For while I have argued that a distinction between Erklären and Verstehen is not obviously part of the best solution to the problem of the unity of experience, it remains entirely possible that it can help us to solve that problem if suitably clarified and motivated. Those of us who wish to understand how our experience relates to scientific knowledge would do well to keep all options on the table and to continue to study the wide range of attempts offered by philosophers, both in the past and the present.

## References


Allard J (2003) Idealism in Britain and the United States. In: Baldwin T (ed.) The Cambridge History of Philosophy 1870–1945. Cambridge University Press, Cambridge, pp 43–59

Anderson RL (2003) The debate over the Geisteswissenschaften in German philosophy. In: Baldwin T (ed.) The Cambridge History of Philosophy 1870–1945. Cambridge University Press, Cambridge, pp 221–235

Bradley FH ([1874] 1968) The Presuppositions of Critical History. Reprint edited by L. Rubinoff. First edition by Oxford: J. Parker. Quadrangle Books, Chicago

Bradley FH ([1883] 1922) The Principles of Logic, 2nd edn. First edition by London: K. Paul & Co. Oxford University Press, London

Bradley FH ([1886] 1935) Is there any special activity of attention? In: ibid., Collected Essays, vol 1. Originally published in Mind 11:305–323. Clarendon, Oxford, pp 181–202

Bradley FH ([1893] 1969) Appearance and Reality: A Metaphysical Essay, 2nd edn. Originally published by Swan Sonnenschein, London. Oxford University Press, London

Bradley FH ([1900] 1935) A defence of phenomenalism in psychology. In: Collected Essays, vol 2. Originally published in Mind n.s. 9:26–45. Clarendon, Oxford, 364–386

Brett GS ([1921] 1965) History of Psychology. Abridged by R.S. Peters. Originally published by G. Allen & Co., London. MIT Press, Cambridge, MA

Dilthey W ([1886] 1989) Introduction to the human sciences. Makkreel RA, Rodi F (eds) Selected Works, vol 1. Princeton University Press, Princeton, NJ

Dilthey W ([1894] 1977) Ideas concerning a descriptive and analytic psychology. In: Descriptive Psychology and Historical Understanding. Nijhoff, Den Haag, 23–120

Dilthey W (1959) Gesammelte schriften. Teubner, Stuttgart

Feest U (2007) Hypotheses, everywhere only hypotheses: On some contexts of Dilthey's critique of explanatory psychology. Stud Hist Philos Biol Biomed Sci 38:43–62

Griffin N (1991) Russell's Idealist Apprenticeship. Oxford University Press, New York

Leiter B (2004) Introduction. In: The Future for Philosophy. Oxford, Oxford University Press, pp 1–24

Lotze H (1888) Logic. 2 vols. Translated by Bernard Bosanquet. Originally published as Logik, 2nd edn., S. Hirzel, Leipzig

Makkreel R (1969) Wilhelm Dilthey and the neo-Kantians: The distinction of Geisteswisssenschaften and Kulturwissenschaften. J Hist Philos 4:423–440

Mander WJ (1991) F. H. Bradley and the philosophy of science. Int Stud Philos Sci 5:65–78

Mill JS ([1843] 1973) A System of Logic. Collected Works (1963), vols 7 and 8. Robson JF (ed) Originally published by J.W. Parker, London. University of Toronto Press, Toronto

Rickert H ([1902] 1986) The limits of concept formation in natural science: a logical introduction to the human sciences, 5th edn., edited, abridged and translated by G. Oakes. Originally published as Die Grenzen der naturwissenschaftlichen Beriffsbildung, 5th edn. originally published 1929. Mohr, Tübingen
Rubinoff L (1964) Critical notice: historicism and the a priori of history. Dialogue 3:81–88
Ryckman T (2005) The Reign of Relativity: Philosophy in Physics 1915–1925. Oxford University Press, New York
Simons P (2001) Whose fault? The origins and evitability of the analytic-continental rift. Int J Philos Stud 9:295–311
Soames S (2003) Philosophical Analysis in the Twentieth Century, vol 2: The Age of Meaning. Princeton University Press, Princeton, NJ
Soames S (2006) What is history for? Reply to critics of the dawn of analysis. Philos Stud 129:645–665
Stout GF (1896) Analytic Psychology. 2 vols. Macmillan, New York
Ward J (1887) Mr. F. H. Bradley's analysis of mind. Mind 12:564–575
Ward J (1925) A theistic monadism. In: Muirhead JH (ed) Contemporary British Philosophy. Second series. Allen & Unwin, London, pp 27–54
Wollheim R (1959) F. H. Bradley. Penguin, Harmondsworth, UK

# Chapter 11
# Individuality and Interpretation in Nineteenth-Century German Historicism

**Jacques Bos**

The methodological controversy in the humanities and the social sciences between the advocates of an explanatory approach similar to that of the natural sciences (*erklären*) and the proponents of an interpretative perspective (*verstehen*) has its roots in a wide-ranging cultural transformation that took place in Europe around 1800. Traditionally, this transformation has been described as the shift from Enlightenment to Romanticism, involving, among other things, the rise of a new, expressivist conception of art and the substitution of a universal notion of rationality by an emphasis on the incommensurability of individual ages and cultures. A different account of the cultural transformation of the early nineteenth century is given by Foucault (1966, 314–354). According to Foucault, a fundamental epistemological rupture took place in the period around 1800, which he describes as the shift from the classical to the modern *épistémè*. A crucial aspect of the rise of the modern *épistémè* is the discovery of man as a transcendental subject that can also be the object of empirical knowledge. Furthermore, in contrast with the emphasis on stable taxonomies of the classical age, the modern *épistémè* perceives the order of things as essentially historical.

Although the traditional and the Foucauldian view address different aspects of the cultural and intellectual reorientation of the early nineteenth century, they should not be seen as radically contradicting each other. In both perspectives historicity and individuality are essential elements of the new discourse that emerged around 1800. The fundamental historicity of Foucault's modern *épistémè* is mirrored in the Romantic notion that the past can only be understood in its own terms, and the modern subject, of which Foucault describes the birth, is essentially a unique individual, endowed by Romanticism with the right to express its individuality in artistic achievements or otherwise. The central place of historicity and individuality in modern thought explains why interpretation has become such an extensively debated philosophical topic. When persons are regarded as distinct individuals and historical periods as distinct epochs, they cannot be straightforwardly understood. In other words, the notions of individuality and historicity emerging after 1800 necessarily involve the problem of interpretation.

---

J. Bos (✉)
Department of Philosophy, University of Amsterdam
e-mail: J.Bos@uva.nl

U. Feest (ed.), *Historical Perspectives on Erklären and Verstehen*, Archimedes, Vol. 21,
DOI 10.1007/978-90-481-3540-0_11, 

A field in which the cultural and intellectual transformation of the nineteenth century is eminently visible is that of historiography. Especially in Germany, historians developed a new conception of their craft, which they explicitly contrasted with the approach to the past of their Enlightenment predecessors. This perspective on historiography is generally known as historicism (*Historismus*). Historicism is not only characterised by a new conception of historicity, but also by a new, and rather complicated notion of individuality. This notion of individuality will be the central theme of this paper. It will be examined by looking at three authors: Wilhelm von Humboldt, who was one of the first to give a philosophical account of the new conception of history, Leopold von Ranke, the "father of modern historiography", and Johann Gustav Droysen, the main theorist of nineteenth-century historicism. Before turning to these authors, however, the meaning of the term "historicism" should be examined more closely, since it is not an unambiguous concept.

## 11.1 Historicism: A Disputed Concept

The word *Historismus* was first used in German at the end of the eighteenth century by Romantic authors such as Friedrich Schlegel and Novalis to indicate a view of the past that understood individualities in their specific context without obscuring them by a contemporary theoretical outlook. Yet, the word was far from ubiquitous in the first half of the nineteenth century: typically, the historians who based their work on historicist principles did not use it (Iggers 1995, 130f., 142). *Historismus* did not become a generally accepted term in German intellectual debates before the early twentieth century, when historicism had reached the end of its vital phase and began to be denounced as something in crisis, as something that had to be overcome (Schnädelbach 1983, 51). This "crisis of historicism" – a phrase coined by Troeltsch (1922a,b) – emerged in several disputes on the nature of history that started around 1880. The central problem in these disputes was that of relativism. The belief in the historicity of man and culture embodied in nineteenth-century historiography implied a relativity of values that authors at the end of the century began to experience as a serious threat to civilization. In the context of this debate Nietzsche severely criticised the lack of meaning and vitality of empirical academic historiography, while authors such as Dilthey, Windelband and Rickert tried to evade the problem of relativism by philosophically analysing the nature of the humanities and the methods of historical interpretation (Bambach 1995; Wittkau 1992).

When the term *Historismus* became broadly used in the disputes on the "crisis of historicism", it was primarily associated with the problem of relativism. The historians of the first half of the nineteenth century, however, did not employ the word *Historismus*, nor were they more than marginally concerned with the possible relativist implications of their work. Because of this contrast Otto Gerhard Oexle distinguishes two kinds of historicism, "historicism I" and "historicism II". The first term refers to the philosophical debates in the late nineteenth and early twentieth century, while the second denotes German academic historiography as it emerged in the nineteenth century (Oexle 1984).

One of the first authors who concentrated on the field of "historicism II" was Friedrich Meinecke. Partly in reaction to the debates on the "crisis of historicism" Meinecke turned to the assumptions lying beneath early nineteenth-century historiography. In *Die Entstehung des Historismus* he describes the rise of historicism as one of the greatest intellectual revolutions in European thought. In his opinion, the core of this revolution is the replacement of a generalising perspective on the past by an individualising and developmental approach. The individualities examined by historicism are people, but also *Kollektivgebilde* such as nations or shared ideas. The nature of these individualities reveals itself in their development (Meinecke [1936] 1959, 1–5). Later theorists of history have severely criticised Meinecke's positive evaluation of historicism, but usually without fundamentally challenging his conceptualisation of its intellectual core (Oexle 1996). Iggers (1997), for instance, emphasises the political dimension of historicism that Meinecke tended to neglect, but agrees with Meinecke that the notion of individuality plays a crucial role in the historicist perspective on the past. A central point in Iggers's argumentation is that historicist authors characteristically expanded the applicability of the idea of individuality from persons to nations and states.

While Iggers corrects Meinecke's perspective by focusing on the political dimension of historicism, other authors highlight other aspects. According to Fulda (1996), the rise of historicism should be interpreted as a primarily aesthetic reorientation. Another important strand in recent research on historicism regards it as a disciplinary matrix or a paradigm, which succeeded Enlightenment historiography, was challenged by positivism and Marxism, and eventually lost its dominant position to historiographical approaches inspired by the social sciences in the second half of the twentieth century (Blanke 1991; Jaeger and Rüsen, 1992; Rüsen 1993). Characteristic of these studies is that they analyse historiography as a "scientific" profession (*Wissenschaft* in German), in which one of the key features of modernity, the individualising and developmental perspective on the past, is conceptualised in connection with specific methodologies, institutions and political interests. The scope of this paper is more limited than the approach of these broad-ranging studies of nineteenth-century historicism. According to Kuhn ([1962] 1970, 40f.), an important element of a paradigm is a set of metaphysical commitments. This paper deals with this aspect of the historicist perspective on the past: I will examine the role of the notion of individuality as a fundamental ontological assumption of "historicism II" and its relation to the problem of interpretation. In this analysis the historicist idea of individuality will be presented as a dual notion, involving on the one hand individual persons and individual agency and on the other hand higher-order individualities such as nations.

## 11.2 Humboldt: Individuality and Idealist Organicism

An important factor in the context in which historicist thought emerged is German idealist philosophy, especially Hegel's philosophy of history. Hegel regarded the course of world history as an inherently teleological process. The determining factor in this process is the *Geist*, the general principle that governs the flow of history and will eventually reach its fulfilment. Opposed to the generality of the *Geist* is the individuality of human beings and their goals and desires. Yet, although we experience our strivings as real and believe that we can have a certain impact on the world surrounding us, Hegel does not think that individuals influence the course of history. In the case of some heroic *weltgeschichtliche Individuen* there does seem to be such an individual impact. Hegel argues, however, that even these people, although they think that they realise their own practical and political goals, are nothing more than instruments of the *Weltgeist* of which they have no consciousness. This is the "cunning of reason" (Hegel [1837] 1986, 29–54).

In Hegel's philosophy of history individual agency is illusory; the *weltgeschichtliche Individuen* who seem to shape the course of history only contribute to the unfolding of its inherent teleology. This means that history can only be understood by analysing this teleology. It is this element of Hegel's philosophy that historicism opposes most strongly. This is very clearly visible in the work of one of the early theorists of historicism, Wilhelm von Humboldt – not primarily a historian, but a philosopher, linguist and politician. In Humboldt's work a notion of individuality comes into view that involves a fundamental critique of the Hegelian view of the historical process. This notion of individuality is not only a central aspect of Humboldt's theory of history, but also plays an important role in his political philosophy (Barash 2004, 101–115; Iggers 1997, 62–85). In his political writings of the early 1790s Humboldt expounds a liberal view of the state. In his opinion, the powers of the state should be limited in order not to obstruct the individual development of its subjects. It is desirable that each individual enjoys an unrestricted freedom to develop his or her *Eigenthümlichkeit*, for the true goal of man is "die höchste und proportionirlichste Bildung seiner Kräfte zu einem Ganzen" (Humboldt [1792] 1960, 64, 69). In Humboldt's later political thought, however, the emphasis has shifted from individual persons to nations and states, to which he attributes a kind of individuality that closely resembles the fundamental uniqueness ascribed to human beings in his earlier work (Humboldt [1813] 1964).

The dual notion of individuality that emerges in Humboldt's political thought is also a central aspect of his philosophy of history. In a short text written in 1814, "Betrachtungen über die Weltgeschichte", Humboldt argues that history cannot be understood by relating events to the intrinsic purpose of the historical process. What matters in history is not an abstract perfection that will eventually be attained, but the development of a richness of individual forms. History is an interaction of forces in which the historian should focus on individual persons and individual nations. This does not mean that there is no coherence in the historical process. Humboldt argues that this coherence is organic, and not rational and teleological, as it is conceived by Hegel. The human race can be compared to a plant, which means that an individual person relates to a higher unity, such as a nation, in the same way as a leaf is connected with a tree. The development of individuals and nations is both driven and constrained by certain organic principles, but can also result in surprising novelties brought about by the interplay of natural forces (Humboldt [1814] 1960, 569, 576).

Humboldt elaborates his view on history in the famous lecture he gave at the Akademie der Wissenschaften in Berlin in 1821, "Über die Aufgabe des Geschichtsschreibers". In this text Humboldt argues that history is not meaningless, but that its meaning does not consist in an abstract *telos* in the Hegelian sense. When searching for coherence, the historian should not sacrifice the richness of individualities that can be observed in the past. Humboldt's perspective remains organic, though with the addition of an important new element: his theory of ideas. In the historical process, three layers can be observed: concrete events, the physical and psychological forces that bring them about, and the not immediately visible ideas lying behind these forces and giving them a certain direction. According to Humboldt, ideas are immaterial and timeless, but they nevertheless pervade all aspects of history and they cannot be known without studying their concrete historical manifestations. Examples of such manifestations of ideas are individualities such as persons and nations (Humboldt ([1821] 1960, 600–605).

Humboldt's view on the role of ideas in history has important consequences for his conception of the task of the historian. Historians should describe what happened in the past. Yet, a simple description of events based on critical and objective research is not sufficient. In addition, the historian should try to understand the ideas that shape the course of history. Humboldt describes the interpretation of ideas as *ahnden*, which means that ideas are grasped intuitively rather than analysed rationally. In this respect, the activity of the historian bears some resemblance to that of the poet, although this does not mean that historians do not have the task of giving a true account of the past (Humboldt ([1821] 1960, 585–589).

This points to an important link between Humboldt's theory of history and his views on aesthetics, which are centred around the concept of *Einbildungskraft*. This term denotes the capacity to transform reality into an image, thereby creating an alternative reality alongside the world of empirical experience (Lau 1999, 354–357; Mueller-Vollmer 1967). Historians do a similar thing when they intuitively grasp and express the ideas behind historical phenomena. Humboldt's analysis of historical interpretation is also related to his philosophy of language. According to Humboldt, language is essentially perspectival. People's use of language depends on their historically situated *Weltansicht*. As a result, understanding someone's utterances is a matter of interpretation. Spoken and written words are not meaningful because they refer to an empirical reality, but because they express ideas (Lau 1999, 323–346).

In his investigation of historical writing Humboldt develops an early notion of *verstehen* that anticipates the more elaborated theories of historical interpretation of authors such as Droysen (Wach 1926, vol. 1, 263–266), and can be argued to have played a barely recognised but nevertheless important role in the development of the hermeneutical tradition (Quillien 1990). Individuality is the cornerstone of Humboldt's historicism, in explicit opposition to Hegel's emphasis on the abstract *telos* of history. This does not mean, however, that Humboldt regards individual agency as the decisive factor in the historical process. His notion of individuality does not only apply to persons, but also to entities of a higher order such as nations. People are organically connected to their nation, and should therefore not be seen as atomic individuals. Furthermore, Humboldt regards individualities as historical embodiments of immaterial ideas, which, although unpredictably, shape the course of history. Thus, Humboldt's criticism of Hegel does not result in a clear-cut individualist ontology, but in an idealist and organicist perspective. It could be argued that the prominence of ideas in Humboldt's theory of history does not involve such a major rupture with Hegelian philosophy of history as Humboldt wants to bring about. Yet, by connecting ideas with individualities and raising the problem of historical interpretation he decisively moves away from Hegel's abstract and universalist point of view. Humboldt's combination of an idealist perspective with an emphasis on the interpretation of specific individualities would become characteristic of nineteenth-century historicism (Rüsen 1993, 99).

## 11.3 Ranke: Individuality and Empiricism

In 1824 Leopold von Ranke published his first major work, *Geschichten der romanischen und germanischen Völker*. In the preface he argues that it is not the task of the historian to pronounce judgements on the past, but to show "wie es eigentlich gewesen" (Ranke [1824] 1874, vii; Vierhaus 1977). With this classical statement of historicist empiricism Ranke condemns the explicit moralising of Enlightenment historiography, but also distances himself from the Hegelian approach to the past. In the 1820s the new University of Berlin was divided in two camps, a "historical" and a "philosophical" school. The central theme in the dispute between these schools was the nature of historical reality, the question whether the multiplicity of past events should be seen as a reality in itself or regarded as an epiphenomenon of a unifying rational principle, as argued by Hegel. An important episode in this dispute was the discussion between Ranke and Heinrich Leo, one of Hegel's students, on *Geschichten der romanischen und germanischen Völker* (Baur 1998, 112–123; Iggers 1997, 89–94). Leo (1828) criticised Ranke because of his focus on particular events, which, in his opinion, obscured the true nature of the historical process. According to Ranke, however, his description of particular events is just as much aimed at discovering general truths as Hegelian philosophy. The main difference is that his depiction of "das Allgemeine" is "unmittelbar und ohne langen Umschweif". In Ranke's opinion, the study of particular events is a more direct and more effective way of discovering the meaning of the past than philosophical analysis (Ranke [1828] 1890, 664).

According to Meinecke ([1936] 1959, 589), individuality is the central category in Ranke's work, especially because it establishes a conceptual connection between generality and particularity, between ideas and their embodiment in people or nations. Individuality is – in Ranke's words – "das Real-Geistige, welches in ungeahnter Originalität dir plötzlich vor Augen steht" and which cannot be reduced to a higher principle (Ranke [1836] 1925, 20). The most important individuality is the state. Every state is a distinct entity governed by a unique principle, an "inner life", which expresses itself in politics. According to Meinecke's interpretation,

Ranke tends to explain the actions of statesmen on the basis of "great motives" that emanate from the inner principle of the state. Hence, Meinecke emphasises the collectivist element in Ranke's approach, but not to the extent that he regards it as a totalising and mechanically applied theory. In his opinion, Ranke's collectivism is "eine sehr viel feinere Art von Kollektivismus" than, for instance, the Marxist approach to the past. The primacy of overindividual ideas in Ranke's work does not imply that these ideas do not need the actions of great men for their emergence and their realisation (Meinecke [1936] 1959, 589–592).

Meinecke's interpretation of Ranke's historicism is not undisputed. It has been argued that Meinecke both underestimates the role of individual agency in Ranke's work and misunderstands his conception of general patterns. As we have seen above, Ranke vehemently criticises Hegel's philosophical approach to the past. This criticism has an epistemological component, the argument that the study of particular events is the best way to get to know the truth about the past, and an ontological dimension, which involves the view that the course of history is not a necessary fulfilment of an abstract *telos*, but a set of contingent episodes (Jaeger 1997, 47–49). An important corollary of this antiteleological emphasis on contingency is the idea of individual freedom, which Ranke categorically affirms at many places in his work, though not without an awareness of its limitations (Frick 1953, 70–73).

In this interpretation, Ranke's perspective is individualist rather than collectivist. It should be noted that Meinecke's collectivist analysis primarily refers to Ranke's political tracts of the 1830s, "Die großen Mächte" and "Politisches Gespräch", in which he develops his theory of the state. When we turn from Ranke's political theory to his descriptive historical work, a more individualist image arises. Here, individual persons often have a decisive influence on the course of history. In *Ursprung und Beginn der Revolutionskriege*, for instance, Ranke describes the decision of Louis XVI to double the representation of the Third Estate in the Estates General as a crucial individual choice, made "im Widerspruch mit allen bestehenden Gewalten ... und zwar zugleich mit einem Mangel an Voraussicht, der fast unverständlich ist" (Ranke [1875] 1879, 28). Similarly, in his books on Wallenstein and Hardenberg (Ranke 1869; 1879–1881), he argues that the decisions of these two individuals were a crucial factor in the historical process, despite the undeniable relation of their considerations and actions to general historical tendencies (Frick 1953, 75–79, 88–97).

This raises the important question what conception Ranke has of these general tendencies in history. His most explicit treatment of this theme can be found in the introduction to a lecture series held in 1831, posthumously published under the title "Idee der Universalgeschichte" (Ranke 1975). Ranke starts his argument with sketching the contrast between history, which focuses on facts and individualities, and philosophy, which speculatively understands the past in terms of abstract general concepts. Yet, this does not imply that historians should ignore general tendencies in history: in every individuality lives a higher principle, "ein Ewiges, aus Gott kommendes". This phrase points towards an important aspect of Ranke's work: its religious dimension. In the end, the course of history is governed by eternal, divine principles, and it is the ultimate task of the historian to grasp these principles. This does not mean, however, that human beings are deprived of their agency.

They can act on the basis of their own, sometimes selfish or irreligious, motives and in that way decisively influence the course of history (Ranke 1975, 77, 79f.). This tension between the divine order and human free will is typical of Ranke's protestant beliefs. He resolves it by strictly separating immanence and transcendence, human and divine, in his historical work (Hardtwig 1991, 3f.). First of all, the historian should thoroughly study particular events and discover causal connections. The next stage is the inductive construction of empirical generalisations, which becomes more difficult when the historian tries to analyse a more extensive domain. The true nature of the totality of world history can only be understood tentatively, but never completely: "die Weltgeschichte weiß allein Gott" (Ranke 1975, 85).

Ranke's work is characterised by many tensions and paradoxes and can hardly be seen as the elaboration of a unified theoretical framework. Two tensions, continuously present from the 1820s to the end of his life in 1886, are especially important in his work: the tension between universality and particularity and that between necessity and freedom (Krieger 1977, 347f.). The notion of individuality is crucial to understand the way in which Ranke deals with these tensions. Ranke's antiteleological empiricism involves an affirmation of the agency of individual persons and the ensuing contingency of history. Nevertheless, this individualist perspective does not preclude that history has a universal dimension. On the one hand, historians can discover the universal aspects of the past by inductively combining their knowledge of particular events. On the other hand, Ranke also believes that the historical process has a transcendent dimension, which he finds in the eternal principles through which God is present in the course of history, and which cannot be directly and completely known (Krieger 1975, 6f.).

Ranke distinguishes individualities at different levels, most noticeably persons and states, both seen as embodiments of ideas. An important question – which Ranke does not explicitly address – is how the relation between these different kinds of individualities should be seen. Meinecke's collectivist interpretation suggests a primacy of the state, which implies that the identity and actions of individual persons are shaped by the ideas inherent in the state to which they belong. Yet, this view contradicts Ranke's repeated acknowledgement of individual agency. The crucial problem here is the ontological status of ideas in Ranke's work. A nominalist interpretation of Ranke's conception of ideas, which involves that they are generalisations envisaged by the historian on the basis of his research, would be in line with Ranke's empiricism, and seems more plausible considering his affirmation of individual agency. His religious perspective, however, suggests that ideas are a reality *sui generis*, ultimately emanating from God.

## 11.4 Droysen: Individuality and Methodical Interpretation

In the middle of the nineteenth century, scientistic positivism had become a more important opponent of historicism than Hegelian philosophy. Positivism rejects metaphysical speculation about the historical process and emphasises that the past should be examined in accordance with scientific standards. Historicism emerged against the same background as positivism, and shares its drive to substitute philosophical speculation with methodical research (Fuchs 1994, 323f.). Yet, historicism and positivism chose fundamentally different directions in the elaboration of their shared principles. The primary aim of nineteenth-century positivism was to transform history into a scientific discipline based on methodological precepts similar to those of the natural sciences. The main consequence of this objective was the demand to search for laws in the historical process. The ontological corollary of this methodological claim is the dismissal of individual agency as a determining factor in the course of history. The English historian Henry Thomas Buckle, one of the major representatives of nineteenth-century positivism, argues that the analysis of human actions should not be based on "something mysterious" as a free will. That would amount to the acceptance of an unscientific, metaphysical dogma. Human behaviour can only be explained in terms of certain laws if it is not to be regarded as the result of coincidence or supernatural forces (Buckle [1857] 1858–1861, vol. 1, 7f.).

Buckle's principal work, *History of Civilization in England*, was reviewed in Germany by Johann Gustav Droysen, the main theorist of nineteenth-century historicism. Droysen argues that Buckle's effort to subsume the history of civilisations under general laws fails to appreciate the freedom of the human will. Human actions are not completely determined by external circumstances, but are brought about by the free will of individual persons. This does not imply, however, that Droysen regards historical actors as totally autonomous, atomist individuals: their actions are oriented towards *sittliche Gemeinsamkeiten*, such as families, states and nations. These moral communities endow individual actions with meaning, and, as a consequence, enable the historian to describe the past as a meaningful totality (Droysen [1863] 1937, 396–403).

The rise of positivism was a major challenge to the historicist paradigm. The positivist claim that it was the only truly scientific approach to the past urged historicist historians to reflect more systematically on their own methods. Droysen took up this task in a series of lectures on the theory and methodology of history, which he gave at the universities of Jena and Berlin between 1857 and 1883.[1] The starting-point of Droysen's theoretical considerations is a fundamental distinction between nature and history, "die weitesten Begriffe, unter denen der menschliche Geist die Welt der Erscheinungen faßt". Nature and history do not indicate two distinct domains of reality, but denote two modes of perceiving the phenomena in the world: in the first our experiences are spatially structured, in the second temporally. As a consequence, history and the natural sciences have fundamentally distinct scientific methods (Droysen [1858] 1937, 325f.).[2] While the natural sciences focus

[1]The text of these lectures was published for the first time in 1937 by Rudolf Hübner (Droysen 1937). In 1858 Droysen published an excerpt of his lectures under the title *Grundriß der Historik*, which is also included in Hübner's edition (Droysen [1858] 1937).

[2]Space and time are the two Kantian *Anschauungsformen*, but Droysen uses them as a formal basis to distinguish history from the natural sciences (Schnädelbach 1974, 94).

on the construction of formal generalisations, the task of history is "forschend zu verstehen" (Droysen [1858] 1937, 328).

An important part of Droysen's lectures is dedicated to the further elaboration of this view of the historical method as a combination of research and interpretation. Research consists of *Heuristik* and *Kritik*, of finding sources and critically examining them. Droysen's analysis of these fields is not remarkably different from the prescripts of his Rankean predecessors. His discussion of interpretation, however, involves a major innovation of historical methodology, primarily because he systematically examines a set of operations that earlier historicists had simply described as the intuitive process of *ahnden*. In connection with this idea of intuitive interpretation, Droysen makes a distinction between "der logische Mechanismus des Verstehens" and "der Akt des Verständnisses"; the latter is a direct intuition "als Tauche sich Seele in Seele" (Droysen [1858] 1937, 329). When intuitively and directly understanding something, we are not conscious of the logical operations involved in this act of interpretation (Droysen 1937, 26). The task that Droysen sets himself in his lectures on historical theory is to uncover these underlying logical mechanisms.

According to Droysen, historical interpretation has four modalities. The first of these modalities is pragmatic interpretation, the reconstruction of a course of events on the basis of the available source material. The interpretation of conditions establishes under what circumstances these events took place: it analyses, for instance, the influence of the geographical situation, the availability of technical means, and the effects of the mentality of a certain period. Psychological interpretation tries to uncover the individual intentions (*Willensakte*) that brought about a particular course of events. In Droysen's opinion, this kind of interpretation is a central element of the historical method, since the historical world is a *sittliche Welt* that bears the marks of intentional human actions in all its aspects. Yet, psychological interpretation has inherent limitations, because historians cannot sketch a comprehensive image of a person's inner life on the basis of its external effects. Furthermore, past events are not exclusively caused by individual intentions; other factors also play an important role. Here, we enter the domain of the interpretation of ideas or *sittliche Mächte*. Examples of *sittliche Mächte* are the family, marriage, the state and the law – phenomena that modern social theorists would call "institutions". Each family or state is a specific expression of an idea, realised by individuals collectively acting on the basis of that idea (Droysen [1858] 1937, 339–344; Droysen 1937, 152–186).

In the end, Droysen places the interaction of historical forces in a theological framework (Hardtwig 1991, 5f.; Southard 1979). Individuals and *sittliche Mächte* try to realise their objectives in the course of the historical process, but this process is only meaningful in light of God's purpose with the world, "der Zweck der Zwecke". This ultimate goal, however, cannot be examined empirically (Droysen [1858] 1937, 356). Thus, Droysen disconnects the ultimate meaning of history from empirical knowledge of the past, two things that were not distinguished all that carefully by earlier historical theorists.

Droysen's methodology of history is not only directed against the scientistic postulates of positivism, but also entails a critique of Rankean historicism.

In his opinion, Ranke's empiricist focus on "wie es eigentlich gewesen" results in a tremendous amount of historical knowledge, but at the cost of making historians specialised "factory workers" for whom the past is a dead object.[3] Droysen's detailed theory of interpretation is not only meant as a critical elaboration of the rather unsophisticated theoretical positions of earlier historicists, but also aims to turn the empirical study of the past into a form of historical *Bildung* that is not devoid of contemporary ethical and political significance. As a progressive liberal (Hock 1957; Southard 1995), Droysen considers Ranke's empiricism a quietist defence of the political status quo, which necessitates the development of a new theoretical perspective from which the past can be reinterpreted and given a new critical relevance (Maclean 1982; Rüsen 1969).

Droysen's rejection of the Rankean variety of objectivism in favour of the critical potential of historiography raises an important hermeneutical problem: that of the subjectivity of the historian. In his treatment of this problem Droysen emphasises the historicity of the historian's point of view. There is no timeless objective position from which the past can be examined: a historian cannot detach himself from the historical process that is the object of his research. This means that the truth he discovers is relative to the *sittliche Mächte* that have shaped his viewpoint. It does not imply subjective arbitrariness: when a historian approaches the past from the perspective of his nation, his state, or his religion, he transcends the subjective prejudices and limitations of his individual personality (Droysen 1937, 287). In a way, this analysis completes Droysen's theory of interpretation. Unlike his predecessors, he does not assume that historiography can somehow escape from the historicity of the world, but also applies the categories that he uses to study the past to the position of the historian (Gadamer [1960] 1990, 219–222; Jordan 1998, 156–172).

## 11.5 Conclusion

Individuality is a core category in the modern worldview that emerged around 1800. Nineteenth-century historiography was highly troubled by two philosophical problems connected with the notion of individuality. In the first place, the question how to conceive of the relation between individuality and universality, and secondly, the question how an affirmation of individual agency can be reconciled with the acknowledgement that there are general patterns in the course of history. The main adversaries of historicism, Hegelian philosophy and positivism, addressed these problems by emphasising the general aspects of the past – either in terms of an abstract *telos* or historical laws – and by bracketing individual agency. Historicist authors rejected these solutions, mainly because they implied a disavowal of the idea of individuality.

[3]The remark is from one of Droysen's letters, where he is more explicitly critical of his predecessors than in his lectures and published texts (quoted in Maclean 1982, 351).

This does not mean, however, that historicism had an unequivocal conception of individuality. Humboldt's notion of individuality does not only apply to persons, but also to entities of a higher order such as nations, which organically encompass individual human beings. Humboldt does not regard the agency of individual persons as a decisive factor in the historical process. Instead, history should be understood by intuitively grasping the immaterial ideas lying behind it. Ranke's conception of individuality is ambivalent and can best be understood by distinguishing two strands in his thought, one focusing on the immanent aspects of the course of history, the other focusing on its transcendent meaning. In the first strand of thought, individual agency is affirmed and plays a crucial role in the empirical study of the past, which can result in generalisations based on induction. At the same time, however, Ranke believes that the course of history is shaped by God and by the ideas that emanate from him – individualities of a different order that cannot be empirically studied but should be understood intuitively.

Although Droysen also believes that history has a final objective, determined by God, in his methodology the transcendent dimension of historiography has been discarded. According to Droysen, individual agents have a decisive influence on the historical process, though not without being connected with certain ideas or *sittliche Mächte*. These ideas, however, are not transcendent entities, but constant factors in the variety of individual appearances of social institutions. The corollary of Droysen's dismissal of transcendent individualities is an elaborate theory of interpretation. As argued at the beginning of this paper, the affirmation of individuality poses the problem of interpretation. To Droysen, the solutions to this problem offered by Humboldt and Ranke – based on intuition and mere empirical research – were unacceptable, not only because of their lack of methodological sophistication, but also because they ignored the historicity of the historian as an interpreting subject.

## References

Bambach CR (1995) Heidegger, Dilthey, and the Crisis of Historicism. Cornell University Press, Ithaca, NY

Barash JA (2004) Politiques de l'histoire. L'historisme comme promesse et comme mythe. Presses Universitaires de France, Paris

Baur S (1998) Versuch über die historik des jungen ranke. Duncker & Humblot, Berlin

Blanke HW (1991) Historiographiegeschichte als historik. Frommann-Holzboog, Stuttgart

Buckle HT ([1857] 1858–1861) History of Civilization in England. 2 vols. Parker & Son, London

Droysen JG ([1858] 1937) Grundriss der Historik. In: Hübner R (ed.) Historik. Vorlesungen über enzyklopädie und methodologie der geschichte. Oldenbourg, München, pp 317–366

Droysen JG ([1863] 1937) Erhebung der Geschichte zum Rang einer Wissenschaft (review of History of Civilization in England by H.T. Buckle). In: Hübner R (ed.) Historik. Vorlesungen über enzyklopädie und methodologie der geschichte. Oldenbourg, München, pp 386–405

Droysen JG (1937) Vorlesungen über Enzyklopädie und Methodologie der Geschichte. In: Hübner R (ed.) Historik. Vorlesungen über enzyklopädie und methodologie der geschichte. Oldenbourg, München, pp 1–316

Foucault M (1966) Les mots et les choses. Une archéologie des sciences humaines. Gallimard, Paris

Frick G (1953) Der handelnde mensch in rankes geschichtsbild. Weiss, Zürich

Fuchs E (1994) Henry Thomas Buckle. Geschichtsschreibung und positivismus in England und Deutschland. Leipziger Universitätsverlag, Leipzig
Fulda D (1996) Wissenschaft aus Kunst. Die entstehung der modernen deutschen geschichtsschreibung, 1760–1860. Walter de Gruyter, Berlin/New York
Gadamer H-G ([1960] 1990) Wahrheit und methode, 6th edn. J.C.B. Mohr (Paul Siebeck), Tübingen
Hardtwig W (1991) Geschichtsreligion – Wissenschaft als Arbeit – Objektivität. Der Historismus in neuer Sicht. Historische Zeitschrift 252:1–32
Hegel GWF ([1837] 1986) Vorlesungen über die Philosophie der Geschichte. In: Moldenhauer E, Michel KM (eds) Werke, vol 12. Suhrkamp, Frankfurt am Main
Hock W (1957) Liberales denken im zeitalter der paulskirche. Droysen und die frankfurter mitte. Aschendorff, Münster
Humboldt W von ([1792] 1960) Ideen zu einem Versuch, die Gränzen der Wirksamkeit des Staats zu bestimmen. In: Flintner A, Giel K (eds) Schriften zur anthropologie und geschichte. Werke in fünf Bänden, vol 1. Wissenschaftliche Buchgesellschaft, Darmstadt, pp 56–233
Humboldt W von ([1813] 1964) Denkschrift über die deutsche Verfassung an den Freiherrn vom Stein. In: Flitner A, Giel K (eds) Schriften zur politik und zum bildungswesen. Werke in fünf bänden, vol 4. Wissenschaftliche Buchgesellschaft, Darmstadt, pp 302–322
Humboldt W von ([1814] 1960) Betrachtungen über die Weltgeschichte. In: Flitner A, Giel K (eds) Schriften zur anthropologie und geschichte. Werke in fünf Bänden, vol 1. Wissenschaftliche Buchgesellschaft, Darmstadt, pp 567–577
Humboldt W von ([1821] 1960) Über die Aufgabe des Geschichtsschreibers. In: Flitner A, Giel K (eds) Schriften zur anthropologie und geschichte. Werke in fünf bänden, vol 1. Wissenschaftliche Buchgesellschaft, Darmstadt, pp 585–606
Iggers GG (1995) Historicism: the history and the meaning of the term. J Hist Ideas 56:129–152
Iggers GG (1997) Deutsche geschichtswissenschaft. Eine kritik der traditionellen geschichtsauffassung von herder bis zur gegenwart. Böhlau, Wien
Jaeger F (1997) Geschichtsphilosophie, Hermeneutik und Kontingenz in der Geschichte des Historismus. In: Küttler W, Rüsen J, Schulin E (eds) Die epoche der historisierung. Geschichtsdiskurs, vol 3. Fischer, Frankfurt am Main, pp 45–66
Jaeger F, Rüsen J (1992) Geschichte des historismus. Eine einführung. C.H. Beck, München
Jordan S (1998) Geschichtstheorie in der ersten hälfte des 19. Jahrhunderts. Die schwellenzeit zwischen pragmatismus und klassischem historismus. Campus Verlag, Frankfurt am Main
Krieger L (1975) Experience, theory, and history in Ranke. Hist Theor. Beiheft 14:1–14
Krieger L (1977) Ranke and the Meaning of History. University of Chicago Press, Chicago, IL
Kuhn T ([1962] 1970) The Structure of Scientific Revolutions, 2nd edn. University of Chicago Press, Chicago, IL
Lau V (1999) Erzählen und verstehen. Historische perspektiven der hermeneutik. Königshausen & Neumann, Würzburg
Leo H (1828) Review of Geschichten der romanischen und germanischen Völker von 1494 bis 1514 and Zur Kritik der neueren Geschichtsschreiber by L. von Ranke. Ergänzungsblätter zur Jenaischen Allgemeinen Literatur-Zeitung (17):129–136 and (18):137–140
Maclean MJ (1982) Johann Gustav Droysen and the development of historical hermeneutics. Hist Theor 21:347–365
Meinecke F ([1936] 1959) Die Entstehung des Historismus. In: Herzfeld H, Hinrichs C, Hofer W (eds) Werke, vol 3. Oldenbourg, München
Mueller-Vollmer K (1967) Poesie und einbildungskraft: zur dichtungstheorie wilhelm von humboldts. Mit der zweisprachigen ausgabe eines aufsatzes humboldts für frau von staël. J.B. Metzler, Stuttgart
Oexle OG (1984) Die Geschichtswissenschaft im Zeichen des Historismus. Bemerkungen zum Standort der Geschichtsforschung. Historische Zeitschrift 238:17–55
Oexle OG (1996) Meineckes Historismus. Über Kontext und Folgen einer Definition. In: Oexle OG, Rüsen J (eds) Historismus in den kulturwissenschaften. Geschichtskonzepte, historische einschätzungen, grundlagenprobleme. Böhlau, Köln, pp 139–199

Quillien J (1990) Pour une autre scansion de l'histoire de l'herméneutique. Les principes de l'herméneutique de W. von Humboldt. In: Laks A, Neschke A (eds) La naissance du paradigme herméneutique. Schleiermacher, Humboldt, Boeckh, Droysen. Presses Universitaires de Lille, Lille, pp 69–117

Ranke L von ([1824] 1874) Geschichten der romanischen und germanischen völker von 1494 bis 1514, 2nd edn. Sämmtliche werke, vol 33/34. Duncker & Humblot, Leipzig

Ranke L von ([1828] 1890) Erwiderung auf Heinrich Leo's Angriff. In: Dove A (ed.) Zur eigenen lebensgeschichte. Sämmtliche werke, vol 53/54. Duncker & Humblot, Leipzig, pp 659–666

Ranke L von ([1836] 1925) Das politische Gespräch. In: Rothacker E (ed.) Das politische gespräch und andere schriften zur wissenschaftslehre. Max Niemeyer, Halle an der Saale, pp 10–23

Ranke L von (1869) Geschichte wallensteins. Sämmtliche werke, vol 23. Duncker & Humblot, Leipzig

Ranke L von ([1875] 1879) Ursprung und beginn der revolutionskriege, 1791 und 1792, 2nd edn. Sämmtliche werke, vol 40. Duncker & Humblot, Leipzig

Ranke L von (1879–1881) Hardenberg und die geschichte des preußischen staates von 1793 bis 1813. Sämmtliche werke, vols 46–48. Duncker & Humblot, Leipzig

Ranke L von (1975) Idee der Universalgeschichte. In: Dotterweich V, Fuchs WP, Schieder T (eds) vorlesungseinleitungen. Aus werk und nachlaß, vol 4. Oldenbourg, München, pp 72–89

Rüsen J (1969) Begriffene geschichte. Genesis und begründung der geschichtstheorie J.G. Droysens. Schöning, Paderborn

Rüsen J (1993) Konfigurationen des historismus. Suhrkamp, Frankfurt am Main

Schnädelbach H (1974) Geschichtsphilosophie nach hegel. Die probleme des historismus. Freiburg, Alber

Schnädelbach H (1983) Philosophie in deutschland, 1831–1933. Suhrkamp, Frankfurt am Main

Southard R (1979) Theology in Droysen's early political historiography: free will, necessity, and the historian. Hist Theor 18:378–396

Southard R (1995) Droysen and the Prussian School of History. The University Press of Kentucky, Lexington, KY

Troeltsch E (1922a) Die Krisis des Historismus. Die neue Rundschau 33:572–590

Troeltsch E (1922b) Der Historismus und seine Probleme. J.C.B. Mohr (Paul Siebeck), Tübingen

Vierhaus R (1977) Rankes Begriff der historischen Objektivität. In: Koselleck R, Mommsen WJ, Rüsen J (eds) Objektivität und parteilichkeit in der geschichtswissenschaft. Beiträge zur historik, vol 1. Deutscher Taschenbuch Verlag, München, pp 63–76

Wach J (1926) Das verstehen. Grundzüge einer geschichte der hermeneutischen theorie im 19. Jahrhundert. 3 vols. J.C.B. Mohr (Paul Siebeck), Tübingen

Wittkau A (1992) Historismus. Zur geschichte des begriffs und des problems. Vandenhoeck & Ruprecht, Göttingen

# Chapter 12
# Shaping Disciplinary Boundaries: Scientific Practice and Politics in the *Methodenstreit* Between the German Historical School and the Austrian School of Economics

**Filomena de Sousa**

In view of the great success enjoyed by the natural sciences by the end of the nineteenth century, scholars working in the social field felt the need to highlight the importance of the human sciences as piece and part of the broad scientific scene. Discussions purporting to the limits and status of the sciences devoted to the study of human behaviour, especially in relation to the logic of the natural sciences, led to the articulation of the conceptual pair Erklären/Verstehen as corresponding to the demarcation between *Naturwissenschaften* and *Geisteswissenschaften*. The differentiation between History and Society took central stage in the context of the debates about the scientific parameters shaping disciplinary boundaries and gave rise to the famous *Methodenstreit* opposing Gustav Schmoller the leader of the German Younger School of Economics to Carl Menger, the founding father of the Austrian School of Economics. The relevance of this episode can be measured by its impact not only on economics, but on the broader context of twentieth century social science, as the Methodenstreit turned on the dispute between Methodological Individualism and Holism. In this essay I tell the story of the divergences between German and Austrian scholars and suggest that the gap between the German Historical School and the Austrian School of Economics may be narrower than standard textbooks suggest.

All in all, the debate over methods can be traced back to a complex historical, social and scientific web whose repercussions extended beyond a single episode to shape pathways and procedures that allowed the social sciences to assert their status as a scientific enterprise in its own right, to form a distinctive scientific field where the problems of knowledge and explanation raised particular challenges. This wasn't an unimportant endeavour. We do need particular methods to study nature and society. But the *Methodenstreit* helped shape disciplinary boundaries between the social sciences and humanities as well and helped delineate the contours separating economics from the other social sciences. It thus contributed to raise awareness of the fact that the field of social science is not homogeneous.

---

F. de Sousa (✉)
Technical University of Lisbon, Portugal
e-mail: Filomenas@iseg.utl.pt

U. Feest (ed.), *Historical Perspectives on Erklären and Verstehen*, Archimedes, Vol. 21, DOI 10.1007/978-90-481-3540-0_12, 

I offer a historical narrative of these developments and bring into focus the historical and institutional background, as well as the political agendas under which new scientific ideas and practices emerged. The scientific, cultural and institutional manifold that led to the *Methodenstreit* is scrutinized, while I also explain how this episode brought about important changes in the way scholars involved perceived their own role, which became increasingly one of specialists. A better understanding of the historical, cultural and political factors that contributed to these scientific developments allows for richer insights into the epistemological issues at stake. In fact, scholars involved in the *Methodenstreit* were fully aware that the scientific practices and ideas they promoted reflected the societal, values, conflicts, and trends of the times. Such awareness led them to articulate the question of a *Wertfreiheit* of scientific practice.

## 12.1 The German Historical School and Its Institutional Setting

The term "historicism" has been understood in various ways since becoming part of German academic discourse in the late nineteenth century. Amongst prominent instances of this movement we find the School of Jurisprudence and the Historical School of Economics. I use the term to identify the School of Economics founded by Wilhelm Roscher, a movement representing a national mainstream sceptical of what we now we refer to as neoclassical economics and dismissive of theory that wasn't derived from historical experience.

Surely, historicism wasn't a particular German phenomenon. But in Germany, during the Second Reich, it dominated social life and academia more conspicuously than elsewhere becoming more influential than similar movements in other countries. These trends fostered a context where the notion of a people's destiny, *Volksgeist*, as a descriptive concept and a normative demand of faithfulness to national genius. Both the School of Jurisprudence and the Historical School of Economics were impregnated by this ideology, which resulted from the efforts made by a regime promoting national unity by encouraging identification with a constructed public memory.[1] As I will show in this essay, the notion of *Volksgeist* also carried with it the kernel of the methodological *holism* that was later going to be confronted with the methodological individualism of the Austrian School of Economics.

By the late nineteenth century, Prussia had become the intellectual center of Germany, having developed an advanced and vibrant industrial economy. But despite its transformation into a commercial and industrial hub, in contrast to England, where industrialisation favoured the rise of a middle class and liberal

[1]For a detailed description of how the idea of *Volksgeist* was a central tenet of the doctrines promoted by the German School of Jurisprudence and the German Historical School of Economics see the works of Hodgson (2001); Lindenfeld (1997); Scaff (1995); and Tribe (1988, 2002).

institutions, German society remained riveted to a paternalistic sociopolitical regime. As Germany sought a path of economic development enfranchised from British dominance of world manufacturing and trade, the teaching of economics was given a prominent place as a body of knowledge turned to the resolution of practical problems (Hodgson 2001, 58).

To counterbalance the overwhelming Anglo-Saxon tradition in economics, the Historical School founded by Roscher dominated German economics from the mid-1800s to the mid-1900s. It was a movement connected by common concerns and similar methodological guidelines, but subdivided in different blocs corresponding to different generations. This lack of cohesion led Schumpeter, Kirzner and Pearson (Schumpeter [1954] 1994; Kirzner 1992; Pearson 1999) to question the terminology of a 'German Historical School of Economics'. However, as I shall explain next, members of the School were the first to welcome this designation, and their contemporaries identified these scholars as members of a distinctive 'German Historical School of Economics'.

The Historical School of Economics was characterized by a list of common concerns: a historical account of economic institutions through an inductive methodology and a view of economic phenomena as part of a web encompassing the state, law, politics and ethics. Economic knowledge was envisioned as growing out of historical research, and economics as an empirical discipline. Accordingly, the economist was portrayed as an economic historian whose work would supplement, and be supplemented by, the work of economic historians proper (Schumpeter [1954] 1994, 808). The Historical School of Economics emphasized the particular and non-repeatable element in human affairs so that interpretation of meanings, *Verstehen*, was an important methodological tool of the historicist economists Given their uniqueness, events could be grasped only by being set into their particular historical and social context, into a holistic network. Whatever the nature of the theory developed from the meanings social actors conferred to their actions, it was historically and empirically grounded denoting the reigning *Volksgeist* (Scaff 1995).

But the crucial factor while identifying a distinctive School is the fact that starting from Roscher, conventionally seen as the founder of the School, and ending with Max Weber, these scholars welcomed the designation of members of a 'German Historical School of Economics' as a label appropriate to describe their outlooks (Tribe 1988, 2002). Not to be neglected is the fact that historicist economics was the expression of an academic discourse directed at ameliorating social ills. Concerns with *Sozialpolitik* were therefore carried over since the beginnings of the Historical School of Economics, which was perceived as a state science, a *gesamte Staatswissenschaften*, with an essential historical and ethical dimension, resulting from an attempt to systematise practical knowledge in the service of bureaucratic education at university level. German intellectual life was dominated by debates over the relationship between ethics, political economy, ideology and social reform. The use of politically oriented historical research had some practical effects on Prussian civil servants and economists, as *Staatswissenschaftlers*, became powerful influences on public opinion (Lindenfeld 1997, 3–6).

## 12.2 Historicity as a Criterion of Demarcating the *Geisteswissenschaften:* The Older School of Economics

The Historical School of Economics was divided in two or three subsets and its foundation was marked by Roscher's publication of *Grundriss zu Vorlesungen über die Staatswirtschaft nach geschichtlicher Methode* in 1843. Educated as an historian and having lectured on politics his whole career, Roscher subsumed economics under politics and ethics. Economic phenomena were seen as entangled in an organic whole, a *Volksleben,* and Roscher's goal was to introduce historical methods into economics to uncover laws of development of national organic wholes. The epistemological roots of his program hailed from the German School of Law that focused on the connection between national spirit and legal system. Savigny's School stressed the historical and cultural specificity of legislation, and the idea that different principles applied to different times and places, but dismissed a universal theoretical system.

Likewise, the *Grundriss,* which pointed to the contrast between "historical" and "philosophical" lines of analysis, depicted political economics as a science devoted to uncovering laws governing the origins, development and functioning of socio-economic life of peoples, *wirthschaftliches Volksleben.* The "philosophical" or "abstract" analysis against which Roscher reacted was epitomised by British classical economists, namely Stuart Mill and Ricardo. Stressing that abstract ideas meant little divorced from the socio-political reality they mirrored, Roscher favoured instead simple descriptions of particular phenomena (Roscher 1843, IV). The goal of his program was threefold aiming at: (a) emphasizing historical contingencies, (b) outlining a general lawfulness underlying history, and (c) turning economics into a pure empirical science.

Roscher was very critical of attempts to explain human actions on the basis of goal-oriented optimizing behaviour which, he believed, could explain the disintegration of society, but not the institutions holding it together. To articulate developmental laws Roscher called for a comparative study of different peoples, assuming that common features of varied developments could be categorised through such laws. Like the School of Jurisprudence he turned his attention to ancient peoples, like the Roman civilisation whose development, he believed, could be assessed in its totality. The *Grundlagen* (1858) illustrates his efforts to articulate a cyclical theory drafted through the biological analogy of youthfulness, maturity, decay and death. Envisioning economic laws embedded in moral and cultural rules, Roscher believed that when cultural and moral standards plunge, the economy follows the same path. As Birger Priddat has underscored, Roscher turned economics into a comparative ethnology. Presumably, his intention wasn't to create a new economic theory, but to develop a descriptive or 'factual' science by 'transplanting' the method of historians into economics (Priddat 1995, 15–34).

At first, Roscher's program had a stimulating effect on German economics, but his scheme of cross-cultural comparisons remained fragmentary and incomplete, a consequence of the naive empiricism underlying his method of comparative analogy.

Besides dismissing an objective, universal, categorization, he was unable to conceptualize history as a continuous process of socio-economic change. Roscher, as Geoffrey Hodgson emphasized, held true to the methodology of the Historical School of Economics, collected and interpreted an enormous body of data, but neglected the theoretical framework required to cast an appropriate description and categorisation of facts (Hodgson 2001, 60).

Members of the Young and Old Historical School alike didn't refrain from criticizing Roscher. Bruno Hildebrand and Carl Knies denounced his failure to periodicize history. Weber criticized Roscher both for smuggling morals into his analysis and his attempts to uncover objective developmental standards, whereas the other representatives of the Old Historical School went one step further, investigating the historicity of the standards themselves (Weber 1903–1906).

Hildebrand, like Roscher, begun his career as an historian but switched to economics, sketching a program for the reformulation of the discipline on historicist lines, comprehending the laws of development of peoples in *Nationalökonomie der Gegenwart und Zukunft* (Hildebrand 1848). Appealing to the connection between the history of the economic, political and juridical development of nations he attempted to articulate developmental laws of *Volksleben,* but instead of using historical jurisprudence as a guideline, he turned to the science of language as conceived in the nineteenth century. In line with neo-Hegelianism, Hildebrand sketched a teleological theory of history where each economic system was seen as a product of a particular stage of development. Hildebrand focused on changes in the institutions of money or credit to provide the foundations for his categorisation that, ultimately, didn't prove out either.

With Hildebrand we start witnessing a shift of focus from historicism into the 'ethical-historical' perspective that characterised the Younger Historical School. An apologist of social reform, he believed that each era presented new challenges, so that their resolution depended upon the moral, intellectual and material progress of mankind, in a scheme where the individual was both the product and architect of history. The emphasis given by Hildebrand to the moral-science nature of economics was highly significant in promoting the distinction between *Naturwissenschaft* and *Geisteswissenschaft.* Although John Stuart Mill had already mentioned a distinct area of studies, the Human Studies (Hildebrand 1848), Hildebrand was, to the best of my knowledge, the first within the field of mid-nineteenth century economic academia to differentiate law that are epistemologically analogous to physical laws (*Naturwissenschaften* laws) from those which, including economic laws, belonged to the social sphere. His social utopia didn't flourish, but Hildebrand's categorisation, which was later popularized by Rickert and Windelband, had an immense impact upon the newly forming social sciences.

Carl Knies, hailed by Schumpeter as one of the leading figures of German economics (Schumpeter [1954] 1994, 508), has equally stressed the historical nature of political economics in *Die politische Ökonomie vom Standpunkt der geschichtlichen Methode* (Knies 1883). Knies taught at Heidelberg for over 30 years. Eugen von Böhm-Bawerk and Friedrich von Wieser, members of the future Austrian School of Economics were his students, as well as American economist

John Bates Clark and Max Weber, who eventually succeeded him. Unsatisfied with the static concepts of British neoclassical economics, he hinted at the concept of value that later became central to the Austrian School: a definition of value in terms of people's subjective estimations. Despite conceding the existence of some 'ethically neutral 'economic phenomena, Knies believed that the psychological causes of values were correlated with historical and national variables, a factor distinguishing his theory from the Austrian subjective theory of value.

Knies understood needs as varying not only from person to person, class to class, but from nation to nation. Viewing ethics as central to the historical method, he was deeply concerned by politics and regarding the transformations occurring in Germany with optimism. Rejecting both authoritarian policies and *laissez-faire*, Knies believed that when individuals failed to attain socially favourable results state intervention, *Staatsaktion*, and "charitable action", *Karritativverkehr*, became a necessary corrective measure. His program was characterised by the same concerns as those of the other economists of the Historicist School. Accordingly, he gave great emphasis to the role of the reigning *Volksgeist*, and stressed that theoretical validity is but a product of its own time. But this was not yet sufficient for Knies. His approach was permeated by an apperception of the relative character of theory itself, an insight setting Knies apart from other economists of his generation. He believed that historical political economy was equally a product of historical development. If social institutions were to be regarded as passing through a series of stages correlative with successive stages of a civilisation, economic science was the result of a particular stage of historical development. Conceptualized in vital connection with the social setting of the period and particular national, local and temporal boundaries, particular economic frameworks could at no point be seen as having attained a definitive form. This typology, however, required analytical tools for the study of each system, a task that Knies neglected.

For Knies, to be 'historical' meant to be 'context-bound' a criterion distinguishing economics from the natural sciences (Lindenfeld 1997, 185). Like his friend Hildebrand he fostered the discussion of the distinction between *Natur* and *Geisteswissenschaften,* but went a step further by asserting that historical political economy belonged to the domain of social and political science rather than to the sphere of the humanities. The pivotal idea is that political and social sciences, the fields where economics belong, have an epistemological status distinct from that of other *Geisteswissenschaften*, such as philosophy, theology, and literature (Knies 1883, 6).

Despite significant differences, Roscher's, Hildebrand's and Knies's works expressed awareness of historical contingency and the changes taking place in Germany and Europe. The combination of growing nationalism and industrial development reinforced their interest in historical stages, raising simultaneously questions about the role of economics as ideology and the comparative economic growth of nations (Lindenfeld 1997, 152). It also furthered their mistrust in a "theoretical" approach, which they feared would bind them to a fixed system of practical economy.

Notwithstanding the diverseness of the works of the political economists of the Older School, all centred around a similar project: understanding the organic interdependence of the ethical, cultural, and historical foundations of economic behaviour.

Reacting to the dominant British tradition, denominated today as the "neoclassical tradition", the program of the German Historical School fit well with the Hegelian emphasis on the reining *Volksgeist.* Rather than a *reine Wirtschaftswissenschaft,* a pure science of economics, economics was conceptualized as a *Volkswirtschasftslehre*, the economy of a people (Koslowski 1995). Despite the fact that scholars associated with the Older School of Economics didn't collaborate in disseminating their ideas nor established a formal school, by the 1860s, 'historicist economics' was a doctrine clearly recognizable to their contemporaries.

## 12.3 The Politics of the Younger School

Policy-oriented economics, or *Sozialpolitik*, played a more prominent part in academic affairs as the Older Historical School was succeeded by a new generation of social scientists led by Gustav Schmoller who studied history and *Staatswissenschaften* before moving to economics. The Younger School, included most academic economists of the newly united Germany forming a cohesive movement including prominent figures such as Lujo Brentano, Adolf Held, Erwin Nasse, H. Rösler, Albert Schäffle, Hans von Scheel, Gustav Schönberg, and Adolf Wagner. This was a period of heavy industrialisation, and by linking the empirical study of economic phenomena to political activism, the historical and comparative investigation of the forces of industrialisation could become a basis on which to define the role of the state as mitigating the negative effects of economic progress. Economics was accordingly seen as a normative discipline to be put to use for political ends.

Schmoller's career, in particular, was anathematized by a tension between scholarship and the cause of social reform, given his conviction that ethics gave meaning and direction to research in economics allowing for the integration of the two roots of economy, philosophy and policy, from an historical perspective (Shionoya 1995). Accordingly, his programme fused history, economics, statistics, psychology, institutional analysis, ethics and *Sozialpolitik*. In his *Grundriss*, Schmoller stressed that economic life couldn't be understood apart from the historical development of customs, laws, and morals. But notwithstanding his focus on the institutional framework that conditioned both the macro and micro economic mechanisms, he saw the choice of a policy framework for nation-building as definitively more important than macroeconomic relations within a given framework. Unlike the older historicists, Schmoller was less interested in deterministic developmental schemes than in explaining typical phenomena. In this sense his was a science of the typical rather than the cyclical (Schumpeter [1954] 1994, 236). The role of *Volksgeist*, ideology and social culture played a central role in the categorisation different socio-economic systems, but the main question confronting the Younger School was whether systems were defined through culture or property relations.

Focusing on the socioeconomic forces underlying the development of the German state, Schmoller did research on an unprecedented scale, collecting data on numerous fields including economic policy and administration. Like that of the older historicists, his work was dominated by the accumulation of monographs on historical studies. Absorbed by the collection of materials and the classification of particulars, Schmoller overlooked the need to articulate explicit theoretical principles on which to categorise empirical data and deduce a theory based upon appropriate assumptions. Yet, if Schmoller's *Grundriss* contained a meagre theoretical content, he explicitly stated that the German Historical School didn't reject theory. Theory was not to be neglected, but from an historical perspective, empirical studies came first and the elaboration of theoretical principles was left for a second stage. As a consequence, the *Grundriss* remained an unfinished work, an accumulation of raw materials. Favouring immersion in historical sources, the Younger School dismissed generalisations, criticised the so-called "method of isolation" and endeavoured to study economic phenomena in a cultural and sociological context.

Like his predecessors, Schmoller didn't yet envision economics as a specialized discipline, but as a program linking economics, history, sociology, statistics, and psychology. Schmoller conceived of the forces operating within society, including the acquisitive instinct, rooted in the human psyche. Hence, psychology was the key, *der Schlüssel*, to all social sciences (Schmoller 1900–1904, 107). Introspection was part of this approach, but not its main component. Schmoller believed in a scientifically perfected psychology, dividing the general psychological elements that influenced economics into two categories: direct causes constituted by individual motivational drives and needs; and indirect causes expressed in cultural institutions such as language or law (Schmoller 1900–1904, 109). This project, as Schmoller complained, was nonetheless hampered by the incipient state of empirical psychology, unable to provide the basis for an adequate account of the psychic foundations of peoples and classes.

For Schmoller, statistics was another key analytical tool that he developed in association with Georg Friedrich Knapp, who came to economics from natural science. This connection illustrates Schmoller's belief in the juxtaposition of *Natur-* and *Geisteswissenschaften*. His economics, indeed, shared affinities with the natural sciences inasmuch as it encompassed technology, ethnography, and demography. However, despite his conviction that economics had a foot in the social and natural fields, whose contours couldn't be clearly delineated, Schmoller pointed to affinities between his and Dilthey's conception of the *Geisteswissenschaften*. This slim connection derived from a shared opposition to the positivistic tendency to apply the simplifying methods of the natural sciences to human society. Nonetheless, economic science, according to Schmoller, didn't entirely fall under the rubric of the *Geisteswissenschaften*.

In his desire to contribute to social reform, Schmoller formed the *Verein für Social politik* in 1872–1873, allowing the Younger Generation to gain organisational coherence, stimulating social research in Germany and fostering a methodology that illustrated how the kind of comparative analysis proposed by Roscher could be linked to the social question. The interplay of party politics and scientific research fuelled controversy with profound implications for the view of economic science promoted by the Younger School. The *Verein* was a vehicle for political activism assembling academic economists, policy makers and businessmen, whose objective it was to find a common basis for social reform. They rejected both *laissez-faire* and liberalism, but with time a division emerged between two political factions within the *Verein*. The left wing favoured a gradual modification of the property law in a direction fulfilling socialist aspirations; whereas the majority advocated state reform based on existing juridical institutions. The social reformers were soon accused of defending a system of state socialism, a charge that earned them the appellation of *Kathedersozialisten* (Lindenfeld 1997, 225).

Despite the efforts of the younger generation, Roscher's project of building a historically oriented economics remained unfulfilled. Early treatments of the historical and institutional nature of economic phenomena were linked to a narrow empiricism. But scholars like Knies, Schmoller, and especially Lujo Brentano who developed sophisticated analytical tools, were concerned with finer methodological issues. Connecting all generations was an hostility towards "abstract economics" underpinned by theorems universally valid, and an allegiance to an inductive approach based upon the accumulation of historical data on a large scale which, as Schumpeter outlined, would contribute to the extension of factual knowledge about economic phenomena (Schumpeter [1954] 1994, 810).

The whole historicist project bore other important consequences, such as a substantial contribution to the development of a strong strand of American institutionalist and evolutionary economics. The weight of Schmoller's historical–ethical approach alone is illustrated by two controversies created by two different dimensions of his approach. In the *Methodenstreit* with Menger, and the *Werturteilstreit* with Weber, Schmoller was doubly criticized for banishing theory because of extreme concerns about historical research and ethical evaluation.

By the early 1900s, the Younger School was being replaced by the "youngest" generation including Weber and Sombart, who were open to a theoretical approach and the refinement of analytical tools. They contributed to shape a new kind of historicism. Far from advocating a split between theorizing and practical deliberation, they expressed misgivings about the previous generation's narrow empiricism and excessive focus on ideological questions at the expense of scientific practice.

Concerns about the future of economics led some of the *Verein* members to try to turn it into a scientifically oriented society. During the first decade of the twentieth century, the problem of *Werturteil* caused heated discussions. Vitriolic exchanges culminated in what amounted to a row at the 1909 Vienna meeting of the *Verein*, which otherwise had offered the ideal opportunity for a fruitful exchange with the Austrian School (Schumpeter [1954] 1994, 804f.). Schmoller's ideological-oriented program was sharply criticized in a *Wertfreiheit* campaign led by Weber and Sombart.

## 12.4 The Road Towards Specialization: The Austrian School of Economics

At the end of the day, however, the most consistent opposition to the methodology and policies of the German Historical School of Economics originated in neighbouring Austria, were a dominating German historicist tradition couldn't prevent the emergence of a distinctive Austrian intellectual tradition. The methodological framework of the Austrian School of Economics was articulated by Carl Menger in the *Grundsätze der Volkswirtschaftslehre*. Menger had been immersed in the German economic tradition while studying at the university of Vienna in the 1860s and, despite its novelty, his work was, to a certain extent, steeped in the tradition of the Older German Historical School (Streissler 1990a). Menger dedicated his *Grundsätze* with "respectful esteem" to Roscher, acknowledging profound indebtedness to the "foundation laid by previous work that was produced almost entirely by the industry of German scholars" (Menger [1871] 1981, 49).

Although the *Grundsätze* clearly reflected Menger's wish to attach himself to German economists, in contrast to English and even to his native Austrian economics, it also demonstrates his desire to articulate an approach lying beyond the scope of narrow German historicism. Indeed, since the first paragraph of the first chapter we get hints concerning Menger's intention to provide the universal theoretical framework historicism lacked. Therefore, from the start Menger appears to have seen himself as mending the errors of historicists by articulating the analytic framework they lacked.

Like his German contemporaries, he sought to understand the processes of historical change, but the novelty of his approach consisted in the articulation of causal-genetic explanations of the emergence and evolution of socio-economic phenomena including human action and physical causes. As highlighted by Karen Vaughn, Menger believed that progress implied increased knowledge, better and more diverse products, complexity, and satisfaction of human needs. This could be accomplished if prices reflected knowledge of the causal connection between goods and the satisfaction of needs conjoined with knowledge about available supplies (Vaughn 1994, 27).

To that degree, Menger's approach was very original and contributed decisively to the foundation of modern economics. The *Grundsätze* called for a scientific foundation for economics and upheld the legitimacy of universal laws governing economic behaviour. Whereas the Germans, on the one hand, dismissed a general theory of economic phenomena, and the British classics such as Mill, on the other, lacked a general theory of value that explained the determination of all prices by a uniform principle, Menger articulated a uniform price theory based on the principle of subjective value.

This represented a fundamental shift away from the classical theory of value - which traced economic growth and capital accumulation back to the labour that produced goods. Rather than being conceptualized as simple physical ratios, economic phenomena were now seen as direct or indirect expressions of consumers' needs: human valuations, preferences and expectations (Kirzner 1992, 73). This explanation of value as subjectively goal-directed became the basis for what is known today as the logic of choice. The theory of subjective evaluations conjoined with a "compositive" or individualist approach became the cornerstone of Austrian methodology, setting the guidelines for an approach so familiar to us nowadays. A satisfactory explanation of economic phenomena, as Menger wrote in the preface to the *Grundsätze,* proceeded by reducing complex phenomena to the "simplest elements that can still be subjected to accurate observation, to apply to these elements the measure corresponding to their nature, and constantly adhering to this measure, to investigate the manner in which the more complex economic phenomena evolved from their elements according to definite principles" (Menger [1871] 1981, 46f.).

The purposes and attitudes of individuals became the building blocks from which to set up models of complex economic structures and deduce laws governing markets' functioning. This method was described by Menger as "atomistic" or "compositive" in manuscript notes (Hayek 1992). But he never used the term "methodological individualism" which was coined by Schumpeter to describe Menger's approach. This individualistic procedure focused on rigorous types and typical relations among phenomena, hence representing a departure from historicist empiricism. By connecting methodological individualism to a subjectivist value theory, Menger believed that he could contribute to solve key problems of economic theory. But this link also points to the dichotomy between the fields of economic knowledge and natural science. It would be improper, Menger wrote, to "attempt a natural–scientific orientation of our science", for past efforts to do so "have led to most serious methodological errors, and to idle play with external analogies between the phenomena of economics and those of nature" (Menger [1871] 1981, 46f.).

So, while holding that economic and natural phenomena are both strictly ordered in accordance with laws, Menger called for a methodology setting apart the social and natural sciences. Economics and sciences studying society use a methodology that appeals to our capacity to understand the meanings of human action in a manner in which we cannot understand physical events. Accordingly, the former are conceived of as "subjective" sciences (Hayek 1992). This raises the important issue of how to gain access to agents' subjective preferences. Most significantly, Menger evaded this topic and didn't explicitly advocate introspection, nor the intuitive discernment of the consciousness of others for collecting economic data. While alluding to our capacity to understand the conduct of economic agents he suggested, according to Hayek's interpretation, that we grasp people's intentions through a *Verstehen* procedure similar to that advocated by Max Weber.

Max Alter, on the other hand, reads Menger as an apologist of a psychologistic conception of *Verstehen* given his claim that we have to proceed by reducing economic phenomena to their ultimate constitutive elements or psychological causes (Alter 1990, 102). Given that Menger didn't explicitly allude to *Verstehen,* suggesting instead that he felt the need for sounder psychological grounds for his theory, Alter's assessment is not altogether convincing. Menger, indeed, left this topic in a rudimentary state in part because the *Grundsätze* was intended as an introduction for an extensive treatise. To offer a consistent foundation for his theory of subjective value he immersed himself in the study of philosophy, psychology and ethnography (Hayek 1992). In the meantime, however, he involved himself in an acrimonious time consuming debate with Schmoller and didn't complete his work.

## 12.5 The Methodenstreit and Its Repercussions

In contrast to the *Grundsätze,* Menger's second book, *Untersuchungen über der Sozialwissenschaften und der Politischen* Ökonomie *insbesondere* (Menger 1883) was a forceful critique of what he took to be the misconceived objectives and defective methodology of the Historicist School. Despite his expressions of indebtedness to Roscher, the *Grundsätze* criticized historicism, albeit in conciliatory terms. Menger's intention was to contribute with a theoretical framework for an evolutionist analysis of social processes providing solid methodological foundations for an alternative to the discredited classical orthodoxy. Instead of appreciating his contribution, Menger's book was seen as a dangerous challenge to historicism which, as emphasized by Huerta de Soto (1998), causing him great frustration. The events that followed culminated in the *Methodenstreit.*

According to Israel Kirzner, the cool reception of his work in Germany probably convinced Menger that a strategic alliance was not viable (Kirzner 1992, 58). Menger then set out to vindicate his methodological innovations and the theoretical approach that the historicists abhorred. The dominant theme of the *Untersuchungen* was the idea that the object of social theory is to trace the unintended consequences of individual actions, *unbeabsichtigate Resultante,* following a compositive method. In response to criticism that theoretical economics was ahistorical, Menger, who was particularly interested in the genesis of institutions, appealed to historical examples illustrating the rise and fall of prices. Believing that scientists would benefit from studying the history of the subject-matter of their discipline Menger viewed the historical approach as complementary to, not a substitute for, the development of theoretical principles.

The idea that socio-economic institutions develop as the unplanned product of individual decisions is, as Hutchison underscored, the key in the arch of the architectonic of Menger's thought, linking together his individualism and the historico-genetic 'organic' approach (Hutchison 1973). Evidently, Menger's emphasis on the effectiveness of spontaneous institutions was irreconcilable with the *Sozialpolitik* of Schmoller's generation. Schmoller's support of Bismarck's authoritarian welfare state, and his amalgamation of methodological, ideological and ethical issues, didn't escape Menger's attention. The *Untersuchungen* includes an appendix that briefly, but sharply, criticises the historicist tendency to fuse science with *Sozialpolitik* and personal values.

That Menger was in favour of value-free (*wertfreie*) economics is corroborated by the absence of explicit political statements from his writings. Nonetheless, notes left by Crown Prince Rudolf, to whom he became tutor in 1876, suggest that Menger leaned towards *laissez-faire* practical policy (Streissler 1990b, 107–130). Additionally, his economics were permeated by an intellectual defence of the market system that explains why the Austrian tradition is widely perceived as sharply antagonistic to interventionist policies. Generally, however, Menger and the original Austrian School maintained an apolitical stance (Myrdal 1954, 128). A position that must be interpreted in the light of a context proper to Austria and

Germany, where there were traditionally two chairs of economics in each university: a chair of economic theory and one of economic policy. Early Austrian economists held chairs of theory and therefore were not responsible for teaching economic policy. Menger's work dealt almost exclusively with positive theory, a tendency strengthened by his emphasis on abstract economics.

The corrosive tone of the *Untersuchungen* may equally be interpreted as a defensive one, given the German reception to Menger's first work, but this "direct attack on what was the only approved doctrine attracted immediate attention and provoked, among other hostile reviews a magisterial rebuke from Gustav Schmoller" (Hayek 1992). The exchange of abuses between the two scholars became known as the *Methodenstreit.* Schmoller acknowledged the validity of abstract theory in principle, albeit arguing that economics was not developed enough to engage in rigorous theoretical approach. He dubbed Menger's concepts as schematic phantoms, fantastic Robinsonades (Schmoller [1888] 1968, 283). However, given his ideas, Schmoller put himself in a position difficult to defend. Mises was quite right while commenting that Schmoller's generation assumed that induction was "the specific method of the natural sciences, whereas in fact it was not used by scientists at all, but by antiquarians" (Mises 1976, 71).

The *Methodenstreit* came to an end with Menger's publication of a brochure further emphasising the misconceptions of the Historical School, to which Schmoller did not deign reply. This didn't prevent him from declaring publicly that "members of the 'abstract' school were unfit to fill a teaching position in a German university" (Hayek 1992). Although the formal exchange between Menger and Schmoller ended there, the repercussions of the *Methodenstreit* extended far beyond this episode creating resentment on both camps and a stream of literature that took decades to subside (Schumpeter [1954] 1994, 814).

According to Streissler, whose version of the debate is popular amongst historians of economics, the change of tone between Menger's *Grundsätze* and the *Untersuchungen* reflects a shift of attitude towards the German School correlated to a generational gap. Accordingly, Menger's quarrel was with Schmoller, who was more aggressive and exclusivist in manner than in substance, rather than with the Older School led by Roscher (Streissler 1990b). Be that as it may, the debate had a huge impact on the teaching of political economy in Germany and Austria and significant methodological repercussions on the discipline. Despite a marked nationalistic tone, to achieve a better understanding of the crisscrossing lines of this debate it is essential to realize that these economists weren't simply divided along national lines. It remains that, as a consequence of the debate, German universities came to be increasingly filled by members of the Historical School and Austrian universities by members of Menger's School (Hayek 1992).

Yet, during the *Methodenstreit* members of the German School, such as Adolf Wagner and Heinrich Dietzel, for instance, seized the opportunity to proclaim their antihistorical position openly criticizing Schmoller. On the Austrian side there was equally dissent from Menger's standpoint, a situation that was slowly reverted after 1884 largely as a consequence of publications by Menger's followers that made the School known worldwide and prized amongst Austrian academic circles.

Many economists agree with Schumpeter ([1954] 1994, 814) assessment of the *Methodenstreit* as having been a waste of energies that could have been put to better use to highlight the particular difficulties facing the discipline. The historical and theoretical methods on which the debate turned concerning different problems, interests, and goals made it useless to quarrel about the relative importance of the methods. As Shionoya judiciously outlined, the real issue was over the scope of economic science, the difference in method merely reflecting different conceptualizations of its subject matter (Shionoya 1995, 163–165).

## 12.6 Economics as a Purely Formal Science

The subjectivist approach introduced by Menger was to play itself out in the work of the next generations becoming the distinctive trait of Austrian economics. It was up to Mises to refine that framework by transforming Menger's insights into a detailed methodological framework. Given Menger's difficulties in elucidating how to grasp the intentions of socio-economic agents, in the 1920s the debate over the psychological foundations of economics reached a climax within the community of Austrian economists. Wieser, one of Menger's disciples, believed that the economist's source of data was his own consciousness. He therefore endorsed introspection, but was faced with difficulties when postulating the *a priori* equation of actual and optimal behaviour. This implied that agents chose what they preferred and preferred what they chose (Giocoli 2003, 80). This tautological argument became an easy target for the Historical School who argued that the Austrians lacked a psychological foundation for their subjectivist theory of value. Wieser's project lost ground and the majority of Austrian economists allied themselves with Mises, who tried to establish value theory as a pure logic of choice, moving economics further into the direction of axiomatisation.

Mises began by setting economic theory into a general theory of human action that he termed "praxeology". This approach was based upon the assumption that goal-directed individual behaviour constrained by available means is the subject matter of social science. The principles underlying praxeology were scientifically warranted by their *a priori* character. While a new experience can force us to discard or modify inferences we have drawn from previous experiences, no kind of experience can ever force us to discard or modify praxiology's *a priory* theorems (Mises 1976, 27). The emphasis placed by Mises on apriorism was a distinctive feature of his approach differentiating his work from that of Menger.

Economics, was seen as the best developed branch of the science of human action, starting from a few universal statements derived from reason and centring on the principle that every action aims at replacing a more desired state for a less desired one. Legitimate economic theory was deduced from this *a priori* premise. Knowledge that humans act purposefully, Mises thought, stemmed from our self-knowledge for "we grasp and conceive rational behaviour" by means of the immutable logical structure of our reason", for the "*a priori* of reasoning is at the same time the *a priori* of rational action" (Mises 1976, 134). Some commentators have equated this approach with *Verstehen,* particularly Lavoie (1991). Caution is imperative here, however, for Mises explicitly stated that comprehension of human action by *Verstehen* was irreconcilable with universal theory, and a point missed by most historians of economics is the crucial distinction that Mises drew between "understanding" and "conception". As Koppl rightly emphasized, despite its importance this distinction has been given little attention (Koppl 2002, 30). While conceding that *Verstehen* supplied helpful information, Mises definitively believed that it could not be a tool for theory, and distinguished it from what he termed "conception", reasoning according to logic rules. Conception, according to Mises, is the domain of "strict logical rules: one is able to prove and disprove" (Mises 1976, 133f.) taking "precedence over understanding in every respect", for that "which results from discursive reasoning can never be refuted or even affected by intuitive comprehension of a context of meaning" (Mises 1976, 133). Mises would probably have agreed that *Verstehen* plays a role in the context of discovery, but not of justification.

Mises dissatisfaction with *Verstehen* was also a consequence of the link that he thought existed between that approach and historicism. That is, with the kind of relativism that was at the centre of *Methodenstreit* and against which he had set out to articulate a new science of praxeology in the first place. For Mises, *Verstehen* conveyed an imprecise type of knowledge and suffered from the same insufficiency that rendered artistic, metaphysical, or mystical approaches inappropriate as analytical tools for assessing rational behaviour (Mises 1976, 134). So, despite his fierce defence of methodological dualism, for Mises, the criterion of demarcation between the social and natural fields couldn't be shaped along the lines of the classical pairs of *erklären/verstehen* or nomothetic versus ideographic methods. While bringing to the forefront themes that originated in the *Methodenstreit,* Mises's efforts were directed towards the refinement of a theoretical science, having in view the attainment of universally valid laws of human action. Accordingly, he viewed the sciences of human action as nomothetic. That is, nomothetic sciences of a particular kind, as Mises makes plain enough when remarking that it would be misleading to assume that "natural science and nomothetic science are identical concepts" (Mises 1976, 118f.).

The founding Austrians meant to demonstrate that sciences explaining socio-economic phenomena are as valid as natural sciences. Their originality consisted in setting forth deductive causal explanations placing economic phenomena under a universal pattern, while dismissing the transfer of the methods of the natural sciences to the social sphere in order to make it more scientific. Herein lies the originality of the Austrian approach. For the Austrians the nomothetic approach was not the dividing line between natural and social science. This is a crucial point. Austrians conceived of economics as a nomothetic science, but not as a natural science. Their endeavour was precisely to demonstrate that a nomothetic approach is not synonymous with a natural science orientation. The economists of the Austrian School claimed that a nomothetic approach was not the hallmark of natural science, that natural science did not hold a monopoly over a nomothetic approach.

It is essential to realise that, for the Austrians, economics was entitled to use a nomothetic procedure but that such a methodological approach did not place economics in the realm of natural science. Economics, according to the Austrians was a social science that used nomothetic methods, for what made of economics a social, rather than a natural science, was its subjectivist orientation. The hallmark of economics, what set it apart from the natural sciences was its subjectivism.

According to the Austrian School of Economics, economics was a social science that used nomothetic methods, yet economics did not belong to the field of natural science for it was a subjectivist science (Cubeddu 1997). All the originality of the Austrian contribution to the debates asserting the specific scientific status of social science, all particularly economics, was that subjectivism was the criterion of demarcation from the natural sciences, not ideography. Following Menger, and adopting a position that epitomized the Austrian approach, Mises valued individualistic explanations stressing that "we must start from the action of the individual.... Everything social must in some way be recognizable in the action of the individual" (Mises 1976, 43). For Mises, as for all the members of the Austrian School, economics focused on individual actions and preferences, it was intrinsically an individualistic science. Individualism and subjectivism were the hallmarks of economics, but economics was not an idiographic science, rather a nomothetic one, but a special nomothetic science distinct from the natural sciences as the quotations inserted above demonstrate.

## 12.7 Historicity and the Theoretical Sciences of Society: The Unbridgeable Gap

Mises denounced the relativism associated with historicism as an expression of the romantic revolt against logic and science arguing that "there is a body of economic theorems that are valid for all human action irrespective of time and place, the national and racial characteristics of the actors, and their religious, philosophical, and ethical ideologies" (Mises [1969] 1984, 38). Extending Menger's campaign against historicism, he made impossible an intellectual rapprochement with the Youngest German Historical School, namely, with Weber. The relationship between Mises and Weber was an interesting one. They developed personal ties during the semester of 1918, when Weber taught in Vienna. Furthermore, Weber was keen on promoting conceptual clarity in German economics and was the editor of a handbook of economics, *Grundriss der Sozialokönomik*, a collaborative project with the Austrians, including Wieser and Schumpeter. There were considerable affinities between Weber and the Austrians and he devoted great attention to Menger's *Untersuchungen* whose *Theorie des subjectiven Wertes* became, in his terminology, *subjective Wertlehere* (Cubeddu 1997, 6).

Weber was in agreement with Menger and Mises, endorsing the constructive aspect of social analytical concepts holding that because human social action is purposive and rational the explanations sketched by social scientists must be related to the values and ideals of socio-economic agents. He articulated a sophisticated set of analytical categories, centering on individualist ideal types of action and stressing the understanding of the meanings that each set of alternatives had for the social agent. Weber's ideal type was precisely the leading form of subjectivism at the time Mises was articulating his position (Koppl 2002, 27). Although for Weber sociology and history were not identical, both made use of the same analytical tool, the ideal type. By forging this bond, Mises thought that Weber's ideal type was an historical construct, rather than a tool for theoretical investigation (Mises 1976, 76–78). He was convinced that, like he other members of the Historical School, Weber had failed to distinguish between the historical and the nomothetical sciences of human action. Mises acknowledged that Weber was one of the most brilliant figures of German science at the time, but had remained bound to a view of economics and sociology as historical sciences so that, at best, he merely used more sophisticated conceptual tools (Mises 1976, 74).

Apart from Hayek, who often expressed sympathy for Weber's views, other members of the Austrian school seem to have concurred with Mises's reservations, which may have been motivated, at least in part, by ideological considerations as well. Much of Mises's work was devoted to investigating the political and ethical consequences of historicism. As a staunch opponent of social planning, Mises was persuaded that the historicist's attack against theoretical economics, particularly Schmoller's, concealed political motivations being "directed toward protecting from disagreeable criticism economic policies that could not withstand scientific examination" (Mises 1976, 107). Weber, in Mises's view, remained closely associated with historicism, realising the extent to which the Schmollerian onslaught on theoretical economics was politically motivated only in his later period (Cubeddu 1997). Mises probably erred here, for Weber had vigorously denounced the state metaphysics of Schmoller. Weber, like Mises, was an adept of capitalism, but despite his ideological allegiances launched a strenuous campaign in favour of *Wertfreiheit* becoming an iconic advocate of a value free social science. While acknowledging the right for scientists to express personal values, Weber deemed it improper to transform the former into the object of discussion of a specialised discipline such as economics (Lindenfeld 1997, 317).

Mises also argued for *Wertfreiheit,* but his epistemological arguments openly promoted the politics of *laissez-faire*. In his intransigent defence of liberalism, Mises, as Hayek commented, was isolated amongst German-speaking economists of his generation, involving himself in a running controversy not only with members of the German School, but also with a Viennese group of Marxist intellectuals some of whose had been his schoolmates and who, through Neurath, were influential on the group of neo-positivists of the nascent *Vienna Circle* (Kirzner 1992, 27–29). As a consequence, Austrians were characterized by their contemporaries as scholars engaged not only in vindicating an abstract science of economics against historicist challenges, but radical liberal policies as well (Kirzner 1992, 88).

Until dispersing in the 1930s, the Austrian School, under the leadership of Mises, remained the center of theoretical work in the German-speaking world, continuing to refine its methodological position that contributed to the development of the discipline. As Koppl remarked, "Mises wanted both science and subjectivism. He attempted to construct a scientific subjectivism" (Koppl 2002, 29). Subjectivism conjoined with methodological individualism became the heart of a specific Austrian tradition favouring deductive methods and conceptualizing economics as a science close to logic. Despite attempts by some authors (e.g., Lavoie 1991) to equate Austrian subjectivism with *Verstehen,* it still remains a topic open to discussion, Austrian economists having eluded the subject or dismissed this interpretation of the approach they defended.

The link between methodological individualism and liberalism is a hallmark of Austrianism that became more evident as its members entered the third generation, engaging in a struggle against the feasibility of socialism. Likewise, the ideological divergence between Germans and Austrians found later echoes in the campaign that Hayek and Popper launched against historicism and holism. In short, each school left a deep impact on the social sciences as both camps grappled with the problem of a criterion of demarcation distinguishing the social sciences from other forms of scientific enquiry. The *Methodenstreit* was part of a the broader story for the clarification of a standard of scientificity peculiar to the emerging social sciences and the proper methods for producing economic knowledge. Holding to history as a criterion of demarcation, the German Historical School proposed a sociological and reformist oriented economics. Viewing economics as part of a comprehensive theory of society they articulated a holistic line of analysis, which derived from their reliance on the notion of Volksleben., conjoined with an inductivist and an empiricist outlook. Raising awareness to the role of the institutional context in shaping socio-economic phenomena was the School's greatest achievement. As Shionoya reminded us , those social scientists and economists that have set out to redefine and create an updated agenda for institutionalism tend to underestimate, or simply ignore, the role of the members of the German School as their forerunners (Shionoya 1995, 78).

## 12.8 Concluding Remarks

The goal of this essay was to offer an historical analysis of the events leading up to the *Methodenstreit* between members of the German and the Austrian schools of economics in order to explore the ways in which various topics around the Erklären/Verstehen dichotomy were at stake throughout the nineteenth century.

We learn more about this legacy by exploring both the points of contact and the different approaches adopted by each School to deal with the challenges that the emerging social sciences found themselves confronted with. Rather than focus on the rhetoric surrounding the *Methodenstreit,* it is far more interesting to set in evidence its scientific achievement that consisted in bringing to the forefront not only the problem of a criterion of demarcation between the natural and the human sciences, but issues addressing the specific methodologies of the different sciences dealing with human action, and the challenge of shaping their particular boundaries. Scholars agree that economics emerged as a distinct discipline in the nineteenth century, but the broader picture, the web of historical and scientific episodes leading to this development, of which the *Methodenstreit* is piece and part, is often left untold.

Scientific knowledge is produced through conjectures and refutations a process in which controversy is sometimes more fruitful than consensus. If the *Methodenstreit* apparently ended without a resolution, it doesn't curtail its importance in the history of the social sciences. Furthermore, a closer look at the debate reveals that the issues lurking behind it are perennial: does social science explain differently from other sciences and what are the principles it uses to produce valid scientific knowledge?

## References


Alter M (1990) Carl Menger and the Origins of Austrian Economics. Westview Press, Boulder, CO

Cubeddu R (1997) The critique of Max Weber in Mises's Privatseminar, Cahiers d'épistémologie 9703

Giocoli N (2003) Modeling Rational Agents. From Interwar Economics to Early Modern Game Theory. Edward Elgar, Cheltenham

Hayek FA (1992) The fortunes of liberalism. Essays on Austrian economics and the ideal of freedom. In: Klein PG (ed.) The Collected Works of F. A. Hayek. The University of Chicago Press, Chicago, IL

Hildebrand B (1848) Nationalökonomie der gegenwart und zukunft. Literarische Anstalt, Frankfurt am Main

Hodgson GM (2001) How Economics Forgot History: The Problem of Historical Specificity in Social Science. Routledge, London

Huerta de Soto J (1998) The ongoing Methodenstreit of the Austrian school. Journal des Economistes et des Etudes Humaines 8(1):75–113

Hutchison TW (1973) Some themes from investigations into method. In: Hicks JR, Weber W (eds) Carl Menger and the Austrian School of Economics. Clarendon, Oxford, pp 15–37

Kirzner I (1992) The Meaning of the Market Process. Routledge, London

Knies (1883) Die politische ökonomie vom standpunkt der geschichtlichen Methode. Braunschweig C. A. Schwetschke und Sohn

Koppl R (2002) Big Players and the Economic Theory of Expectations. Palgrave, New York

Koslowski P (1995) Economics as ethical economy in the tradition of the historical school. Introduction. In: Koslowski P (ed.) The Theory of Ethical Economy and the Historical School. Springer, Berlin, pp 1–11

Lavoie D (1991) Hermeneutics and Economics. Routledge, London

Lindenfeld DF (1997) The Practical Imagination. The German Sciences of State in the Nineteenth Century. University of Chicago Press, Chicago, IL

Menger C ([1871] 1981) Principles of economics. Originally published as Grundsätze der volkswirtschaftslehre, Baum, Wien

Menger C (1883) Untersuchungen über die methode der sozialwissenschaften und der politischen ökonomie insbesondere. Duncker & Humblot, Leipzig

Mill JS (1848) The Principles of Political Economy. W. J. Ashley, London

Mises L, von ([1969] 1984) The Historical Setting of the Austrian School of Economics. Arlington House, New York

Mises L, von (1976) Epistemological Problems of Economics. New York University Press, New York

Myrdal G (1954) The Political Element in the Development of Economic Theory. Harvard University Press, Cambridge, MA

Pearson H (1999) Was there really a German historical school of economics? Hist Political Econ 31(3):547–562

Priddat B (1995) Intention and failure of W. Roscher's historical method of economics. In: Koslowski P (ed.) The Theory of Ethical Economy and the Historical School. Springer, Berlin, pp 15–34
Roscher W (1843) Grundriss zu Vorlesungen über die Staatswirtschaft nach geschichtlicher Methode. Dieterische Buchhandlung, Cotta Göttingen
Roscher W (1858), Die grundlagen der national ökonomie. Ein hand- und lesebuch für geschäftsmänner und studierende. Cotta, Stuttgart
Scaff LA (1995) Historicism in German social and economic thought. In: Koslowski P (ed.) The Theory of Ethical Economy and the Historical School. Springer, Berlin, pp 313–331
Schmoller G, von ([1888] 1968) Zur literaturgeschichte der staats-und sozialwissenschaften. Burt Franklin, New York
Schmoller G, von (1900–1904) Grundriss der allgemeinen volkswirtschaftslehre. 2 vols. Duncker & Humblot, Leipzig
Schumpeter J ([1954] 1994) History of Economic Analysis. Oxford University Press, Oxford
Shionoya Y (1995) An appraisal of Schmoller's research program. In: Koslowski P (ed.) The Theory of Ethical Economy and the Historical School. Springer, Berlin, pp 57–78
Streissler E (1990a) The influence of German economics on the work of Menger and Marshall. In: Caldwell BJ (ed.) Carl Menger and His Legacy in Economics. Duke University Press, Durham, pp 31–68
Streissler E (1990b) Carl Menger on economic policy: the lectures to Crown Prince Rudolf. In: Caldwell BJ (ed.) Carl Menger and His Legacy in Economics. Duke University Press, Durham, pp 107–130
Tribe K (1988) The Reformation of German Economic Discourse 1750–1840. Cambridge University Press, Cambridge
Tribe K (2002) Historical schools of economics: German and English, Department of economics working paper series, Keele University, 2002–2002
Vaughn KI (1994) Austrian Economics in America. The Migration of a Tradition. Cambridge University Press, Cambridge
Weber M (1903–1906) Roscher und Knies und die logischen Aufsätze. In: Gesammelte aufsätze zur wissenschaftslehre. J.C.B. Mohr, Tübingen

# Chapter 13
# From Mill via von Kries to Max Weber: Causality, Explanation, and Understanding

**Michael Heidelberger**

In the second part of his "Critical studies in the logic of the cultural sciences" published in 1906, which carries the title "Objective possibility and adequate causation in historical explanation" (Weber 1906, 164-188/266-290)[1] Max Weber (1864–1920) wrote that he feels "almost embarrassed in view of the extent to which here again, as in so much of the preceding argument, I am 'plundering' von Kries' ideas" (Weber 1906, 186/288).[2] Weber thus admits a very strong influence on his approach by the physiologist, philosopher, and theoretician of probability, von Kries (1853–1928), who was for sometime his colleague in Freiburg in southwest Germany. Von Kries had suggested a legal criterion for attributing a deed to an agent that exerted a strong influence on German civil law and was also taken up by the legal system of other countries. This earned him the title of an honorary doctor of the law faculty of the University of Erlangen in 1897.

In the following text, I want to read and interpret Weber as much as possible from the perspective of von Kries. In this way I will show that the methods of the natural and the social or the historical sciences were for Weber much more similar to each other than is widely assumed. Thus, his conception of "interpretative understanding" or *Verstehen* is much more distant from the tradition of Dilthey, Windelband, and Rickert and much closer to the usual conception of "explanation" in the natural sciences than commonly believed. At the time, however, there persisted a deep misunderstanding about the nature of a statistical law that might also have influenced proponents on both sides of the controversy.

Although there have been a few studies of the bearing of von Kries on Weber (Ringer 1997, Ch. 3; Ringer 2002; Ringer 2004, Ch. 3; Nollmann 2006; Eberle 1999,

M. Heidelberger (✉)
Universität Tübingen, Philosophisches Seminar, Bursagasse 1, 72070 Tübingen, Germany
e-mail: michael.heidelberger@uni-tuebingen.de

[1] The numbers after the slash refer to the pages of the German original.

[2] The translators have obviously misunderstood the German original here: "Der Umfang, in welchem hier wieder, wie schon in vielen vorstehenden Ausführungen v. Kries' Gedanken 'geplündert' werden, ist mir fast genant." The last word is taken from French as it was used colloquially among educated Germans at the time: "gênant", meaning "embarrassing!" Shils and Finch translate: "I scarcely mention the extent to which here again…," thus belittling Weber's debt to von Kries!

U. Feest (ed.), *Historical Perspectives on Erklären and Verstehen*, Archimedes, Vol. 21,
DOI 10.1007/978-90-481-3540-0_13, © Springer Science+Business Media B.V. 2010

99–105; Wagner and Zipprian 1986; Turner 1983), the topic seems to have been relatively neglected. The studies mentioned are, in my opinion, either thwarted by bringing in alien viewpoints from the present, or they handle the work of von Kries in too cursory a fashion. My chapter has seven parts: I first set out the claim that for Weber explanations in the natural and the cultural sciences do not differ very much after all, contrary to what is commonly thought, even if the aims are different.

In the second part, I outline how John Stuart Mill viewed causal attribution and how this was taken up by legal practice in Germany from the 1870s onward. The third part deals with von Kries' criticism both of Mill and of the legal "equivalence theory" and shows how he wanted to reform the situation by introducing criteria for an "adequate cause" of an action that allows unequivocal objective imputation of a consequence to an agent. The fourth part then explicates the way in which Max Weber tried to transfer von Kries' criterion from legal science into history and sociology. In the fifth part, two ways of explanation that are used in the social sciences are discussed. Besides nomological explanation, which is standard in science, there also exists ontological explanation, which is typical of disciplines that deal with mass phenomena. In the sixth part, the reasons why Weber insisted on interpretative understanding of causal adequacy in history and social science are considered. The final part spells out the consequences and importance of Weber's 'Kriesian' approach for the contemporary theoretical discussion about the nature and status of social science. It is argued that Weber's theory was a kind of mediating position among the approaches of von Kries, Lexis, von Bortkiewicz, and Tschuprow who moved probability into prominence and the attitude of Knapp, von Oettingen, and others who denied the value of statistics and called for a social science that allowed for analyzing developments in history or society as lying within the responsibility of individuals.

## 13.1 The Different Aims of the Natural and Social Sciences

Weber writes that the "the historian's problem of causality also is [always] oriented towards the imputation of concrete effects to concrete causes, and not towards the fathoming of abstract 'uniformities' (*Gesetzlichkeiten*)" (Weber 1906, 168/270; translation amended). At another place, however, he declares that if

> "the causal knowledge of the historians consists of the *imputation* of concrete effects to concrete causes, a *valid* imputation of any individual effect without the application of 'nomological' knowledge – i.e., the knowledge of recurrent causal sequences – would in general be *impossible*" (Weber 1904, 79/179).

At first glance, this seems to be a blatant contradiction. On the one hand, Weber affirms that the historian does not need any laws in order to do his work; on the other hand he says (as if in the same breath) that historians must rely on causal regularities or nomological knowledge. On both occasions, Weber uses the term "imputation" or *Zurechnung*, which very often appears in his work as "causal imputation." To answer the question regarding the role that laws really play for Weber in historical and sociological explanations, we have to comprehend what he means by "causal imputation."

'To impute an effect to someone' means 'to make a person liable or responsible for some happening or "consequence" (*Erfolg*),' as the jurists say. The lawful (physical) connections, the "recurrent causal sequences" that exist between the act of the person and the outcome are normally taken for granted or even regarded as trivial.

"Imputation" is a legal term (introduced into jurisprudence by Samuel Pufendorf in 1660) that has been understood in the double sense of *imputatio facti* and *imputatio iuris* or *legis* since at least the eighteenth century. Factual or causal imputation concerns the question whether someone has caused a certain event, and legal or normative imputation deals with the further problem of whether a person having caused an event can be held responsible for it in the light of some legal norm. Therefore, the juridical imputation *presupposes* causation and is applied only after additional criteria are fulfilled. Normally, the outcome of an action is physically caused by the agent except if he or she acted by omission or induced someone else to act or the like. It is the task of a criminal or civil court to decide who is legally liable for some (mostly harmful) consequence. So, e.g., a carriage driver might be imputed for reckless driving that caused harm to someone. The concept of imputation can also be figuratively applied to impersonal circumstances: a rusty bolt was responsible for the carriage falling apart and the latter effect might thus be imputed to the bolt.

This consideration, I think, can resolve the contradiction: Weber says that the historian's aim is not to *establish* a (historical) law or to *explain* an event by a law or laws, but to *use* nomological knowledge that is already available to *causally attribute* an event, fact, or circumstance to a person or a set of conditions. But then the question arises as to what is the difference between an explanation given by a scientist who gets to the bottom of a causal law and an explanation given by a historian. The first kind of explanation asks for the law that was causally effective in the happening, whereas the second kind of explanation looks for the constellation or event or action to which another event or action or set of circumstances can be imputed on the basis of presupposed nomological knowledge. In both cases, the explanation is given by completing fragmentary information: depending on the context, either the general law establishing the connection between a cause and a consequence is looked for, or the action of an agent that led to a concrete outcome whose operative causal history is already known is singled out.

Even if explanation of outcomes by imputation dominates history and the social sciences in general, and even if the explanation of outcomes by natural law is the central concern for natural science, the two fields are not fundamentally different: History cannot do without alluding to recurrent causal sequences even though it has the tendency to take them for granted, and the natural sciences cannot do without concrete individual initial conditions when explaining an outcome although they are inclined to make them "disappear" behind general laws.

Following von Kries (1888, 195 f.), Weber describes the difference between the two kinds of explanation – or better, between the two manifestations of explanation – also with the pair of opposing concepts as either "concrete" or "abstract." The historian searches the concrete causes for singular concrete consequences while keeping as close as possible to everyday causal knowledge. The scientist, however, wants to identify the general law that governs the connection of cause and effect in order to explain an event in a fashion that is as abstract as possible, i.e., disregarding the concrete character of the effect and instead taking it as a general case.

According to Weber, both ways of explaining have their origin in the same scheme and are therefore not necessarily restricted to one class of sciences. The difference between the natural and cultural sciences cannot be grounded in a fundamentally different method, but only in the different goals that are followed in explanation. At one point, Weber writes that it is not, as Dilthey thought, the difference in ontology that demands a difference in method but the difference in "logical peculiarities of the objective of knowledge [*Erkenntnisziel*]" (Weber 1903/1906), 12 fn.).[3] It would be possible to concentrate on the concrete singular case also in the natural sciences or to ask for the general laws in the cultural sciences, but this change of perspective would not carry high epistemic value (ibid., 126).

The different estimation of laws and concrete initial conditions in the natural and cultural sciences thus results from their different objectives: In the cultural sciences, the establishment of causal laws "is not the *end* but rather the *means* of knowledge. It is entirely a question of expediency, to be settled for each individual case, whether a regularly recurrent causal relationship of everyday experience should be formulated into a 'law'" (Weber 1904, 80/179). Conversely, in the natural sciences, the individual case is a means to a general knowledge of laws, but its concrete registration is generally not the aim. Whereas the cultural sciences regard the general as the fortuitous and try to neglect or eliminate it as much as possible (Weber 1903/1906), 5), the natural sciences deem the historical and individual as incidental, which is to be eradicated wherever it is possible (ibid., 8; Weber 1904, 85/185; Weber 1906, 117 f./219 f.).

## 13.2 Mill on Causality and German Legal Practice

How does factual or causal imputation actually work in history and the cultural sciences? In order to answer this question, Weber went back to von Kries who developed a new theory that resulted from a close analysis of Mill's work. In his *System of Logic* of 1843 (first German translation in 1849), John Stuart Mill (1806–1873) expanded the regularity view of causality as conceived by David Hume, and analyzed it as the set of all necessary conditions that are jointly sufficient for the effect. It is true, he maintained, that an event is caused by another one if an "invariability of succession" can be observed between the two. Yet it is rarely, if ever, "between a consequent and a single antecedent, that this invariable sequence subsists. It is usually between a consequent and the sum of several antecedents" (Mill 1843, III, v. 3). This means that we have to take the whole complexity of a situation into account. "For every event," Mill wrote, "there exists some combination of objects or events, some given concur-

[3] Quotes in English not taken from published translations and not indicated by a double reference to page(s) are by Michael Heidelberger.

rence of circumstances, positive and negative, the occurrence of which is always followed by that phenomenon" (III, v. 2).

Mill's example is a person who eats a particular dish and subsequently dies because it was poisoned.

> "There needs not, however, be any invariable connection between eating the dish and death." To find such a connection, the act of eating the dish must be "combined with a particular bodily constitution, a particular state of present health, and perhaps even a certain state of the atmosphere; the whole of which circumstances perhaps constituted in this particular case the *conditions* of the phenomenon" (III, v. 3).

In other words, when asking for the cause of the death we have to take into account the whole set of conditions on which this particular death depended. "The cause, then, philosophically speaking, is the sum total of the conditions, positive and negative taken together; the whole of the contingencies of every description, which being realized, the consequent invariably follows" (ibid.).

Mill's proposal can be formulated in the following way (where $C$ is a conjunction of conditions and $E$ a fact):

$C$ is the cause of an effect $E$ in Mill's sense if and only if

(1) $C$ and $E$ are actual.
(2) $E$ is later than $C$.
(3) $C$ is the set of all conditions $c1, \ldots, cn$ (for some positive integer $n$), which together are sufficient for $E$ (i.e., it is always the case that: if $C$, then $E$).
(4) Each $c_i \in C$ is necessary for $E$ (for all $i \in 1, \ldots, n$) (i.e., if not-$ci$, then not-$E$)

(This definition assumes that an effect has only a finite set of conditions and that a "negative condition" for which Mill allows can be stated as a positive one. The question of causal asymmetry between $C$ and $E$ is also neglected.)

If "cause" is defined in this way, Mill gets into serious trouble with everyday intuitions – a difficulty he describes as follows: "It is very common to single out one only of the antecedents under the denomination of Cause; calling the others merely Conditions" (III, v. 3). In the example with the poisoned dish, "people would be apt to say that eating of that dish was the cause of his death" and they would drop or forget all the other conditions. The everyday usage of "cause" thus regularly leads to a conflict with the philosophical and scientific use.

Mill considers several reasons why we are inclined in everyday life to prefer a single condition over the rest: the more immediate connection of one of the conditions with the effect could play a role, it being the only instantaneous change on the background of stable continuous conditions, or it having a stronger connection with the effect than the other conditions. Yet, he eventually comes to reject all this because he realized that, with a little imagination, each of the conditions could be made to play the role of the salient condition or "real" cause in the everyday sense.

Mill concludes that it is just a matter of rhetoric or subjective perspective and preference which of the many conditions onc chooses over the others as *the* cause of an outcome. It really depends on "the purpose of our immediate discourse" and is not a matter of objective fact: "[I]n practice, that particular condition is usually styled the cause, whose share in the matter is superficially the most conspicuous, or whose requisiteness to the production of the effect we happen to be insisting on at the moment." Yet it would be erroneous to suppose, as it is done in everyday life, that one of the conditions has a "closer relation to the effect than any of the other conditions has." In strict philo-sophical and scientific discourse we must, according to Mill, stick to the view that "the real cause is the whole of these antecedents, and we have, philosophically speaking, no right to give the name of cause to one of them, exclusively of the others" (III, v. 3).[4]

In addition, he reaffirms that for the scientific view "all the conditions were equally indispensable to the production of the consequent; and the statement of the cause is incomplete, unless in some shape or other we introduce them all" (III, v. 3).

For Mill, however, this was not yet the end of the matter. He realized that the set of causes one reaches if one follows considerations like these would become unmanageable and confusing. The analysis of large systems like the society or the economy as a whole would have to deal with too many conditions at once. There is also the threat of a potential infinity of the antecedent conditions. He therefore introduced the concept of "ultimate laws" that act behind the "observed" and "derivative uniformities." Each causal condition of a mere regularity, like the one between the eating of the poisoned dish and the death of the eater, is conceived to be itself the result of a composition of ultimate causes. It is these causes that Mill wanted to discover and investigate with his famous inductive methods of experimental inquiry. Much of the *System of Logic* is committed to the problem of analyzing the "assemblage of conditions" or "intermixture of effects" in terms of "ultimate laws" (esp. III, x–xviii).

In the 1870s, Mill's theory of causality became a point of departure for many German legal authors who dealt with causation in the law (compare von Bar 1871, 6–8; Mayer 1899, 20–22; Traeger 1904, 17; Hart and Honoré 1959, 391 ff.).[5] This was a time when the legal system of the just united Germany had to be reorganized and brought into harmony with the former local systems. Mill's treatment provided the legal experts with the most authoritative and scientifically respectable notion of causality available to date. It was commonly referred to as the "philosophical theory." In addition, its empirical nature and high flexibility made it easily applicable to cases of everyday life.

Hart and Honoré mention several respects in which Mill's theory was closer to life and thus more attractive for legal practice than available alternatives, e.g., Hume's or Kant's: Mill's conditions include not only events as points in time but also persistent states, thus allowing for structural conditions to contribute to an

---

[4] Compare also "If we do not, when aiming at accuracy, enumerate all the conditions, it is only because some of them will in most cases be understood without being expressed, or because for the purpose in view they may without detriment be overlooked. For example, when we say, the cause of a man's death was that his foot slipped in climbing a ladder, we omit as a thing unnecessary to be stated the circumstance of his weight, though quite as indispensable a condition of the effect which took place" (Mill 1843, III, v. 3).

[5] Sometimes the philosophers Johann Friedrich Herbart and Hermann Lotze, but also Thomas Hobbes, are regarded as independent forerunners of Mill..

effect. In addition, Mill's theory does not only take account of positive happenings but also of negative conditions as possible causes. This makes room for dealing with causation by prevention or omission, and assimilates negative conditions to positive acts. By identifying the antecedents with a "concurrence" of many variable conditions and not just a single one that is nomologically related to the effect, Mill also recognizes the complexity of real-life situations. And finally, if one admits that there are causes that can counteract other causes, as Mill does, a condition does not have to be necessary for its effect, but only sufficient. This opens up the possibility for a "plurality of causes," as Mill called it: there may be several independent causes for a certain type of event after all (see esp. III, x). In short, the fact that Mill's theory can easily be turned from a theory of causal conditions into a theory of attributing causality to agents – that is, into a theory of causal imputation – proved to be of great advantage in German legal contexts.

For all the advantages Mill's theory then presents to legal practice, the problem of the divergence of the common usage of "cause" from the scientific one has to be dealt with also in the courtroom: Which of the conditions is to be regarded as decisive for attributing responsibility for an event to someone? If the cause of an event is a whole set of conditions including not only a particular human act but also a whole lot of other circumstances, how can an agent ever be held responsible for his or her deed? How can we say that Mr. X caused the death of Mr. Y because he mixed poison into Y's dish? True, he might have had his share in the "whole of the contingencies" – but could not every other condition that was present be taken as *the* cause of death as well? Following Mill, we might as well blame Mr. Y himself for his own death because if he had not opened the door to let Mr. X in, his dish would not have been poisoned and he would not have died. If we blame Mr. X for having committed the crime, however, are we not just following a subjective or rhetorical, in short: an unscientific and irrational inclination to prefer one factor over another? There seems to be no general method of reasonably narrowing down causality and attributing it to an agent in an impartial and objective way. The vagaries of human action as they become subject to legal considerations seem to require, as much as the strict philosophical or scientific outlook, taking the complete set of necessary conditions that are together sufficient for an event as its cause.

One of the first authors, perhaps even the first one, to have posed Mill's "problem of the salient condition" (as one could call it) in this way was Ludwig von Bar who proposed an answer to the problem that already lies in the direction of von Kries' later solution: An agent who produces a condition for an effect is a cause of it in the legal sense if he or she thereby changes the regular and habitual course of events (von Bar 1871, 11). In 1873, however, Judge Maximilian von Buri bit the bullet and maintained that *each* single necessary condition of an event is cause of the effect, and therefore to be regarded as "equivalent" from the legal point of view to any other necessary condition (von Buri 1873). His doctrine was therefore called the "equivalence theory," and also the "sine qua non theory" or "theory of conditions." In line with Mill's doctrine, it regarded a condition to be a cause in the legal sense "if one cannot imagine it as being absent without the result not occurring," as one can still read in juridical literature today. Von Buri argued in the fashion of Mill that it would be much too vague and arbitrary, and therefore unscientific, to isolate one element among the antecedent complex of conditions as *the* cause, and neglect the other conditions as irrelevant (as von Bar had done). It would be equally vain to grade the conditions in their amount of contribution to the outcome as other authors have tried. Each necessary antecedent condition is causally connected with the effect in the same way as any other because it is equally required for the production of the effect and thus just as relevant and efficient. Von Buri concluded, in this case against Mill, that *each* necessary condition of an effect must be called the "cause" of the outcome. There is thus no difference between the cause of an outcome and a condition for it. (See von Kries 1888, 399 ff. for a criticism of von Buri's theory.)

The deeper reason for von Buri's approach lies in his special concept of action that was new at the time: He distinguished the objective, external side of action that is thought to be unrelated to the will of the agent from the subjective, inner side that is assumed to be linked solely with guilt (*culpa*), i.e., the violation of a norm. Von Buri's view was directed against the popular Hegelian view of the time that the subjective and objective side of an action are inseparably intertwined in the will of the agent, which has to be understood solely in a teleological way. Von Buri's proposal has been associated very often directly with Mill and is followed by German criminal courts until today (Hart and Honoré 1959, 393). Many other continental European and Latin American countries have also incorporated this theory into their legal system.

The biggest problem for the equivalence theory is not so much the fact that it gets too many 'synchronous' necessary conditions and therefore causes of a deed, as one could say, but too many diachronic ones. This is because "being a necessary condition of" is a transitive relation: If the deed of an agent A is a necessary condition of an effect E, and if the procreation of A by A's father is a necessary condition of A's existence, then the procreation of A is a necessary condition of E and A's father is responsible for E! Followers of the equivalence theory have thus to show why some necessary conditions are not really causes or must limit the ensuing liability by other means. The most popular strategy to deal with this problem was and is to appeal to defeat by subsequent events or by abnormal coincidence or to the closeness of an act to the effect: The causal connection between A and his father is defeated by A's act which is more proximate to E. (It would lead us too far afield here to discuss the other methods available to criminal law in order to deal with the problem.)

## 13.3 Johannes von Kries and "Adequate Causation"

Johannes von Kries' original intuition now was that there *must* be an *objective* reason for singling out in "common parlance" just one factor from the set of antecedent conditions as *the* cause. His arguments thus originate from dissatisfaction both with Mill's theory of causality and with von Buri's legal equivalence theory: they both attribute to each element of the antecedent conditions of a consequence the same effectiveness and thus the same logical status. In addition to this, von Kries felt challenged to show that his newly developed philosophical interpretation of probability could deal with such an unsatisfactory situation. In 1886, in his philosophical work *Principles of Probability Theory*, which developed the so-called range theory (or *Spielraum* theory) of probability, von Kries had considered many applications of probability in a variety of different areas: in physics (especially statistical thermodynamics and error theory), moral and social statistics (in close connection to the work of his Freiburg colleague, the economist and mathematician Wilhelm Lexis), medical statistics, and the constitution of legal juries (von Kries 1886; on von Kries in general see Heidelberger 2001). Therefore, it was only natural that he tried to expand his approach to legal problems as soon as he learned about the unsatisfactory situation in jurisprudence (von Kries 1888; 1889). He suggested an objective, nonarbitrary, and scientifically acceptable criterion that allows to single out one element (or a small amount of elements) of the "assembly of conditions" as crucial and to distinguish it (or them) from the other conditions that play only a subordinate role. He called the crucial condition in question the "adequate cause," and distinguished it from the rest of the conditions that are only "accidental" or "coincidental." Although Mill is scarcely mentioned by name, there is no doubt that von Kries' discussion is centered on Mill's problem of the salient condition, that is, on the problem of rationally choosing one of the antecedent conditions as crucial for the situation in question (von Kries 1888, 203–210).[6]

It is very likely, by the way, that von Kries took the expression and also partially the concept of "adequate causation" over from Spinoza (1677). In his *Ethics*, Spinoza distinguished between "adequate" and "inadequate" or "partial" cause in a similar way as von Kries did between "adequate" and "accidental" cause. Spinoza wrote, "I call a cause 'adequate' if its effect can be clearly and distinctly perceived through it. I call it 'partial' or 'inadequate' if its effect cannot be understood through it alone" (III, D 1). This terminology is then applied to action, in a way similar to von Kries:

> "I say that we 'act' when something happens, either within us or externally to us, whereof we are the adequate cause – that is (by the foregoing definition) when through our nature something takes place within us or externally to us, which can be clearly and distinctly understood through our nature alone. On the other hand, I say that we are acted on when something happens within us, or something follows from our nature externally, of which we are only a partial cause" (III, D 2).

Quite in accordance with Spinoza, von Kries wanted to find an objective criterion for isolating the antecedent condition that is, as one could also say, *intrin-sically* accountable for the causal outcome and separate it from those conditions that are only *contingently* or *externally* (or, as Spinoza says, partially) so. He wanted to identify the requirements that have to be fulfilled for a condition (from a whole set) to be an efficacious cause and not merely a contingent factor of an outcome. Von Kries' decisive idea was that there is usually only one condition that *generally*

[6] Weber also realized this: "Von Kries himself has shown the contrast between his theory and John Stuart Mill's [...] in a way which is entirely convincing to me" (Weber 1906, 168/270).

increases the objective *possibility* of an effect in a significant way, i.e., that favors an effect *in general.*

Consistent with this idea, von Kries arrives at the following definition: An adequate cause is a necessary antecedent condition of an event whose realization generally and significantly *favors* the outcome, i.e., *increases the chances* of the outcome, when compared to the other conditions. Von Kries writes:

> "Where it is established that a factor has a causal impact on an effect, we still want to distinguish whether its connection with the effect can be generalized or whether it is only a [contingent] peculiarity of the case in question; i.e., whether the factor, as we usually say, is generally suited – has a tendency – to produce such an effect, or whether it has occasioned the effect only accidentally" (von Kries 1888, 200 f.; compare also 199, 188; von Kries 1889, 531 f.).

It is thus the increase of the general tendency or objective possibility of the outcome in relation to the rest of the conditions that makes a certain necessary condition an adequate cause. Estimates of this possibility drawn from general probability statements allow choosing the right causal condition as the objective cause of an event. To identify an adequate cause is to consider the relation of a particular causal condition to the effect in a general, or, as von Kries also writes, in an "abstract" way (von Kries 1888, 201). Von Kries compares the situation to the use of a loaded dice. In such a case we would say that a certain condition (the eccentric center of gravity of the dice) is a "favoring condition" for a certain outcome, e.g., the falling of a six. Since the realm of social phenomena is also characterized by mass repetitions of similar cases (like in the game of dice), we are justified to use the same terminology for the social sciences.

If we apply von Kries' criterion to Mill's example of the man who died because he ate a poisoned dish, the condition of the eating of the poisoned food is not just chosen as *the* cause of the incident because of our subjective inclination, as Mill maintained, or of our contextual interest. The probability to die after having eaten a poisoned dish is generally higher than after having eaten an uncontaminated one if all the other conditions (like the state of the atmosphere, the man's bodily constitution, his state of health, etc.) stay the same. The reason of the choice lies, therefore, in the *increase of the possibility* of the man's death by one condition to the exclusion of the other necessary conditions present. "Very often," writes von Kries, "we can state beyond any doubt that a causal condition increases the possibility of an effect, that is, that the existence of this condition produces the consequence for a much bigger variety of circumstances than without" (von Kries 1888, 202). The concrete singular case is to be considered, as von Kries says, in a "generalizing manner"; i.e., a general idea of the process must be prescinded. The discussion shows that von Kries does not want to introduce a foreign technical element into our everyday understanding that would be available only to legal experts. On the contrary, he claims to have shown that (and how) our everyday causal and legal intuitions in attributive contexts are guided by an objective criterion that can be made explicit.

In order to make von Kries' definition of the "adequate cause" more conspicuous and to explain the connection with von Kries' ideas on probability, we must deal more closely with his terminology and his views on probability theory. Suppose, there is a set of conditions $c_1, \ldots, c_n$ to be considered. These conditions can form different "formations" or "configurations" (*Gestaltungen*) of antecedent conditions. A "formation" of $c_1, \ldots, c_n$ is a conjunction that contains either *ci* from the set of conditions or its negation ¬ *ci*, but not both. So $c_1, \neg c_2, \neg c_3, c_4$ would be a possible forma-tion, or $\neg c_1, c_2, \neg c_3, c_4$ another one. If there are *n* conditions forming the necessary part of a complex that is sufficient for the outcome *E*, we get $2^n$ different possibleformations. A specified outcome occurs only in a certain genuine subset of these possibilities. Von Kries calls the class of all formations of $c_1, \ldots,$ *cn*, for which the outcome *E* occurs the "range" (*Spielraum*) of the outcome, i.e., the leeway or scope that is given by a certain part of the formations for the effect (see Heidelberger 2001, 38 f.).

Let us assume that *E* obtains in the following four formations of $c_1, \ldots, c_n$, but not in the remaining 12 outcomes. These formations then constitute the "range" of *E*:

$$c_1, c_2, c_3, c_4$$
$$c_1, \neg c_2, \neg c_3, c_4$$
$$\neg c_1, c_2, \neg c_3, \neg c_4$$
$$\neg c_1, \neg c_2, c_3, \neg c_4$$

Note that $c_1$ appears three times in this range of four formations, $c_4$ and $c_3$ appear twice and $c_2$ appears only once.

In conformity with elementary probability theory, we can then define the objective possibility μ of an outcome relative to one of the factors $c_i$ of the antecedents in the following way:

$$\mu(E|c_i) = \frac{\mathrm{Prob}(ci \,\&\, E)}{\mathrm{Prob}(E)}$$

In the words of von Kries: "The size of the possibility which a condition has for an effect can be designated by the ratio of the *Spielraum* bringing about the effect to the entire *Spielraum*" (von Kries 1888, 184, fn. 2). In other words, given a certain condition, the probability of an event is the share this condition would have among the outcomes of all possible formations resulting in *E*. In the example chosen above, the objective possibility of $c_1$ for *E* would be 3/4, because it appears in three of the four cases of the range of *E*. (Unfortunately, von Kries did not express his results in a formal way himself. In my symbolic rendering I tried to stay as close as possible to von Kries' verbal formulations. For the connection of von Kries' definition with Wittgenstein's, Waismann's, and Carnap's conceptions of logical probability that make use of formal means see Heidelberger 2001.)

The concept of the "range" or *Spielraum* of an outcome now allows us to give a more precise and more conspicuous definition of the concept of an adequate cause:

A condition $c_a$ is an adequate cause of an effect $E$ if and only if

(1) there are conditions $c_1, \ldots, c_n$ which are a cause $C$ of $E$ in Mill's sense.
(2) $c_a$ is among $c_1, \ldots, c_n$.
(3) $\mu(E \mid c_a) >> \mu(E \mid \neg c_a)$; i.e., the range of $c_a$ is considerably greater than the range of any other $c_i$ ($i \in 1, \ldots, n$).

In other words, $c_a$ is an *adequate cause* of an effect $E$ if and only if it generally and considerably increases $E$'s objective possibility. Accordingly, a condition $c_a$' is an *accidental* (or *coincidental*) *cause* (or a *mere condition*) of $E$ if and only if it is *not* an adequate cause of it, i.e., if the relevant objective possibilities are approximately equal. We can also say that $c_a$' merely *occasions* or *triggers E*, without actually being a cause of it.

Von Kries illustrates his definition with the case of a coachman who gets drunk ($c_a$) before he drives a passenger in a horse carriage ($C$). As a result he falls asleep and misses a turn (von Kries 1888, 201; compare also 1889, 532). To make matters worse, the passenger in the carriage subsequently gets struck by lightning ($E$). According to Mill and the legal equivalence theory, the fact that the driver is drunk and asleep must be taken as the cause of the fate that hit the passenger. The drunken behavior of the driver must be taken as an effective condition of the accident because if the driver had not been drunk he would not have made the detour and the passenger would not have been struck by lightning. According to von Kries, however, the driver's behavior is only *coincidental* or *contingent* for this event, although the accident would not have occurred if the driver had not been drunk and as a result of it had stayed on his normal route. The reason is that missing a turn as a carriage driver does not in general increase the possibility of getting struck by lightning (or, if you insist, only in a negligibly small way). According to the notation introduced above, this would mean that $\mu(E \mid c_a) \approx \mu(E \mid \neg c_a)$. The behavior of the driver was therefore *not* an adequate cause for the death of the passenger but only a contingent one.

The matter would look different, however, if the driver had led his carriage into a ditch and the passenger subsequently got injured or killed by the falling carriage ($E$'). In this case, we could blame the driver on the account that drunkenness generally increases the objective possibility for a carriage driver to cause an accident. (This would mean that $\mu(E' \mid c_a) > \mu(E' \mid \neg c_a)$.) In such a case, the driver's drinking ($c_a$) would be the *adequate cause* of the accident. "It is true that the negligence of the driver will not necessarily lead to an accident like the overthrow of the carriage, but one will insist that it is generally apt, or that it has the tendency, to let happen such an accident – that it increases its possibility or probability" (von Kries 1888, 201).

Von Kries' conception thus easily solves the problem that is left open by Mill and the equivalence theory. It makes possible to identify the objective factor among the many necessary conditions of the antecedent complex of an event playing the salient and effective role in bringing about the conclusion. In contrast, the equivalence theory has to use artificial and twisted principles in order to free the driver from the responsibility for the death of the passenger due to lightning (in the first example). Adherents of the equivalence theory must take rescue in strange constructions, e.g., in the idea that a causal chain is "broken" or the causal efficacy of the conditions is "counteracted" by another influence: accordingly, in the case of the coachman, one would have to say that drunkenness is not a causal factor of being struck by lightning, although it seems to be so, because it has been "cancelled out" by the causal efficacy of the interfering lightning or because the causal effectiveness of the antecedent is *preempted* or *defeated* by an interfering event (lightning).

Von Kries, however, does not have to resort to such problematic ideas: Drunkenness, although causally effective, does not increase the probability of getting struck by lightning, and is thus not an adequate cause of the event; the condition and the outcome are not *generally* connected with each other, although the condition is a necessary element in the sufficient set of conditions of the event and therefore part of the concrete cause. "The specificity of the case," von Kries argues,

> "lies only in the fact that, because of the very special and not foreseeable configuration, an action that is generally suited to the highest degree to lead to a certain outcome did not produce it. [...] No causal connection is interrupted, but the culpable act has not fully developed its causal effectiveness in the usual way, because of the special formation of the singular case" (von Kries 1888, 210).

How does Mill's example of the person eating the dish compare with von Kries' example of the carriage driver? Mill's treatment is (at least preponderantly) guided by the aim to find a scientific explanation of the mortal incident by subsuming it under an unexceptionable regularity. In von Kries' case of the driver, however, we already *have* the explanation of the passenger's death, namely having been struck by lightning or having been crushed by the carriage. What von Kries is after in his examples concerns the decisive factor among the complex of antecedent conditions of the outcome that allows attributing causal responsibility in an impartial, unforced, and objective way. In Mill's context, we search for a causal law that, together with the "given concurrence of circumstances," allows subsuming the effect under it, whereas in the legal context of von Kries, we want to learn to which of the circum-stances causal responsibility is to be attributed. In both cases, knowledge of general regularities is required, but it is of a different kind: For Mill it is knowledge of natural laws, or regularities in Hume's sense, whereas in von Kries' case it is knowledge of objective possibilities, or more precisely, it is comparative knowledge of the differ-ence a condition makes to a set of objective background possibilities. Consequently, von Kries finds Mill's theory of causation much too narrow: "Any theory of causation for which there is no other causal connection than the constant ordered sequence of A and B – which bases a causal connection only on an unexceptionable regularity of succession – will, ultimately, prove sterile" (von Kries 1888, 201 f.). This can be interpreted as meaning that to make explanatory use of causality by causal laws is not the only use there is. There is also an attributive one that does not ask for a regular uniformity but for a statistical tendency.

## 13.4 Weber's "Causal Analysis of History" in von Kries' Manner

In 1906, Weber applied von Kries' theory to history in order to give an adequate analysis of the special way in which historians explain historical events:

> "We ask first, in common with juristic theory, how in general is the attribution [*Zurechnung*] of a concrete 'effect' to an individual 'cause' *possible* and realizable in principle in view of the fact that in truth an *infinity* of causal factors have conditioned the occurrence of the individual 'event' and that indeed absolutely *all* of those individual causal factors were indispensable for the occurrence of the effect in its concrete form?" (Weber 1906, 169/271; slightly amended.)

Weber felt justified in applying von Kries' method in the same way to history as it was applied to human actions in criminal law at the time, because asking for the cause of an historical event is like asking in a legal context whether someone caused an effect through his or her action:

> "It is natural that it was precisely the jurists and primarily the jurists specializing in criminal law who treated the problem since the question of penal guilt, insofar as it involves the problem: under what circumstances can it be asserted that someone through his action has 'caused' a certain external effect, is purely a question of causation. And, indeed, this problem obviously has exactly the same logical structure as the problem of historical 'causality'. For, just like history, the problems of practical social relationships of men and especially of the legal system, are 'anthropocentrically' oriented, i.e., they enquire into the causal significance of human 'actions'. And just as in the question of the causal determinateness of a concrete injurious action which is eventually to be punished under criminal law or for which indemnity must be made under civil law, the historian's problem of causality also is oriented towards the correlation [*Zurechnung*] of concrete effects with concrete causes" (Weber 1906, 168/270).

In order to find an adequate cause of an historical event, the historian has to carry out, according to Weber, a series of abstractions: First he or she has to isolate from the infinity of the conditions those that are interesting, essential, and relevant from the historian's point of view. It would actually be impossible and meaningless to consider *all* the factors of an event in their individual totality. Next, one has to compare the causal components of an historical episode that are actually present with a fictitious course of events where one or more components are imagined to be absent, i.e., to compare objective "possibilities" with each other in a counterfactual way. As Weber famously said: "In order to penetrate to the real causal interrelation-ships, we *construct unreal ones*" (Weber 1906, 185 f./287). Such an "imaginative construct" presupposes the isolation of the different components, which is given in "ontological" knowledge. However, it also requires "nomological" knowledge of how human agents normally act in, or react to, given situations, i.e., one has to know the general impact a condition has. Thus, judgments of possibility entail "the isolation and generalization of the causal individual components for the purpose of ascertaining the possibility of the synthesis of certain conditions into adequate causes" (Weber 1906, 177/279).

Weber gives several examples in order to illustrate his application of von Kries' method to historiography (Weber 1906, 171/273 f., 174/276 f., 184 f./286). The first one is the Battle of Marathon, the victory of the Athenians over the Persians in 490 BC. Although the dimensions of the battle were tiny, it is of great historical importance, because so much of later history depended on the actual course of events. The decisive question to be asked by the historian is what would have happened if the Persians had won this battle. Ontological knowledge resulting from isolating the important elements of Persian culture together with the nomological knowledge that a theocratic–religious development would have been possible in Hellas at the time allow objective considerations of the possibilities involved. As a result, the historian can say that the Battle of Marathon was the adequate cause of the creation of the Attic fleet, which in turn enabled the salvation of the independence of Greek culture and thus of the autonomous development of the occident in later times.

A second example regards the "March Revolution" of 1848 in Berlin (165 f./268, 181/283, 185/287). The question to be answered concerns the historical significance of the shots that were fired during the night of the revolution in March 1848. These firings – although they triggered the relevant revolutionary events – are historically insignificant and only coincidental, because if the shots had not been fired (= fictitious course), the decisive street fighting would still have broken out. Therefore, the historian must conclude that factors other than this shooting were the cause of the outbreak at the time. The implied ontological knowledge concerns the actual social and political situation at the time and the nomological knowledge that is applied in this case concerns the readiness of the population for street fighting given the concrete situation of the society.

## 13.5 Two Kinds of Explanations in Natural and Social Science

It would still be much too hasty at this point to conclude that von Kries' theory of adequate cause is not concerned with what is at the center of Mill's concern: scientific explanation. An adequate cause is indeed able to provide an explanation of an important feature of reality which cannot be given by Mill, and which is to be distinguished from the deductive–nomological explanation that we have considered above. In order to understand this kind of explanation, we have to take a closer look at the way in which von Kries' distinction between "nomological" and "ontological" features of reality fares in an idealized game of chance: Take a roulette disc that is straightened out on an infinite stripe and divided into infinitesimally small sections of black and red. If a ball were pushed along this stripe it would land on red or on black according to the initial force with which it is pushed. Any variation of this force – even if it were infinitesimally small – would lead to a change in the outcome. This "push-game" or *Stoss-Spiel*, as von Kries calls it, is an ideal chance device with maximal instability. The chance character of the game, von Kries maintains, does neither result from our limited knowledge of the conditions nor because of an alleged suspension of the deterministic nature of the natural laws involved (von Kries 1888, 180). The reason for calling it a chance device is its aleatory character, the peculiar objective setup of the game, or, to put it into the context of von Kries' view of probability, the special production of the elements of the range through maximal instability. (One could point, of course, to other chance mechanisms playing a similar role, e.g., the Galton Board (quincunx) or the simple throw of a dice. Von Kries chose to illustrate his approach by the push-game because it also emphasizes certain other important features more as opposed to the other devices.)

If we slightly change a parameter in the push-game, say, if we increase the size of the black sections – even if it were only by an infinitesimal amount – the tendency of the ball to fall on black would also increase in the long run. It is now very interesting to have a closer look at this kind of tendency: It can certainly be expressed as a regularity. Is this regularity *derivable* from the laws of mechanics that govern the behavior of the ball? Or do we have to assume the reality of an *additional* law over and above the laws of mechanics? The answer must be: neither! It is rather the outcome of a special and contingent distribution of alternatives in the range of the game, or, to put it in Mill's terms, the result of the special "complexity of the conditions" (von Kries 1886, 87; von Kries 1888, 189). It is exactly this special complexity that von Kries calls an "ontological" feature of reality. He distinguishes it from the "nomological" ones that concern the laws of nature. Whereas nomological features relate to "lawful necessities," ontological ones have to do with the contingent properties and the general setup of the world in which these laws are effective. The laws of mechanics are the necessities governing the chance mechanism, whereas the equality or inequality of the red and black sections of the device concern its boundary conditions; they have to do with the setup's contingent features: with the "dimensions of the ranges" or, as von Kries also calls it, with the "proportion" of the "formation" of its ranges.

Now, according to von Kries, the two basic kinds of properties of the world, its ontological and its nomological character, lead to two different kinds of scientific explanation: There is the explanation of a fact by subsuming it under general laws in virtue of the realization of certain antecedent conditions. But there is also explanation by reference to the given objective yet contingent possibilities, as in the explanation of the frequency with which the ball hits a red stripe in the long run. "An explanation," von Kries writes,

> "satisfying our intellectual needs is produced, first and foremost, as soon as the observed facts are recognized as the necessary consequences of empirical laws of nature. An explanation can, however, also be given by referring to the dominating range of the ontological arrangement. Take the almost equal frequency distribution of black and red in the outcomes of a roulette game in the long run. It is not possible to give an explanation of this observation in the usual sense by reducing it to reality's own lawfulness. We can, however, convince ourselves that, in a case like that, to demand an explanation of this kind would be unreasonable anyway" (von Kries 1925, 159 f.; compare von Kries 1886, 167 f.).

What is the relevance of von Kries' distinction for his examples of the carriage driver? In the case in which the carriage turned over in the ditch, we explain the acci-dent through the action of the coachman: He is the originator of the accident because he increased the possibility of a danger by his behavior. The explanation we give here does not operate by invoking a law of nature, but by comparing the ontological features of two cases with each other: driving the passenger in a sober state and driving him in a state of drunkenness. "Asking for the causality of a certain object amounts to the question what would have happened had this real condition ([i.e.] a particular part) of the complex of conditions been missing, but everything else had stayed exactly the same" (von Kries 1888, 198). The behavior of the driver is thus judged in respect to whether he changed some ontological feature of the world. So, the explanation is not completely analogous to the case where we explain by reference to the dominating range of the ontological arrangement. It is rather an explanation which is given by *contrasting* two configu-rations of ranges with each other (driving in a drunken state implies a higher possibility of accident than driving in a sober state), that is to say by reference to a *change* of on ontological arrangement (increasing the risk of an accident).

Ideal chance games do not only illustrate the distinction between ontological and nomological features of reality, they are also the only means with which, von Kries claims, we can objectify statements of probability. Von Kries regards probability as a subjective magnitude that signifies the degree of rational expectation of an event in the light of the information that is available to the subject (of its "intellectual state," as he says). There are, however, as we have already seen, "objective possibilities" for the occurrence of events. In many cases, the objective possibilities can be divided into symmetric alternatives and a range, a *Spielraum*, can be attributed to them. In a game of dice, any number from one to six has an equal chance of being thrown so that the probability for each number is one-sixth. Strictly speaking, an objective quantitative measure of the probability of an event only makes sense if the possible outcomes, the ranges, are symmetrical. This is only the case for ideal chance devices.

These setups have a further property which we have not yet considered that concerns their foreseeability: The expected probability of an ideal chance result will not change if we increase our information about the initial conditions, i.e., the game's ontological constitution as a result of the maximal instability of the system. Von Kries calls such probabilities "universally valid" and maintains that in such cases probability takes on an objective meaning. There are so many different ways of how the push-game can realize its outcome ("equally extended configurations of conditioning circumstances," as von Kries expresses it) that no increase of information about the exact initial conditions will give us a reason to prefer one outcome (the falling of the ball on red) over the other (falling on black). In the case of an ideal dice, each of the six outcomes would be equally probable and thus would have the same range and therefore the same *objective* possibility. (I am passing over other conditions that, according to von Kries, have to be fulfilled for probabilitiesin order to become objective possibilities; see Heidelberger 2001, 179 f.)

An ideal game of chance allows us, so to say, to standardize and normalize ranges in the realm of objective possibilities: we are justified in these cases in assuming "that the (universally valid) probability of an effect is as big as its (objective) possibility, i.e., that the same numerical value indicates probability and possibility" (von Kries 1888, 190). A possibility statement is thus not based on a *lack* of information, as in Laplace's classical conception of probability theory, i.e., it is not based on a subjective "intellectual state," but on an objective circumstance in the outside world.

One could object that there are not enough objective possibilities known to us according to von Kries' criterion in order to build a science on them. There are, as a consequence, very few cases where we can really be sure of the extreme instability required. Von Kries sees a way out of this problem in the *estimation* of how much a certain type of event diverges from an ideal chance setup. In order to judge the possibilities of actual cases, we have to compare them with an ideal type: the ideal chance setup. A game of chance that offers us objective knowledge about the size of the ranges involved can serve as a measuring stick that can be compared with other cases and thus reach at least a qualitative estimation of objective probability.

## 13.6 Ideal Type and "Interpretative Understanding" in Weber

At first glance, von Kries' theory seems to be intuitively very attractive and an effective way out of the impasse created for legal practice by Mill's concept of causality. A closer look reveals, however, that there is a threat looming for von Kries' approach from an unexpected direction. Although von Kries rightly talks of "objective possibilities," these possibilities are *epistemic* ones, i.e., they depend on a (consciously or unconsciously chosen) *epistemic base*, a set of information available to the knowing subject that is taken for granted. In order to elicit what is relevant here, let us vary von Kries' example with the carriage driver a little.

Suppose that there was a rusty and brittle bolt in the carriage that broke when the driver drove into the ditch. Let us further assume that an expert found out that if the bolt had been sturdy, the carriage would not have been overturned when the driver drove it into the ditch even when taking into account his drunkenness. In such a case we would say that it was not the drunkenness of the driver that increased the probability of the accident, but the brittleness of the rusty bolt and that, in consequence, the bolt (or rather the person responsible for it) is to be regarded as the adequate cause of the event. Therefore, an increase of information about the conditions involved can change our belief about the adequate cause.

This example demonstrates three things: first, that the adequacy of a cause for a consequence can only be justified *relative to available knowledge of the circumstances*. Second, it shows that imputation of a consequence to an agent *depends on the knowledge available to the agent or on the knowledge that can rightly and justifiably be expected* from him. If the driver had received the carriage from the company owner before he started to transport the passenger and if the driver assumed in good faith its normal functioning and did not know of the rusty bolt, he cannot be made responsible for its breaking and thus of the accident even if he were drunk. Third, the example also proves that a condition being an adequate cause depends on the *description* of the condition in question. "Driving a carriage" and "driving a carriage with a dangerously rusty bolt" makes a difference!

There are some (slight) indications that von Kries was conscious of this objection (see esp. von Kries 1888, 393 ff.). Yet he treated this lightly and thought it would suffice to point to the vagueness and indeterminacy of knowledge in general; this, however, was unsatisfying. It was the philosopher of law, Gustav Radbruch (1878–1949), later a colleague of Weber in Heidelberg for some years, who criticized the approach of the "adequate causation" in this respect in order to defend the theory of the *conditio sine qua non* as superior (Radbruch 1902, 348 ff., 382–384). He distinguished three possibilities to deal with the limited knowledge situation in the theory of adequate cause: either one takes the ontological knowledge of the agent as decisive for judging the adequate cause (knowledge *ex ante*, i.e., from the perspective of the time of the act) or the ontological knowledge of the judge (*ex post*, i.e., retrospectively) or that of the "normal" or "ideal human being" (under no special temporal perspective) (348, 353). The first possibility Radbruch identified with von Kries' own version; the second one with the "standpoint of the objective retrospective prognosis" developed by the jurist Max Rümelin, where it must be decided from the knowledge of the conditions after the action has taken place whether the consequence would have been foreseeable at the time of the action (Rümelin 1896); and the third possibility was recognized as standing in the tradition of the *bonus vir*, the average right-thinking man, whose intuition and common sense is taken as the basis of the decision. It is remarkable to see how much von Kries' terminology of "ontological" and "nomological" knowledge has been taken over by Radbruch (and many other jurists at the time). One of the objections raised by Radbruch against von Kries' theory was that it shares all the errors that the equivalence theory was usually blamed with, i.e., that by varying the knowledge base and the perspective one could declare almost any condition as the adequate one (Radbruch 1902, 350 = 26; doubly paginated).

It would lead us too far afield to follow the legal discussion any further. If we ask now for the relevance of the foregoing for Max Weber, it seems plausible to assume that Weber introduced the category of *Verstehen* or interpretative understanding in order to solve the problem of the knowledge base that was posed by Radbruch. If the historian asks for the adequate causation of historical events, which epistemic basis does he have to choose? From which perspective does he have to judge the development? If one believed von Kries that grasping the causal meaning of a historical process is comparing it with the absolute chance of an ideal chance mechanism one would risk getting causal relations that are irrelevant from a historian's point of view. Therefore, it seems that Weber's category of understanding is a means to limit the number of causal relations that could come into play and, by their sheer quantity, devalue the criterion of adequate causation. In order to illustrate this situation we can return to the modification of the example with the carriage driver. If we want to find out whom we have to blame for the accident of the carriage overturning in the ditch we would not be satisfied by hearing about the rusty bolt. We would insist on learning about whose responsibility the functioning of the bolt would fall under. Weber had the idea of introducing the perspective of the *ideal agent* into the picture instead of an ideal chance process. One can then sift the causal processes and only retain those for which human agents are responsible and thus limit the number of candidates for adequate causes in a way that excludes irrelevant causal aspects. Thus, I can wholeheartedly agree with Turner and Factor when they write:

> "But *Verstehen*, or more precisely the requirement of *Sinnadäquanz* [adequate understanding], seems to shrink the field of descriptions, and does so drastically. If one requires that

the description be 'adequate at the level of meaning', one eliminates the bulk of possible descriptions, and does so in a way which is entirely consistent with the 'interest' of the historian in the subjectively meaningful" (Turner and Factor 1981, 21).

We have seen that for von Kries rigorous knowledge of objective possibilities in the world can only be achieved by a comparison of reality with an ideal case. An ideal case functions, as Weber later put it, as

> "a conceptual construct (*Gedankenbild*) which *is* neither historical reality nor even the 'true' reality. It is even less fitted to serve as a schema under which a real situation or action is to be subsumed as one *instance*. It has the significance of a purely ideal *limiting* concept with which the real situation or action is *compared* and *surveyed* for the explication of certain of its significant components" (Weber 1904, 93/194).

Or, a little later: ideal types are "of great value for research and of high systematic value for expository purposes when they are used as conceptual instruments for *comparison* with and the *measurement* of reality. They are indispensable for this purpose" (Weber 1904, 97/198 f.). Although Weber only talks about the cultural and social sciences here, his statements are equally valid for the natural sciences.

How can the idea of an ontological explanation that compares a concrete process or event with an ideal type be translated into sociology and history? It cannot be done by taking an ideal chance mechanism as the standard of comparison but only "a conceptually pure type of rational action," Weber says (Weber 1921, 2/544). In order to justify this, Weber discusses (at another place) the analogy to throwing a dice:

> "Consider a concrete throw of a dice. The fact that the number 6 turns up – assuming that the dice are not loaded – simply cannot be the object of any causal account [*Zurechnung*]. It appears to be 'possible,' in the sense of being consistent with our nomological knowledge. [...] The fact that – assuming the dice are not loaded – given a very large number of throws, each of the six sides turns up at an approximately equal rate, appears 'plausible'. We 'comprehend' the empirically verifiable validity of the 'law of large numbers' in such a way that the contrary case – certain sides turning up at a higher rate in spite of the increasingly large number of throws – would force upon us the question of the causes to which this difference in frequency could be attributed" (Weber 1903/1906), 126/ 68; translation slightly amended).

The individual event of throwing a six, for example, remains irrational and only after the occurrence of many throws can we reach justified probability statements as, for example, the proposition that the dice is loaded or that another condition is responsible for leading to a change in the equal distribution of the cases.

In history, a behavior can be regarded "not only as nomologically 'possible,'" as Weber continues, but also as

> "'teleologically' *rational*. Not in the sense that we can establish, as a result of the ascription of causes, a statement of *necessity*. But rather in the sense that his conduct has an 'adequate cause.' I.e., given certain intentions and (true or false) beliefs [...] [of the acting person], and given also a rational action determined thereby, a 'sufficient' motivation can be identified" (ibid., 127/ 68 f.).

In other words, the assumption that an event is an action and thus imputable to a person as the cause of it, i.e., that it has "a concrete 'motive' or complex of motives 'reproducible in inner experience'" (Weber 1903, 67) enables us to ask for potential deviations from the logic of a rational means-end analysis. This happens in analogy to processes in nature where we assume that, in the normal case at least, objective possibilities are uniformly distributed (i.e., that a chance process is underlying it that does not favor a special outcome) and that any deviation from this ideal must find an explanation.

## 13.7 Consequences for the Social Sciences

The above-mentioned analysis calls for a consideration of the consequences of introducing von Kries' methods into the social sciences: As the most immediate consequence, social science is now considered to be dealing with the ontological features of society instead of with the nomological ones. The proper way of explanation is by referring to dimensions of ranges (i.e., to objective possibilities or general tendencies) rather than by subsuming events under general laws. As a result, von Kries' approach can be understood as the core of a research program for the social sciences that mediates between two rival programs that were prevalent at the time: the school of Adolphe Quetelet (1796–1874) and the opposed "historical school." I would like to argue that Weber, by taking over von Kries' approach, consciously took part in such a mediating paradigm. The authors to be considered in this respect are Ladislaus von Bortkiewicz (1868–1931) and Alexander Tschuprow (1874–1926) to whom Weber refers as standing in a "very close relationship to von Kries' theories" (and *a fortiori* Wilhelm Lexis (1837–1914) who stands in the background for both these authors) (Weber 1906, 167/ 269 f.). As representative of the Quetelet opponents one can take someone like the economist Georg Friedrich Knapp (1842–1926) who is treated both by von Bortkiewicz and Tschuprow as the major contemporary opponent to Quetelet.

Both parties of the controversy developed their standpoint from a close consideration of the recent history of statistics. Their viewpoint is almost identical as regards the judgment of a long stretch of this historical period (Tschuprow 1905, 12–14; von Bortkiewicz 1904, 246 ff.; Knapp 1871/72, 1872). Both sides agree that everything began with Quetelet, who developed a determinist "social physics" or *physique sociale* (thus the subtitle of Quetelet's work of 1835) that took the "average man" or *l'homme moyen* and his different penchants (among them most famously his *penchant au crime*) as the subject matter of statistics. In addition, Quetelet wanted to connect these factors with the actual condition of the *système social*. Statistical regularities in society, especially those of moral statistics that show the temporal constancy of the *budget du crime* and the like in a country, were therefore of utmost importance to Quetelet and seen by him as scientific laws with the same status as the laws of physics. The role an individual can play in this is at most to represent a small perturbation that is ultimately canceled out by counteracting processes. Freedom of the will is thus only an illusion. Knapp wrote that according to Quetelet "the acting human being is either wholly comparable to a falling stone, or to an enchained dog to whom is dictated with mathematical inexorability the location in which he is allowed to freely bop around" (Knapp 1871/72, 241).

A generation later, both sides continue their report, Quetelet's approach was criticized by perceptive German professors, spearheaded by the Leipzig Herbartian, Moritz Wilhelm Drobisch. This criticism eventually led to the defeat of "Queteletism." There followed a period of almost theory-free activity of collecting vast amount of data when statistics was little more than a narrative enterprise.

Up to this point, the two sides agree completely in their appreciation of the history of statistics, but from here on they start drawing different conclusions. In a way, both sides express a certain disillusionment that leads to an eclectic and synthetic approach choosing elements both from the Quetelet as well as the Drobisch tradition and molding them into something new. Tschuprow sees the tide of the history of statistics turning with Wilhelm Lexis, who proposed a criterion of statistical dependency by comparing statistical series with chance processes. Normal games of chance with complete independence of the individual results of the game with each other have what he calls a "normal dispersion." If there is a causal dependence between the results for the different rounds, it is called supernormal or subnormal dispersion depending on whether there is an influence of result A on result B favoring the similarity of B to A or to not-A. Therefore, according to Tschuprow, Lexis has shown that statistical data can be regarded as empirical manifestations of objective probabilities and that one can gain insight through these probabilities into relations between social phenomena.

Tschuprow summarizes his standpoint with the following words:

> "The method of statistics must support the inductive method, which was created for investigating unbreakable causal connections, but which is helpless in respect to the looser connections that are more important in practical contexts than the former ones. I have tried to show how, in pursuing this aim, the statistician unveils the chance character of the frequencies coming to the fore when singular phenomena are concentrated in statistical masses and he reduces them to the underlying objective probabilities in order to decide by comparing statistical series with each other whether there is a dependency between the observed phenomena or not" (Tschuprow 1905, 68 f.).

For the historical school, however, statistical regularities lack a deeper meaning and there is no reasonable way to regard them as "external" scientific laws in the sense of physics coercing the individual. It follows, for this school, that social science can and must do without probability theory. Such an approach is held irrelevant for a true analysis of society and does not say anything about the particular individual that forms a constitutive part of the whole. Social science is in the end a narrative enterprise and the role statistics can play is only marginal. Georg Friedrich Knapp sees in moral statistics a tool to supplement the qualitative description of society in a more exact and quantitative way. It could help to determine in which degree and to what extent the freedom of the individual is limited by belonging to a certain social system. The school of the historian Thomas Henry Buckle, wrote Knapp, who radicalized Quetelet's approach, "explains from the outside to the inside; it sees the constancy of the whole und limits therefore the individual. The German school, however, explains from the inside to the outside; it takes the individual as he is and looks for reasons of the constancy of the whole" (Knapp 1871/72, 243). For the "inside," that is for the motives of the individual person, one can also assume a kind of an "inner law," which is influenced by the motives of the environment into which the individual is "thrown through the fate of his birth". Instead of serving a "social physics," statistics can serve a "social ethics" in the sense of the Dorpat theologian Alexander von Oettingen and return responsibility to the individual in place of society (245–247). Von Oettingen does not only argue against Quetelet's view, which he sees guided by "analogies from physics," but also against those who, like the "Manchester school of national economy," assume man's freedom, but "disconnect man from all his interrelationships with the community [*Gemeinschaft*]" (247). Knapp's emphasis on the "inner law" and the "approach from the inside" seems to be contrary to von Buri's (and thus also von Kries') approach, where the inner aspect is separated from the causally effective aspect of action and considered only in respect to guilt, i.e., the individual infringements of norms.

How does Max Weber's modification of von Kries' views fit into this picture? It seems to be some kind of compromise or mediation between the school of Lexis and von Kries on the one hand and the standpoint of someone like Knapp and von Oettingen on the other. As von Bortkiewicz stresses, there are two fundamental conceptions Quetelet and Lexis share with each other, and I would say, also with Weber: The first is the interest in the constancy of statistical series, and the second is the interest in the "average man" or, as it is called by Lexis, the "abstract man" (von Bortkiewicz 1904, 248 f.). In social science one has to compare with types. For Weber, as for Lexis and von Kries, statistical regularities are thus important in social science and have to be analyzed with the help of probability theory, i.e., the "real analogy" between "mass phenomena" and "games of chance" has to be exploited (ibid., 234). These uniformities cannot, however, as all statisticians (the Lexis–von Kries side and the Knapp–Oettingen one) agree, be regarded as laws of nature in the sense of Quetelet; they do not represent nomological knowledge, but pertain instead to the ontological character of society, as von Kries has shown. Ontological knowledge holds information about causal interrelationships that cannot be given otherwise. This is directed against Knapp's contention that statistics is only a perspicuous and rigid way to present results that can be found and known already qualitatively.

Yet Weber is also on the side of Knapp who criticizes and rejects the tendency of statistics since Quetelet to let the individual disappear behind mass uniformities. People like Knapp criticize, on the one hand, that constructs like the "average man" are taken to say something about a particular individual, but on the other hand they condemn the tendency in statistics to forget the original motives of the acting individual and to dissolve the contribution of the individual in the mass of data. In this respect, Weber's transfer of von Kries' method of imputation to history and sociology was doubly clever: it does recognize a certain part of Knapp's approach, yet it does this with means that were already available in other domains of the Lexis–von Kries view. The requirement of interpretative understanding can be seen as a move to secure responsibility of the individual in history and sociology and to block the way of imputation of historical and social developments to mere impersonal powers and processes.

## References

von Bar L (1871) Die lehre vom causalzusammenhange im rechte, besonders im strafrechte. Bernhard Tauchnitz, Leipzig

von Bortkiewicz L (1899) Die erkenntnistheoretischen grundlagen der wahrscheinlichkeitsrechnung [review of Ludwig goldschmidt, die wahrscheinlichkeitsrechnung. Versuch einer kritik. Hamburg & Leipzig 1897]. Jahrbücher für Nationalökonomie und Statistik (3rd series) 17 (=72), pp 230–244

von Bortkiewicz L (1904) Die theorie der bevölkerungs- und moralstatistik nach Lexis [review of Wilhelm Lexis, abhandlungen zur theorie der bevölkerungs- und moralstatistik, Jena 1903]. Jahrbücher für Nationalökonomie und Statistik (3rd series) 27 (=82), pp 230–254

von Buri M (1873) Ueber causalität und deren verantwortung. Gebhardt, Leipzig

Eberle TS (1999) Sinnadäquanz und Kausaladäquanz bei Max Weber und Alfred Schütz. In: Hitzler R, Reichertz J and Schröer N (eds) Hermeneutische wissenssoziologie. Standpunkte zur theorie der interpretation. Universitätsverlag: Konstanz, pp 97–119

Hart HLA, Honoré AM (1959) Causation in the Law. Clarendon, Oxford

Heidelberger M (2001) Origins of the logical theory of probability: von Kries, Wittgenstein, Waismann. Int Stud Philos Sci 15(2):177–188

Knapp GF (1871/72) Die neuern Ansichten über Moralstatistik. Jahrbücher für Nationalökonomie und Statistik 16:237–250

Knapp GF (1872) A. Quetelet als Theoretiker. Jahrbücher für Nationalökonomie und Statistik 18:89–124

von Kries J (1886) Die principien der wahrscheinlichkeits-rechnung. Eine logische untersuchung. Mohr, Freiburg

von Kries J (1888) Ueber den Begriff der objectiven Möglichkeit und einige Anwendungen desselben. Vierteljahrsschrift für wissenschaftliche Philosophie 12(2–4):179–240, 287–323, 393–428

von Kries J (1889) Über die Begriffe der Wahrscheinlichkeit und Möglichkeit und ihre Bedeutung im Strafrechte. Zeitschrift für die gesamte Strafrechtswissenschaft 9:1889, 528–537

von Kries J (1925) Johannes von Kries. In: Grote LR (ed) Die medizin der gegenwart in selbstdarstellungen. Meiner, Leipzig, pp 125–187

Mayer ME (1899) Der causalzusammenhang zwischen handlung und erfolg im strafrecht. Eine rechtsphilosophische untersuchung. Heitz, Strassburg

Mill JS (1843) A system of logic ratiocinative and inductive, being a connected view of the principles of evidence and the methods of scientific investigation. In: The Collected Works of John Stuart Mill, Robson JM (ed.), vol VII (Books I–III) & vol VIII (Books IV–VI). University of Toronto Press, Toronto; Routledge, London, 1974. (Cited according to book, chapter and paragraph.)

Nollmann G (2006) Max Webers Vergleich von Rechts- und Sozialwissenschaft. Archiv für Rechts- und Sozialphilosophie 92(1):93–111

Radbruch G (1902) Die Lehre von der adäquaten Verursachung. (Abhandlungen des kriminalis-tischen Seminars an der Universität Berlin. N. F. Bd. 1 (3)) Guttentag, Berlin, pp. 325–408 (=1–84)

Ringer F (1997) Max Weber's Methodology: The Unification of the Cultural and Social Sciences. Harvard University Press, Cambridge, MA

Ringer F (2002) Max Weber on causal analysis, interpretation, and comparison. Hist Theor 41(2):163–178

Ringer F (2004) Max Weber: an Intellectual Biography. The University of Chicago Press, Chicago

Rümelin M (1896) Der Zufall im Recht. Mohr, Freiburg & Leipzig

Spinoza B (1677) Ethica ordine geometrico demonstrata. rieuwertsz, amsterdam

Traeger L (1904) Der kausalbegriff im straf- und zivilrecht. Zugleich ein beitrag zur auslegung des B.G.B. Elwert, Marburg

Tschuprow AA (1905) Die Aufgaben der Theorie der Statistik. Jahrbuch für Gesetzgebung, Verwaltung und Volkswirtschaft im Deutschen Reich 29:421–480 (=11–70)

Turner SP (1983) Weber on action. In: Lassman P (ed) 2006, Max Weber. Ashgate, Aldershot, pp 179–192. (First in Am Sociol Rev 48:506–519)

Turner SP, Factor RA (1981) Objective possibility and adequate causation in Weber's methodological writings. Sociol Rev 29(1):5–28

Wagner G, Zipprian H (1986) The problem of reference in Max Weber's theory of causal explanation. In: Lassman P (ed) 2006, Max Weber. Ashgate, Aldershot, pp 157–178. (First in Hum Stud 9(1):21–42)

Weber M (1903/1906) Roscher and Knies and the Logical Problems of Historical Economics. Transl. by Guy Oakes. Free Press, New York, 1975 (German original: Roscher und Knies und die logischen Probleme der historischen Nationalökonomie, in: Jahrbuch für Gesetzgebung, Verwaltung und Volkswirtschaft im Deutschen Reich 27( 3):1181–1221; 29(4):1323–1384; 30(1):81–120. In: Weber M 1988, pp 1–145)

Weber M (1904) 'Objectivity' in social science and social policy. In: Weber M 1949, pp 50–112 (German original: Die 'Objektivität' sozialwissenschaftlicher und sozialpolitischer erkenntnis. In: Weber M 1988, pp 146–214)

Weber M (1906) Critical studies in the logic of the cultural sciences. In: Weber M 1949, pp 113–188 (German original: Kritische studien auf dem gebiet der kulturwissenschaftlichen logik. In: Weber M 1988, pp 215–290)

Weber M (1921) Basic Concepts in Sociology. Transl. and introd. by Secher HP. Citadel Press, New York, 1962, repr. 2002 (Parts I & II of German original in: Weber 1988, pp 541–581)

Weber M (1949) The Methodology of the Social Sciences. Transl. and ed. by Shils EA, Finch HA (eds) Free Press, New York [Translations of a selection of articles from Weber 1988]

Weber M (1988) Gesammelte aufsätze zur wissenschaftslehre. Winckelmann J (ed) 7th ed. Mohr, Tübingen

# Chapter 14
# Social Science Between Neo-Kantianism and Philosophy of Life: The Cases of Weber, Simmel, and Mannheim[1]

**Daniel Šuber**

*"In the human and social sciences there are different types of mediation between 'understanding' and 'explanation.'"*

Apel (1987, 132)

## 14.1 Introduction

In recent years, a group of social scientists have credited Wilhelm Dilthey with the status of a "classical sociological theorist" (Bakker 1999) and a key figure with regard to the establishment of the social sciences since the last decades of the nineteenth century.[2] Such evaluations stand in distinct contrast to Dilthey's reputation as a firm critic of sociology on the one hand and his dubious standing within his proper field, philosophy, on the other, where he is perceived as a failed epistemologist. Generally, his influence on social and cultural science is associated with his notion of *Erleben* and understanding as fundamental categories for the interpretive sciences and their unique relatedness to their particular subject. On the basis of this starting point, he eventually established a division between *Verstehen* and *Erklären* and, correspondingly, human and natural sciences.

In the following chapters, I intend to give a different reading of Dilthey's impact on diverse conceptions of social science that does not rest on displaying conceptual affinities and/or homogeneities between Dilthey and sociological theorists, but instead follows a more complicated strategy. Starting from the (meta-theoretical) assumption that the "meaning of an individual concept is rooted in its whole context" (Mannheim 1969, 18), I attempt to indicate that the pervasiveness of Dilthey's think-

D. Šuber (✉)
University of Konstanz, Department of History and Sociology,
P.O. Box: 5560 D 28
D-78457 Konstanz

[1]For very helpful comments I am indebted to Katherine Arens and Uljana Feest.

[2]See also Acham (1985, 1992) and Hahn (1999).

U. Feest (ed.), *Historical Perspectives on Erklären and Verstehen*, Archimedes, Vol. 21, DOI 10.1007/978-90-481-3540-0_14, © Springer Science+Business Media B.V. 2010

ing in social theory is rather due to a comprehensive theory of structure (*Strukturlehre*) that allows for connecting certain types of categories and phenomena, which could not be linked on different theoretical grounds. Hence, to display Dilthey's influence on social theory, we must reconstruct – to apply the language offered by Karl Mannheim once again – the "logical structure of the prior systematization" (Mannheim 1969, 25) upon which a particular concept of social science is built. In simpler words, our main target of attention shall be the meta-theoretical and epistemological presumptions, which we can delineate in Weber's "Wissenschaftslehre," Simmel's "theory of a priori" of the historical sciences as well as Mannheim's "theory of culture and understanding." Such meta-theoretical claims expound specific views of reality (*Wirklichkeit*), give priority to particular methodological strategies to grasp these phenomena, determine evaluative criteria for scientific assertions, and also reflect upon the relation between science and reality in general terms.

Before we turn to the social sciences, we have to define and discriminate the elementary principles that frame Dilthey's conception of knowledge. For this purpose, it seems appropriate to contrast it with the (Badian) neo-Kantian system of knowledge, which had been widely perceived as its anathema.

## 14.2 Dilthey's *Critique of Historical Reason* and Rickert's *Philosophy of Values:* Two Versions of a General Theory of Knowledge

Although Dilthey and neo-Kantians are regularly referred to as philosophical opponents, they both share a close relation to Kant and developed their foundational theories encroaching upon the Kantian epistemological guidelines. With regard to the outcome of our protagonists' endeavors, we can also detect some general resemblances like the assertion that natural sciences must be separated from human respectively cultural sciences, that there are two distinguishable methods, and that modern philosophy must, in general, overcome the fetters of metaphysical speculation, which had infused philosophical thinking up to the end of the nineteenth century. As is widely known, Dilthey and Rickert used quite divergent methodological strategies and key concepts to finally come up with a solution to their central issue of providing a new theory of knowledge. Likewise, their results bore very different epistemological consequences. Thus, to assert that they had divergent impacts on the establishment of modern social science is not seem a very surprising. However, with the notable exception of Max Weber, social scientists have only recently invested efforts in contextualizing the concrete philosophical environment of such figures as Simmel, Mannheim, Schutz, and others. But even in the case of Weber, as will be outlined below, commentators conveyed very inconsistent interpretations of his position with regard to Dilthey and Rickert. Instead of toiling with the pointless task of coming to grips with the vast amount of commentaries on Weber, it appears more appropriate to spell out the systematic diversity in the works of Dilthey and Rickert in the first place. Through this operation we can extract helpful heuristic criteria by which we can evaluate the philosophical implications of the "social epistemologies" (Fuller) of the sociological classics.

For lack of space, we have to confine ourselves to an extrapolation of the respective architectures of Dilthey's and Rickert's conception of knowledge. We use the terminology of a *holistic* versus a *dualistic* foundational strategy in order to mark the most crucial dissimilarities between both approaches. Since those terms are usually used for varied purposes, a few words cannot be omitted here.

In the following context, dualism and holism shall denote the basic *ontological* premises of the neo-Kantian and Diltheyean foundational systems. However, as we will see, those presumptions do, in fact, structure their subsequent theoretical arguments. Hence, they are not used, as is commonly done, as oppositional concepts to depict different *methodological* starting points.[3] While Dilthey's theory of knowledge is based on the ontological position according to which "inner" and "outer experience" cannot be separated but both pertain to one and the same "psychic nexus," Rickert sought to establish a novel ontological dualism by introducing the "sphere of values" (*Wertsphäre*) as an autonomous domain of philosophy. By separating "values" and "reality" (*Wirklichkeit*) as two distinct areas, he also introduced an original line of demarcation between philosophy on the one hand and empirical sciences on the other – while Dilthey was famed for founding both, human and social sciences, upon the "nexus of lived experience, expression, and understanding." (Dilthey [1910] 2002, 109) Rickert himself denounced such ontological premises as uttered by Dilthey as a "metaphysics of immediate experience" (Rickert 1999, 24) because it would not only reduce the world to a single substance, but, moreover, declare it unintelligible at the same time (Rickert 1999, 23). Rickert instead wrote: "Science has to be, at least, dualistic" (Rickert 1999, 24).

Dilthey's holistic notion of *life nexus* or *psychic nexus* as constituent of all forms of experience and knowledge on the one hand and Rickert's dualist scheme of value and reality on the other exemplify two distinct vantage points for philosophical foundation. In varied version, this opposition still stands in the background of contemporary debates on *Verstehen* in the social sciences.[4] Some important features and epistemological consequences shall thus be examined in more detail.

### *14.2.1 Dilthey and Holism*

Dilthey is widely identified as the philosopher who inaugurated an *ontological* demarcation criterion from which he conclusively derived a separation between humanities and natural sciences.[5] A closer look would, however, reveal that Dilthey

[3]A similar coining of the dualism–anti-dualism divide has been provided by Bertram and Liptow (2001) with reference to opposition philosophical formulations of the mind–world-relation.

[4]Soeffner (2004, 8), for instance, associated the various methodological controversies in the social sciences with the opposition between hermeneutical approaches and Cartesian standpoints.

[5]Among contemporary critics Rickert ([1899] 1986a, 12, 29, 70–72) and Weber (1968, 12) are the most prominent.

himself rejected such a position and insisted that "it is clear that the human sciences and natural sciences cannot be logically divided into two classes by means of two spheres of facts formed by them." (Dilthey [1910] 2002, 103) His ontological vantage point, as indicated above, was not dualistic in nature but rather holistic. What a holistic postulation presupposes can be grasped from Dilthey's "principle of phenomenality": "Whatever is there for us – because and insofar as it is there for us – is subject to the condition of being given in consciousness" ([1883] 1989, 246f.) From this perspective, it becomes clear why Dilthey's rejected an ontological demarcation criterion for the distinction between human and natural science. The difference between objects of "outer" and "inner experience", Dilthey continued, was only relative to their distinctive *mode of givenness* in our consciousness. While objects of "outer experience" were represented in our consciousness as mere "phenomena" (Dilthey [1894] 1977, 27), objects of "inner experience" would be given "immediately" and constituted what Dilthey referred to as "lived experience" (*Erlebnis*) or simply "life" (*Leben*). Dilthey added that only with the latter objects we were *involved* with the *whole totality* of our mind's structuring activity whereas the former objects are merely the result of rational thinking activity so that it would be inappropriate to talk of *involvement* in its pure meaning. On that basis, Dilthey introduced the method of *explanation* for the analysis of objects of outer experience and the method of *understanding* for objects of inner experience. He attributed to inner experience a superior status against outer experience. This becomes evident in his famous dictum that "thought cannot go behind life" (Dilthey [1931] 1968, 180). In one of his later sketches, this idea of primacy of *Erleben* over *Erkennen* is justified in a more concrete manner: "It is the nexus of life itself which brings about knowledge. We cannot go behind it" (Dilthey 1990, 312).

It was Dilthey's ambition in his psychological writings to analyze and objectify the complex psychic nexus in which *all* forms of human knowledge were grounded. But likewise his hermeneutical approach that he had developed by the end of his life supported the holistic starting point. In the *Aufbau* we find the following portrayal of the life nexus upon which Dilthey proceeded to ground the whole of the human and social sciences:

> The first givens are lived experiences. But, as I have attempted to show previously, they belong to a nexus that persists as permanent amidst all sorts of changes throughout the entire course of a life. [...] it encompasses our representations, evaluations, and purposes, and it exists in the connection of these constituents. And in each of these constituents, the acquired nexus exists in distinctive connections – in relationships of representations, in assessments of values, in the ordering of purposes. We possess this nexus; it operates constantly in us. [...] Thus it is always present and always efficacious, but without being conscious (Dilthey [1910] 2002, 102).

As had been widely perceived, Dilthey, in reaction to Ebbinghaus's critique of Dilthey's psychology, finally abandoned the vexatious attempts to expound the psychological mechanism and interrelations within the human cognitive apparatus. Instead, he installed a life-philosophical version of hermeneutics to illuminate the complex structure of human existence. He thus reformulated his original psychological impetus in recognition that "[t]he psychophysical life-unit is familiar with itself by means of the same double relationship of lived experience and understanding" (Dilthey [1910] 2002, 108).

The road to acquaintance of human existence would therefore lead via a detour of an interpretive account of the existing cultural forms of life (*Ausdruck*) that, in turn, relates back to common human activities and forms of perception (*Erleben*). Hermeneutic understanding (*Verstehen*) becomes a reliable method of acquisition of the human sciences because all three categories – *Erleben*, *Ausdruck*, and *Verstehen* – are grounded in the same existential substructure: life. Again, we find a holistic model at the basis of the human and social sciences:

> Thus the nexus of lived experience, expression, and understanding is everywhere the distinctive procedure by which humanity is present for us as an object of the human sciences. These sciences are founded upon this nexus of lived experience, expression, and understanding (Dilthey [1910] 2002, 109).

An often neglected but radical implication of Dilthey's holistic presumption is his dismissal of the very idea of epistemology as the primary foundational discipline.[6] Since the process of thinking was conceived of as a *secondary* phenomenon relative to the immediate forms of perception where the "the whole person" was involved, any attempt to separate a single part of our cognitive apparatus would detract us from the right track of *grasping life as it is*. Consequently, he introduced a novel discipline that would not only account for the reflexive capacities of human consciousness but also for the other forms of experience:

> This analysis of the total content and nexus of the facts of consciousness, which makes possible a foundation for the system of the sciences, we call "self-reflection," in contrast to "theory of knowledge" ([1883] 1989, 268).

A valid and full-fledged foundation of scientific knowledge would afford much more profound philosophical inquiry than had been undertaken so far, Dilthey suggested.

From this rough picture, we can derive that the tendencies of Dilthey's psychology as well as his hermeneutical foundation of the human and social sciences can only be comprehended from his holistic conception of psychic respectively life nexus, which led him to conclude that the function of *Erleben* is prior to reflexive cognition.

### *14.2.2 Rickert's Dualism*

The following picture of Rickert's theory of knowledge will probably differ from previous renderings. In most of them, we are exclusively confronted with his logic of the cultural sciences which he had offered in his "*Grenzen*" and, in a popularized version, in *Kulturwissenschaft und Naturwissenschaft*. Such a restricted view, however, omits that Rickert's theory of cultural science was actually couched in a particular concept of philosophy that was directed at overcoming the cultural and, at the same

[6] For the history of epistemology as a foundational discipline see the important description by Köhnke ([1986] 1991, 36–43).

time, scientific crisis of his times. Similar to Dilthey and many other representatives of the German mandarins (Ringer 1969), Rickert felt that a reconciliation of "science" and "life" (Rickert 1999, 9) – an often used counter-position to indicate the mainspring of the troubled situation – could only be achieved through a "*rehabilitation* of philosophy" (Krijnen 2001, 77). This momentum explains why even such an earnest philosophical character as Rickert did not hesitate to publish polemic pamphlets against any philosophical tendencies he considered harmful. Hence, to do justice to his genuine philosophical ambition, we would have to recontextualize his famous theory of concept formation within his broader architectural framework. Furthermore, on these grounds, we can also better grasp the philosophical reasons that made Weber, Simmel, and Mannheim repudiate some of the theoretical implications of the Badian version of neo-Kantianism.

In comparison to Dilthey, Rickert had built his theoretical edifice upon *dualistic* ontological assumptions. In a programmatic article that opened the first issue of the journal *Logos*, Rickert laid down the principles upon which his further intellectual development was grounded. Here, we encounter one of his most central philosophical teachings which reads: "Science has to be, at least, dualistic" (Rickert 1999, 24). This enigmatic statement becomes more profiled in contrast to a particular form of philosophical reasoning that Rickert had been wrestling with throughout his entire career. That position he termed "intuitionism" or "metaphysics of immediate experience" (Rickert 1999, 24). Such accounts, among which Rickert probably had subsumed Dilthey's theory, betrayed a "mystical monism" that would not only reduce the world to a single substance, but, moreover, declare it unintelligible (Rickert 1999, 23). In Rickert's eyes, this position would do away with science after all, since it declared scientific reflection incapable of grasping its proper object. Against monism Rickert declared that we could bring down our original experiences (*Erlebnisse*) to either the realm of *values* or the realm of *reality* so that "there is no problem that would be, put in adequate terms, a problem of philosophy" (Rickert 1999, 24).

By pointing at values, Rickert, in the first instance, intended to introduce a novel category of scientific objects into philosophical reflection which had so far been neglected by his predecessors. He formulated: "Apart from reality there are values whose validity (*Geltung*) we seek to understand" (Rickert 1999, 13). Secondly, and more generally, Rickert deemed it necessary to enlarge the philosophical concept of the *world* for the sake of establishing a well-grounded "Weltanschauungs*lehre*" (Rickert 1999, 323) as a solution to the current crisis. "Reality" and "world" are thus not equivalent concepts in Rickert's understanding. He reserved the latter term for marking out the special domain of philosophy while he asserted that empirical sciences would only be dealing with "reality." The properties of values were hence meticulously distinguished from the qualities he attributed to *Wirklichkeit*. Both categories, accordingly, belonged to different *ontological* provinces. While real things *existed*, values would either *be valid* (*gelten*) or not (Rickert 1999, 14). Values, as Rickert concluded, belong to "a realm of its own, beyond subject and object" (Rickert 1999, 14).

Rickert's version of dualism did not only contrast with the classical metaphysical notion of Descartes but also with Kant's division between "things-in-themselves" and their "appearance." We can characterize Rickert's theoretical shifting of these conceptions as an *ontologization*. He himself hinted at this particular nature of his taxonomy:

> It is within the domain of experiential reality itself that we have to affirm an ontological dualism, which is different from the traditional psychophysical dualism in principle (Rickert 1999, 377).

However, Rickert took a great deal of pain to prove his adaptation in line with Kant's transcendental idealism (Rickert 1999, 377). Virtually, he presented Kant as an analyst who was not only concerned with epistemological issues, but, to the same extent, also with "ontological problems" (Rickert 1999, 353).[7]

After this detour to Rickert's concept of philosophy, we can now summarize what I have referred to above as Rickert's dualistic version of a general theory of knowledge. Rickert has taught us, firstly, that the world did not consist of *real* things only, but also of entities that were *valid* and had meaning. Thus, he, secondly, rejected any scientific effort that would reduce objective things to unintelligible, subjectivist categories as faulty as well as any objectivist approach that would not even admit *the unintelligible*. The dualism of value and reality indicated to him a new starting point from where to pursue a legitimate theory of knowledge. According to this basic ontological division, Rickert went from *epistemological* projects to investigating the realm of *meaningful* qualities that were still a matter of ontological inquiry to him.

It is now left to analyze Rickert's logic of historical science and to deal with the question whether epistemology and ontology were compatible in the shape that Rickert had given them. Since his epistemological masterpiece, "The limits of concept-formation in natural science" ("*Grenzen*"), was reedited and also revised by Rickert several times, it also would seem compelling to look into the motives of Rickert's amendments. Within the limits of this study, we will have to confine ourselves to give only a slight impression of the direction that Rickert steered his epistemology. The main focus shall therefore concern the role he had advised the concept of value to play within the context of epistemology.

Similar to Dilthey in the famous "Vorrede" to his "Introduction to the human sciences," Rickert also recognized that Kant's epistemology left behind a "methodological gap" (Rickert 1999, 361). Since Kant did not develop a clear image of human and social science, his epistemology had merely been leaned on and carried out for natural scientific knowledge. Yet, Rickert perceived that an expansion of

---

[7] Be reminded that Rickert ([1899] 1986a, 27) had attacked Dilthey exactly for his ontologization of "nature" and "spirit." Symptomatically, contemporary adherents of Rickert aim at belittling his tendencies which, in fact his disciple Rudolf Zocher (1939, 35–37) already portrayed as an "ontologization of critical idealism." See for instance Krijnen (2001, 156) and Bohlken (2002, 13).

Kant's "Critique" to the realm of historical sciences would not be impeded by in principle obstacles: "The soul [*das Seelische*] can be treated with the same method as the body [*das Körpersein*]" (Rickert 1999, 361).[8] He finally came up with the strategy of introducing a methodological, or rather logical, criterion that would make both sciences discernible. A logical principle, in Rickert's eyes, could be offered by the respective modes of "concept-formation" (*Begriffsbildung*) prevalent in the natural and historical sciences. While the former would aim at *generalizing* single observations for the purpose of subsuming them to general laws, the latter would, alternatively, go for the *individual*, in the sense of singular, content within the empirical material. Still, both methods of concept formation could be applied in the natural as well as human sciences, Rickert added.

With that strategy, which he saw perfectly in accordance with Kantian transcendental idealism, Rickert shifted the epistemological focus away from questions about the qualities of the sciences onto the processes of subjective concept formation. Eventually, it was on the level of subjective experience of reality that Rickert did, explicitly, define an *ontological* difference between historical and natural scientific cognition (Rickert [1899] 1986a, 36; [1926] 1986b, 137f.; 1999, 362). At this point, the concept of value is imbued with important theoretical functions. Rickert continued to determine the difference between *nature* and *culture*. Accordingly, they could be distinguished on the grounds of the role they fulfill for human practice. Natural things around us, for instance trees, become "cultural goods" (Rickert [1926] 1986b, 138) as soon as they are bestowed with meaning, or rather value (in a neutral sense). Hence, it is this moment of "value-relatedness" (*Wertbezogenheit*) that made natural and historical objects separable.

Here, we can rediscover the dualistic opposition between reality and value that we have discussed above. Rickert claimed an "indifference of the real to the concept and to value" (Rickert [1926] 1986b, 214). On the one hand, this was Rickert's way of restating Kant's distinction of "thing-as-it-is" and its "appearance." Thus, a rift between reality and human perception was established. But, on the other hand, Rickert did not subscribe to the Kantian view that *all* cognition was, to a certain extent, determined by universal a priori. Symptomatically, he did not maintain the Kantian transcendental subject as his epistemological starting point, but replaced it, as was already noted by Iso Kern, by a "real and empirical subject" (Kern 1964, 396). On the footing, that there was a "*hiatus irrationalis*" (Lask), i.e., an unbridgeable gap between reality and cognition, values were supposed to regulate the process of concept formation on the side of the subject. Thus, only those sections of our experienced reality would become objects of our scientific efforts which conveyed those meanings that were of "interest" for us, i.e., that expressed values that are essential to us (Rickert [1899] 1986a, 198f.). However, by this "solution"

[8] This statement was launched as an attack against any position that would maintain an ontological difference between the objects of natural science and those of the human sciences. Of course, Rickert settled upon Dilthey as the most prominent upholder of such a viewpoint who is, consequently, marked as Rickert's adversary and even "whipping boy" (Oakes 1990, 161) until the day.

the problem was in a certain sense even aggravated, because values had so far been identified as the roots for nonobjective, relativist consequences in the philosophy of science which Rickert throughout his academic career had fought against. Naturally, he could not help recognizing the precarious implications of his theoretical move, at least in the first instance, when he wrote that "it is to be expected that our epistemological standpoint will be called 'subjective' in the pejorative sense" (Rickert [1926] 1986b, 216).

Rickert planned to alleviate the unsound constellation by declaring the act of relating to values, or "value relationship" (*Wertbeziehung*) as a formal and neutral process that would not imply any "valuations" (*Wertungen*) (Rickert 1929, 649). He was well aware that it was on this matter that his colleagues would probably turn away from him (Rickert [1899] 1986a, 168). Yet, he continued to assert that confusion about the scientific status of values was the source for many irritations in contemporary philosophy (Rickert 1999, 14). Considering the potential threat to his entire work, Rickert only invested little philosophical energy to confirm his hypothesis that relating to values would really not touch the objective status of historical knowledge. Instead, he confined himself to giving mere illustrations of it.[9] For this reason, he is even today called the "father of relativism" (Eliaeson 2002, 25).

From this overview we have seized the centrality of Rickert's dualism of reality and value on the field of philosophy of science. It served a twofold function: first, Rickert established an ontological criterion for the distinction of the objects of natural and human sciences: they are distinguished by the *relatedness to values* which only pertains to "culture." Second, Rickert referred to (objective) values for the purpose of securing the objective status of the historical sciences. In conclusion, we have observed how Rickert's entire philosophical thinking was organized around his dualistic starting point, namely the reality–value – distinction. The philosophical problems he deals with on distinct theoretical spheres are altogether related to the moment of disconnectedness of reality and value. To account for the essence and methodological significance of values must thus be judged as Rickert's most eminent ambition.

## 14.3 Sociology Between Neo-Kantianism and Philosophy of Life: Three Case Studies

With very few exceptions, the epistemological contributions by nonphilosophers or, as Weber put it, a "specialist in a single discipline" (Weber [1903] 1975, 209), have not been recognized by academic philosophers. This statement seems to be valid for then as for the present. Authors like Simmel, Durkheim, Mannheim, or Schutz are practically ignored even in textbooks on the philosophy of social sciences. Against this tendency, the following offering is aimed at reconstructing a glimpse

[9]See Oakes (1990, 90f.) and Bohlken (2002, 72).

of what could be specified as the "cognitive identity" (Lepenies 1981, I) of modern sociology by inquiring into some of their (explicit as well as implicit) arguments vis-à-vis the rival foundational strategies introduced in the chapter before. For reasons of space, we have to confine ourselves to very few key arguments, which tell us about the epistemological frameworks upon which these authors intended to ground the social sciences.[10]

Already hinting at our result, we can finally pick up Apel's statement that opened this contribution. It alludes to a particular predicament that is caused by a certain *tension* between the two elements of *Erklären* and *Verstehen*. Similarly, the upcoming (three) cases display different syntheses of holistic and dualistic interpretive figures.

### *14.3.1 Weber's Wissenschaftslehre*

Weber's methodology represents the most familiar case in point of mediation between *Erklären* and *Verstehen*. Both methodological strategies were amalgamated in his eminent definition of sociology as "eine Wissenschaft, welche soziales Handeln deutend verstehen und dadurch in seinem Ablauf und seinen Wirkungen ursächlich erklären will" (Weber [1922] 1980, 1). Still, there is disagreement with regard to the question of whether Weber's conception of *verstehende* sociology rested upon neo-Kantian or hermeneutical groundwork. While Burger, among many others, claimed that Weber's "theory is practically identical with that of Rickert" (Burger 1976, xii), there are other prominent interpreters like Weiß (1992, 355) who defended that "die Webersche 'verstehende Soziologie' [...] genau die durch die Diltheysche 'Kritik der historischen Vernunft' hindurchgegangene Gestalt dieser Wissenschaft (repräsentiere)."[11] A closer look at Weber's actual mediation between these two philosophical counter-positions shall not be bypassed here, despite the fact that this field being already comparatively well charted.

Weber, as cannot be ignored, drew heavily upon conceptualizations and theoretical models which were submitted by Rickert.[12] It is not necessary to retell this story in detail, since it belongs to the stock knowledge of sociological theory.[13] We will rather stress the points where Weber seems to leave behind the grounds of neo-Kantian philosophy.

---

[10] For a more extensive analysis see Šuber (2007).

[11] A similar position has already been taken by Wanstrat (1950, 31).

[12] Weber ([1903] 1975, 213) himself indicated: "One of the purposes of this study is to test the value of his ideas for the methodology of economics."

[13] Rickert's definition of "reality" as "an extensively and intensively infinite multiplicity of phenomena" Weber ([1903] 1975, 55), his exposition of "culture" (Weber 1949, 76) and, last but not least, the most essential parts of Rickert's concept of value-relation are the most significant adoptions by Weber.

The first point to mention where Weber (1994, 228) apparently went beyond the confines of Rickert's conception of *Kulturwissenschaften* is manifest in his definition of social science as *verstehende Wissenschaft*. From the first to the fifth (and so far last) edition of the *Grenzen*, Rickert was evidently averse to leaving any significant methodological function to *understanding*.[14] Against Rickert's firm assertion of "fundamental inaccessibility of other minds" (Weber [1903] 1975, 217), Weber rebutted that "both the course of human conduct and also human expressions of every sort are susceptible to a meaningful interpretation" (Weber [1903] 1975, 217f.). In another substantial passage, Weber – quoting Bernheim – more generally denoted the importance of *Verstehen* with regard to the possibility of historical knowledge: "history and the properties peculiar to it are possible because and only insofar as we can 'understand' men and 'interpret' their conduct" (Weber [1903] 1975, 260). This affirmative articulation toward the importance of *Verstehen* which we encounter in Weber's essay on Knies somewhat corresponds formally to the often quoted edict in the "*Objektivitätsaufsatz*":

> The transcendental presupposition of any *cultural science* is *not* that we find one or any "culture" to be of *value*, but that we are cultural *beings* endowed with the capacity and the desire to adopt a position with respect to the world, and lend it *meaning* (Weber [1904] 2004, 380f).

The last dictum has mostly been cited as verification of Weber's close affiliation with Rickert. However, taken together with the previous quotes we can also decode it as a confirmation of the human capacity of understanding as the primary precondition for the cultural and social sciences.

The next question we have to consider is whether Weber's notion of *Verstehen* was actually compatible with his neo-Kantian axioms. To begin with, it must be emphasized that Weber did not pursue to explore the nature of *Verstehen* either psychologically or philosophically, but contented himself to apply it to the logic of scientific forms of concept formation. What Weber centrally argued for was the possibility of "the sort of 'interpretation' which produces knowledge of *causal* relations" (Weber [1903] 1975, 149). He consequently ignored the propositions of hermeneuticists like Schleiermacher and Boeckh "since they do not pursue epistemological aims" (Weber [1903] 1975. Against contemporary proponents of a strong *Verstehen thesis*, like Knies, Münsterberg, and Gottl, Weber objected to the possibility to base the singular quality of the human sciences on *Verstehen*. Especially against Münsterberg, who had advanced a dualist conception of science according to which only the natural sciences could provide causal explanations, Weber objected that "individual human conduct is in principle intrinsically less 'irrational' than the individual natural event" (Weber [1903] 1975, 125). He thus denied that explanation and understanding, as the two principle forms of scientific method, could simply be attributed to natural respectively human sciences. On that account, he continued to argue (now against Simmel and Gottl), that scientific

[14] For an illustration of Rickert's (1913, 422–424) aversion to the concept of *Verstehen*, read the chapter on "explanation and understanding" in his "Grenzen."

interpretation must not be reduced to the immanent disclosure of inner states and relations, but that it could in fact be controlled objectively. He eventually declared unmistakably

> Every putatively valid piece of knowledge concerning the concrete complexes of immediate experience rests on "observation" of exactly the same logical structure as the "observation" employed in any analysis of the "objectified" world (Weber [1903] 1975, 161).

In this outlook, scientific interpretation would not only not be reliant on *Erleben*, but to the contrary, be feasible "only after the 'experience' itself has elapsed" (Weber [1903] 1975, 162).

To summarize, Weber on the one hand rejected the position that the phenomenon of *Verstehen* could be employed for the purpose of discriminating the natural and cultural sciences ontologically, but, on the other hand, he asserted that it fundamentally sustained the specific method of knowledge acquisition of the cultural sciences[15]. Weber's *Verstehen* thesis, in contrast to traditional hermeneutics, aimed at displaying the *logical* autonomy of interpretation and thus marked a line between *scientific* and *actual* understanding. Here, we finally recognize an overt familiarity to Rickert who had delivered a logical description of historical concept formation. Nonetheless, it can be demonstrated that Weber implemented Rickert's theory of value relation for distinct purposes and within a theoretical context that differed from its origin.

We must see that Weber turned to Rickert's theory of value relation not for its own sake, but for reasons that stand in relation to the problem of understanding. As already indicated, it had been Weber's main focus to confirm the practicability and scientific validity of causal interpretation. A striking difference between Weber's and Rickert's vision of historical epistemology lies therefore in the fact that the former attributed the task of reconstructing the meaning of individual actors to his "*verstehende*" sociology, while the latter primarily cared about the comprehension of objective values.[16] In fact, value relation was employed as a means to the bigger aim of the reconstruction of meaning. Weber declared that valuations were most helpful for producing causal knowledge.[17] Neither had values been assigned to control the practice of concept formation in the cultural sciences nor to secure their epistemological validity like in Rickert's attribution. Their methodological function was instead limited to the task of acquiring "*causal* status" (*kausale Durchsichtigkeit*) (Weber [1903] 1975, 63) and, thus, "communicability" (Weber 1968, 123).

---

[15] In *Economy and Society*, Weber (2004, 322) wrote: "In the case of 'social forms' (and in contrast to 'organisms') we can rise above the mere registration of functional relationships and rules ('laws') typical of all 'natural science' (where causal laws are established for events and patterns, and individual events then 'explained' on this basis) and achieve something quite inaccessible to natural science: namely, an '*understanding*' of the behavior of participating *individuals*."

[16] On this occasion we can again quote Weiß (1992), 52), who remarked that for Weber the concept of meaning was of more import than that of the concept of value.

[17] Since these passages are not yet translated we present its original formulation: "eminent leistungsfähige Geburtshelfer kausaler Erkenntnis" (Weber 1968, 124).

Traditionally, Weber's theory of the ideal type is taken for the core of his methodology. His incessant warnings of blurring the difference between real experience and constructed types of experience in the "*Objektivitätsaufsatz*" have frequently been perceived as a document of Weber's reliance upon Rickert. Indeed, *Verstehen* as a key problem of interpretative science is rarely referred to in this particular methodological piece of Weber's at all. As I want to indicate, it was the reference to the notion of "*Erfahrung*" that finally collapsed Rickert's strict dualistic distinction between *concept* and *reality*.

*Verstehen*, Weber once reasoned in his essay on Knies (1903), was in certain respects dependent on experience:

> "Verstehen" – im Sinne des evidenten "Deutens" – und "Erfahren" sind auf der einen Seite keine Gegensätze, denn jedes "Verstehen" *setzt* (psychologisch) "*Erfahrung*" *voraus* und ist (logisch) nur durch Bezugnahme auf "Erfahrung" als *geltend* demonstrierbar (Weber 1968, 115).

In this statement Weber points at a twofold function of experience that we shall outline briefly. First, we must portray the way in which *Erfahrung* is constitutive of concept formation as indicated in the last citation. In passages which are often overlooked, Weber laid out that only such ideal constructions would be valid that were in accord with so-called rules of experience (Weber 1968, 111). In his "*Soziologische Grundbegriffe*" (Weber ([1921] 2004)), Weber defined with reference to sociological concept formation:

> Only those statistical regularities which correspond to the understandably intended meaning of a social action are in the sense used here understandable types of action, i.e. "sociological rules." Only such rational constructions of meaningfully intelligible action are sociological types of real events observable in reality to some degree Weber ([1921] 2004), 319).

In the second place, Weber attributed to those *rules of experience* another role within his epistemology, namely to evaluate the results of historical knowledge. Again in the "*Grundbegriffe*," Weber had introduced "evidence" as an evaluative criterion for the *verstehende* approaches to science. Evidence eventually referred to rules of experience again. Weber had used the concepts of "objective potentiality" supplied by von Radbruch and the notion of "adequate causation" by Kries as theoretical instruments to define and to bring under methodological control the experiential rules. The form of validation of scientific results in the cultural sciences, according to Weber, would not have to be dependent on a theory of values, but could be supplied by the before mentioned categories.

To bring this to a close, Weber, by means of the concept of *Erfahrung*, had offered alternative solutions to problems that Rickert had already struggled with. With regard to the process of concept formation, Weber had, against Ricker, linked the scientific constructions back to *Erfahrung* and thus replaced Rickert's solution to relate them to transcendental values. Also, on the level of validation, Weber again relinquished the proposition provided by Rickert who inferred that this question had to be left to a philosophical theory of value. Instead, Weber insisted on criteria that arose from within the confines of empirical science and would be derived from *Erfahrung*.

In conclusion, by reassessing some of the writings of Weber's *Wissenschaftslehre*, we finally detected some indications at the ontological basis on which Weber grounded his methodology of the cultural sciences. His adoptions of the concepts of *Verstehen* and *Erfahren*, which we found at the bottom of Weber's theory of interpretation as well as his theory of ideal type, display a telling affinity to Dilthey's architectonic. In the latter's theory, they indicated the structural nexus upon which human sciences would rest. Notably, Weber's theory of ideal-typical concept formation testified to the fact of scientific concepts being directly contingent on everyday experience. This hypothesis, however, rebuts Weber's continual declamation of his theory being in line with Rickert's as well as Weber's rhetorical insistence on the gap that separated practical and scientific concepts and knowledge. He thus integrated philosophical ideas that originated from incompatible ontological presumptions and theories of knowledge. It is exactly this particular tension that elicited the recurrent laments about the inconsistency of Weber's *Wissenschaftslehre*.

## 14.3.2 Simmel's Theory of Historical Knowledge

Similar to Weber, Simmel has also been identified as a delegate of neo-Kantianism. Even Rickert adopted him as a disciple of his school. However, it often tends to be disregarded that Simmel had put forward an original account of "*Kantwissenschaft*" already in 1892 – 2 years before Windelband ([1894] 1998) outlined the principles of Badian neo-Kantianism in his famous address.[18] Hence, any attempt to reconstruct his theory of knowledge would have to start from an interpretation of his studies on the "Problems of the philosophy of history." What makes it difficult to determine a clear-cut position to Simmel, though, is the fact that he revised this work completely several years later in 1904 and started to work on a another (third) revision around 1913.

Similar to Dilthey, Windelband, and Rickert, Simmel developed the key categories of his theoretical approach to the historical sciences in close dialogue with Kant. Thereby, he was especially drawn by Kant's refutation of sensualism and empiricism. He took up the Kantian distinction between *knowledge* and *experience*, respectively *form* and *content*, and applied it to historical knowledge.[19] Yet, a simple transposition of the Kantian dualism to the field of human sciences would not be practicable as Simmel added. Through all the editions of his theory of historical knowledge he remained most sensitive to the central question of how it is possible to draw inferences from external factors given in *reality* to inner psychic states (Simmel [1892] 1989a, 422). Still, on a formal level he maintained, against Kant's notion of a priori, that it had installed too rigid a line between form and content. In fact, Simmel argued,

[18]According to Köhnke (1996, 102), Simmel chose the term "*Kantwissenschaft*" for strategical reasons, i.e., not to be rated under the prominent strands of neo-Kantianism.

[19]From this point on, as he summarized in an autobiographical sketch, he enhanced the distinction between form and content to a universal "metaphysical principle" (Simmel 1958, 9).

> Nicht scharfe, systematische Scheidungen, sondern allmählichste Übergänge bestehen zwischen den allgemeinsten. [...] Formen und den speziellen, selbst empirisch gewonnenen und als A priori nur für gewisse Inhalte anwendbaren (Simmel [1892] 1989a, 305).[20]

In other words, the a priori must not be fixed and, even less, universally defined, since every form of knowledge would rest on a specific combination of *real* and *epistemic* elements. Simmel thus appended to the "absolute a priori of the intellect" so-called relative a priori (Simmel [1892] 1989a, 304) that would give rise to various *practical* forms of knowledge. Finally, he generalized and transposed Kant's epistemological strategy to all forms of knowledge production. In that manner, he felt that Kant's separation of theory and practice was too strict:

> Die Praxis wie die Theorie machen in jedem Augenblick von Verbindungsformen für das empirische Material, von jenem eigentümlichen plastischen Vermögen des Geistes Gebrauch, das jeden gegebenen Inhalt durch die Art, ihn anzuordnen, zu stimmen und zu betonen, in die mannigfaltigsten definitiven Gestalten gießen kann (Simmel [1892] 1989a, 306).

Not by chance, we feel reminded at Dilthey's refutation of Kant's concept of the a priori.[21]

In the first edition of the *Problems*, Simmel affirmed that a theory of historical knowledge could not dispense with psychological inquiries that would account for the "unconscious and unproven preconditions" (Simmel [1892] 1989a, 308f.) on the side of the knowing subject. Psychological descriptions were, according to Simmel, also presumed to shed light on the enigma of historical understanding. From behind this formulation of the core problematic of historical epistemology, it is plausible to assume a positive impact of Dilthey on Simmel. Against such prejudgment, we shall prove that Simmel, in this respect similar to Weber, did not support a *hermeneutical* notion of *Verstehen* after all.

Although in the first instance Simmel paraphrased Dilthey's famous emphatic dictum: "How different is our knowledge of psychic life!" Simmel ([1905] 1997), 53) by stating: "In ganz anderer Weise, könnte man sagen, sei uns die Geschichte zugänglich, wie die Natur" (Simmel [1892] 1989a, 319), he instantly expressed his doubts that a "natural uniformity of the soul would create a bridge over the gap between the I and the non-I" (Simmel [1892] 1989a, 319). *Verstehen*, in Simmel's eyes, was not a process that would rely upon emphatic reexperiencing of other person's states (Simmel [1892] 1989a, 360). Instead, he believed that *Verstehen* necessarily implied a deviation from original motifs, emotions, and beliefs (Simmel [1892] 1989a, 320). These passages must be read as a critical comment to Dilthey's psychological foundation of historical knowledge. Even though it had been his initial aim to detect some of the rules that governed the syntheses of individual and objective aspects in the formation of historical knowledge, the final results of

---

[20] Unfortunately, only the second edition of *Problems* has been translated into English so far. For that reason we prefer to quote the original language.

[21] It is confirmed that large parts of Simmel's epistemological works have to be read as a (mostly hidden) remonstration with Dilthey. See the important data presented by Köhnke (1989, 1996, 358–360, 380–383, 420–424).

Simmel's first account remained unsystematic and opaque. In the end, he even admitted that he had not yet developed a "positive picture" of the Verstehen process (Simmel [1892] 1989a, 318).

On the basis of the *Probleme I*, we can thus identify Simmel's middle position between a straight Kantian position and Dilthey's hermeneutical psychology: on the one side he rejected a drastic form of separation between formal categories of thought and reality in the tradition of Kant; on the other hand, he also disbelieved that an objective understanding between different persons was available. With regard to the problem of objectivity, Simmel's final chapter gave some suggestions, but he did again refrain from offering clear-cut criteria. Of course, he rejected the view that there would be a royal road to objectivity for the historical sciences. The results of historical interpretation would remain on subjective grounds, since understanding was, according to Simmel's starting point, conditional on psychological properties.

In the preface of the second edition of his book *Einleitung in die Moralwissenschaft* in 1904, Simmel hinted at the direction to which his thought had meanwhile changed: from "psychological" to "objective (*sachliche*) method" (Simmel [1892] 1989b, 9). What he actually meant can be grasped from a gaze at the revised edition of the *Probleme* which was issued the same year.

In the introductory chapter, Simmel reformulated the problem of psychological understanding by bringing in another important distinction. He demanded a clear differentiation between the "content, i.e. the conceptual significance of psychic dynamics from those dynamics per se" Simmel ([1905] 1997)). On this account he came to a more precise separation between psychology and history than he had in 1892. Psychological understanding would be of different sort than historical understanding (Simmel ([1905] 1997), 272).

In an added chapter Simmel filled several examples for the sake of demonstrating how the given historical material itself would shape the construction of historiography so that he could conclude that there is, besides a history of the individual, a "history of the mere content of events" (*Geschichte der Geschehensinhalte*) (Simmel ([1905] 1997), 285). Compared to the first edition, Simmel's new position thus reinforced a stricter distinction between actual history and the form of its scientific representation. Because Simmel accentuated the autonomy of the historical forms against their content, Rickert was not astray to see Simmel switching sides and supporting his proper position. Nevertheless, Simmel did not sustain Rickert's value-philosophical solution to the problem of objectivity at all. In opposition to Rickert, he maintained that since historical constructions could not be measured against reality, they, consequently, would only "provide a truth corresponding to their proper aim" (Simmel ([1905] 1997), 289).[22]

[22] On this ground, we can easily bridge the gap which leads to Simmel's ([1900] 1990, 117) formulation of *relativism* as a novel principle of knowledge that he had launched in his *Philosophy of Money*. From the vantage point of logical autonomy of cognitive forms, there was no way to keep up objectivity in the traditional sense. Thus, Simmel was looking for an evaluative criterion that would "not claim exemption from its own principle."

Still we have not reached the end of our story. While the problem of *Verstehen* had been cast aside in the revised edition of the *Probleme* or, to be more precise, replaced with the logic of concept formation in the historical sciences, we observe that it was accorded an important place again in Simmel's last contributions to the philosophy of history. This time, it had not been the psychological backdrops of understanding, but the connections between scientific and "quotidian understanding" (Simmel 1980, 102) that were of central interest to him.

Starting from the basic presupposition that "*Erleben*" was our "most primitive mode in which a content of consciousness [...] becomes accessible to us" (Simmel 1980, 145), he drew attention to the question of the relatedness of scientific knowledge to *life*. This project again displays familiarities to Dilthey's hermeneutical works. Even from the title of one of his final essays, entitled "Vom Wesen des historischen Verstehens," it becomes evident that the *old* Simmel had meanwhile developed a clearer image of *Verstehen* than he did in 1892. How did he, then, 25 years after the *Probleme I* conceive of the interrelation between life and knowledge?

Each form of cognition, Simmel began, would start from *Erleben*. He added that, despite this origin, history as a form of knowledge would still reside in an ideal sphere independent of its origin (Simmel 2000, 322). Simmel had developed this assumption in the *Probleme II*, as we have just indicated. Yet, he went beyond this stage by establishing the "form of life" (Simmel 2000, 322) as a particular mode of perception that would be constitutive for the form of history, after all. He gave various illustrations of how both forms did, in fact, link up. Instead of delving into the fine details of the argument, we can restrict ourselves to the conclusion that Simmel, in the last years of his life, acknowledged that cognition itself could not be fully explained independently from its origin in life.

In his *Vom Wesen des historischen Verstehens*, he conceptualized these ideas in more general terms. In the course of his explanations he inferred "that we observe *the whole person*. The isolation of his corporality is a product of subsequent abstraction" (Simmel 1980, 102). Here we encounter argumentative figures that had been central in Dilthey's holistic approach to scientific foundation. They remind us of Dilthey's prominent conclusion according to which *Erleben* was prior to *Erkennen*. With regard to a definition of the nature of understanding, Simmel also defined it "an irreducible, primitive phenomenon in which a universal relationship between man and the world is expressed" (Simmel 1980, 124). This again reads like another concession to Diltheyean hermeneutics and a radical revision of his early standpoint. In the same direction, we have to interpret his depiction that scientific understanding was "a variant of our contemporaneous, thoroughly quotidian understanding" (Simmel 1980, 102). Eventually, he defined *life* as being

> the ultimate authority of the spirit, its court of last resort. Therefore the form of life ultimately determines the forms in which life itself can be intelligible. Life can only be understood by life (Simmel 1980, 124).

This passage reveals a line of argumentation that resembles Dilthey's description of the particular variant of a hermeneutic circle as it pertained to all human sciences almost to the letter. The old Simmel accounted for a close interrelatedness of forms of knowledge on the one hand and the form of life on the other. Thereby, he accorded primacy to the formational function of life. Simmel thus restated Dilthey's preeminent epistemological idea of the primacy of life over knowledge.

As our cursory overview of the development of Simmel's ideas on the epistemological preconditions of the historical sciences have indicated, he was, very similar to Weber, struggling with the question whether scientific knowledge constituted a categorical realm beyond, and autonomous from, ontological properties or not. While, at the first stage, he maintained that the actual grounding of historical knowledge was psychological in its nature, he, since 1900, strove to confirm the *logical* autonomy of historical forms. At this point, he was approximating Rickert's position. Symptomatically, *Verstehen* was not bestowed with any crucial epistemological relevance for the construction of historical images any more. Only in his final writings, he took up philosophical principles that display many structural similarities to Dilthey's life-philosophical assumptions. The doctrine of logical autonomy of scientific forms was demoted in favor of that of a close connection between scientific forms and *life*. Now, the *quotidian* understanding was seen as fundamental for *scientific* understanding.

Therefore, in Simmel's as well as in Weber's methodologies we can observe a more or less explicit resort to a foundational structure upon which the logic of scientific concept formation is conditional. Although its influence is not spelled out in detail and occasionally contradicted by superficial (self-) characterizations, the tendency to deny the neo-Kantian dualist separation of science and life is predominant. Unlike Rickert, neither Weber nor Simmel took refuge to transcendental tenets in order to save objectivity for the cultural and historical sciences. Their conceptualizations of social science can be considered as being holistic in that they presuppose that scientific methods and practical modes of perception are interrelated and must not be separated.

### *14.3.3 Mannheim's Theory of Cultural Sociology*

In the final chapter, we pursue to interpret the early epistemological writings of Karl Mannheim as one of the rare versions of philosophy of social sciences that was set on holistic parameters.[23] For several reasons, it is often passed over the fact that he had conceived sociology of knowledge not only as an answer to the difficult political constellation in Germany during the Weimar Republic,[24] but also, and even more so, to the foundational crisis in the human sciences. Mannheim's work, similar to Simmel's, has regularly been criticized for being unsystematic and ambiguous

---

[23] The influence of Dilthey and the historicist tradition in Mannheim's work is obvious and frequently stated in the secondary literature.

[24] The important comprehensive analysis of Mannheim's work provided by Kettler et al. (1989, 14f.), which today has a status as a standard, has claimed to behold Mannheim as a "political theorist" and his *Wissenssoziologie* as a "critique of the predominating tradition of political theory."

(Loader 1985, 3). Before we begin to make sense of Mannheim's theory of cultural sociology, we shall therefore recontextualize the specific motifs that originally stimulated his turn to conceptual questions.

Mannheim's earliest scientific footsteps render a clear picture of the troubles that concerned the student. They confirm that, although raised in Hungary, he was fully aware of the intellectual topics informing the scientific discourse in central Europe. In his first public address, in 1918, he presented himself as a spokesman for the young generation that was confronted with resolving the division of objective and subjective culture (Mannheim 1970, 66–69). Here, the 25-year-old Mannheim not only supplied an impressive sketch of the cultural–historical background of the development of Western thought, but also suggested a solution that was oriented toward Simmel's analysis of cultural phenomena (Mannheim 1970, 77). We can thus conclude, with Wolff, that already the student Mannheim was concerned with finding a "synthesis of idealism and Marxism respectively mind and society" (Wolff 1964, 13). The profundity of Mannheim's knowledge about the systematic philosophies of his time is well documented in his dissertation thesis from 1917 where he looked out for a new principle of logical classification of the predominant philosophical systems. From this outlook, we must neither be surprised nor irritated by the announcement that he was striving for "a novum organon in the humanistic sciences" (Mannheim [1980] 1982, 150). He introduced this formula in the second of two manuscripts that were probably written in 1922 and edited posthumously under the heading "Structures of thinking" in Mannheim ([1980] 1982), in an English translation). Both works were designated to explore the properties of "cultural sociological knowledge." The question why Mannheim had actually chosen a formula that was alluding to Francis Bacon's masterpiece *Novum Organon* instead of restating Dilthey's *Critique of Historical Reason* cannot be decided. But, for reasons that will be elucidated soon, we can assume that he deliberately disowned the Kantian heritage. What is evident, though, is that Mannheim really sensed the relevance of a (re)formulation of a general theory of historical knowledge. He occasionally referred to it as "a theory of knowing the qualitative" (Mannheim [1980] 1982, 160).

How did it come about that, even four decades after Dilthey's "Introduction," the problem of knowledge still was not, at least in Mannheim's eyes, satisfactorily resolved? According to the latter, this curious occurrence was

> due to the circumstance that the new disciplines of human scientific research have grown out of a philosophy different from that in which the still dominant methodological theory originated (Mannheim [1980] 1982, 151).

Since Mannheim, as we have indicated, was not ignorant of the various positions among the foundational discourse, it will be instructive to listen to his comments on the alternative philosophical schools.

Against the neo-Kantian system in the versions of Windelband and Rickert, Mannheim betrayed a stubborn adherence to "rationalistic premises and prejudices which had come into it out of the rationalistic tradition of the Enlightenment" (Mannheim [1980] 1982, 161). He continued by spelling out that the

> starting point alone, from which the problem is approached, closes off, in our opinion, the possibility of a fundamental solution to the questions relevant here (Mannheim [1980] 1982, 161).

As the unproductive starting point on the side of Badian neo-Kantianism, Mannheim disclosed the grounding of the distinction between natural and humanist sciences on a *formal* criterion. Mannheim categorically defined that any "attempt to construct an epistemology without ontological presuppositions must today be regarded as having already failed (Mannheim [1980] 1982, 283).[25] We notice that Kant, on whom the young Mannheim had commented quite neutrally, if not benevolently, in his doctoral thesis, figured as critical foil in the later works.

Apart from neo-Kantianism, Mannheim also remarked upon the phenomenological school around Edmund Husserl. On the one hand he positively professed to be an adherent of the broader phenomenological movement himself ([Mannheim [1980] 1982, 281), on the other hand he seemed to be reluctant to engage in a true examination of the basic arguments of Husserl's works. He justified his hesitation by hinting at the fact that "the results of pure phenomenology" were still "unfinished, problematic, and unstable" (Mannheim [1980] 1982, 110). From different side remarks it can be concluded that he conceived of phenomenology as still reliant partially on Kantian conceptual figures. For instance, when he wrote: "But while phenomenology continued to operate with a timeless consciousness, in this respect resembling Kantianism [...]" (Mannheim [1980] 1982, 173).

In Dilthey's theory, Mannheim detected a sounder philosophical base for the foundation of cultural and social science. To be more concrete, here he found the "philosophical grounds more congenial to the humanistic sciences." (Mannheim [1980] 1982, 162) According to Mannheim, Dilthey sprang from a very different philosophical tradition of thinking, i.e., historicism or romanticism, which would be open to such preconditions which have been commonly neglected. He especially deemed Dilthey's theory of worldviews as the most eminent theory of history. However, the most eminent idea that Dilthey had forged with regard to a foundation of humanistic science was summarized as follows:

> Dilthey had made everyday experience or "general experience of the world" a problem of history. We consider it to be one of the most important problems of cultural sociology (Mannheim [1980] 1982, 134).

On these premises, Mannheim developed the primal quest not only for his account on cultural science but also, as shall be seen, for his formulation of sociology of knowledge, namely to grasp the "fact of inner connection between thinking and existence" (Mannheim [1980] 1982, 163). This connectedness, as he continued, would be typical of humanistic knowledge, but not for the natural sciences. He thus derived a division between the "total consciousness" from the "theoretical consciousness" (Mannheim [1980] 1982, 187), and in this manner defined an *ontological* demarcation that he further delineated.

---

[25] Mannheim (1922) had already come to this conclusion in his dissertation.

The resemblance with Dilthey's concept of the psychic nexus is apparent already at this point. If we listen to Mannheim's portrayal of "total consciousness," the correspondence becomes even more striking. He asserted that it involved subjective aspects and modes of experience such as "'loving', 'acting', 'wanting to change'" (Mannheim [1980] 1982, 186). He, thus, based his novel conception of cultural science on

> the fact (and the methodological consequences of this fact) that the subject of cultural-scientific knowledge is not the mere epistemological subject, but the "whole man" (Mannheim [1980] 1982, 50).

Another blatant analogy comes to the fore when Mannheim spells out "that every act of knowledge is only a dependent part of an existential relation between subject and object" and then adds that in this "existential relationship [...] knowing is only one side" (Mannheim [1980] 1982, 186). In this argument we immediately discern Dilthey's life-philosophical key idea according to which knowledge could not be explained isolated from the overall structure of the whole psychic nexus.

Given these conditions, Mannheim carried on to define their consequences with regard to the methodology of understanding in the cultural sciences. Thereby, he came to a most significant conclusion that would eventually become the vantage point for his sociology of knowledge. Because

> this kind of knowledge is always anchored in far-reaching foundations, and that products of knowledge of this kind are valid only for the circles whose existential attachments they express and in the form in which they present themselves [...] This kind of knowledge, then, has conjunctive validity only, not objective validity (Mannheim [1980] 1982, 193).

The foremost idea that is indicated in this passage is the idea of perspectivism of knowledge, and consequently the main justification for a sociology of knowledge as a "theory of the social determination of knowledge" (Mannheim [1929] 1968, 239). In 1922, he adopted the formula "conjunctive knowledge" from Freiherr von Weizsäcker to signify the genesis of that particular type of knowledge.

The next thing we learn is that understanding is not immediately given, but that it had to be gained through interpretation. Moreover, understanding in this sense would only be likely on the premise that subject and object shared similar experiences to some degree. In the end, he defined: "interpretative understanding means penetration into an existential space, bound to a community, into its formations of meaning and their existential bases" (Mannheim [1980] 1982, 243). Here, we can perceive Mannheim reiterating Dilthey's theory of *Verstehen* of the "Aufbau" where he had determined that an interpretation (*Verstehen*) of cultural expressions (*Ausdruck*) would rest upon the givenness of the foundational structure of life (*Leben*). Dilthey, indicating the movement of cyclical *Verstehen*, had put it in the simple terms: "Human spirit can only understand what it has created" (Dilthey [1910] 2002, 148). In Mannheim we read:

> Conjunctive knowledge thus everywhere comes upon nothing but spiritual realities. It takes in institutions, works, and collective creations – in short, nothing but formations of meaning [...]. It always takes in things spiritual filled with the same spirit (Mannheim [1980] 1982, 237).

With Dilthey's theory and methodology of historical knowledge we have thus tracked down the philosophical foundations upon which Mannheim, since 1925, started to build his sociology of knowledge. For him, this new discipline represented a specific technical device for the interpretation of a particular type of knowledge for which it is assumed that it "can only be relationally formulated" (Mannheim [1929] 1968, 270). Daring a more provocative articulation, we can even conclude that Mannheim developed and transposed Dilthey's holistic theory into a sociological methodology.

The foundational potential of this approach has rarely been acknowledged so far, mainly because of the unhappy reception of the sociology of knowledge since World War II.[26] In comparison to Weber and Simmel, Mannheim resolutely discharged the neo-Kantian presumptions of any solemn relevance with regard to a theory of cultural scientific knowledge. He went further in forging conceptual pathways that have been opened by Dilthey's critique of historical reason.

## References

Acham K (1985) Diltheys Beitrag zur Theorie der Kultur- und Sozialwissenschaften. Dilthey-Jahrb. f. Phil. u. Gesch. d. Geisteswiss. 3:9–51

Acham K (1992) Das Verstehen und die Wissenschaft von der Gesellschaft bei Wilhelm Dilthey. Annali di Sociologia 8(1):85–111

Apel K-O (1987) Dilthey's distinction between 'explanation' and 'understanding' and the possibility of its 'mediation'. J Hist Philos 25(1):131–149

Bakker JI (Hans) (1999) Wilhelm Dilthey: Classical sociological theorist, Quart J Ideol 1&2:43–82

Bertram GW, Liptow J (2001) Zu einer antidualistischen Rekonstruktion sprachlicher Bedeutung: Robert B. Brandom und Wilfrid Sellars. Phil. Rundschau 48(4):273–300

Bohlken E (2002) Grundlagen einer interkulturellen ethik. Perspektiven der transzendentalen kulturphilosophie heinrich rickerts. Trierer Studien zur Kulturphilosophie, Bd. 6. Königshausen & Neumann, Würzburg

Burger T (1976) Max Weber's Theory of Concept Formation. History, Laws, and Ideal Types. Duke University Press, Durham

Dilthey W ([1883] 1989) Introduction to the human sciences. Selected works vol. I. Translated by Rudolf A. Makkreel and Frithjof Rodi. Originally published as Einleitung in die geisteswissenschaften. Versuch einer grundlegung für das studium der gesellschaft und der Geschichte. (Duncker & Humblot, Leipzig). Princeton U.P., Princeton

Dilthey W ([1894] 1977) Descriptive psychology and historical understanding. Translated by Richard M. Zaner and Kenneth L. Heiges. Originally published as Ideen über eine beschreibende und zergliedernde psychologie, and das verstehen anderer personen und ihrer lebensäußerungen. (Sitzungsberichte der Berliner Akademie) Nijhoff, Den Haag

Dilthey W ([1910] 2002) The formation of the historical world in the human sciences. selected works, vol. III. Translated by Rudolf A. Makkreel and Frithjof Rodi. Originally published as "Der Aufbau der geschichtlichen Welt in den Geisteswissenschaften." In: Abhandlungen der preußischen akademie der wissenschaften. (Verlag der Königlichen Akademie der Wissenschaften, Berlin). Princeton U.P., Princeton

[26] Since World War II, the dominant schools of sociology of knowledge have systematically tended to discount the epistemological surmises and implications of their subdiscipline.

Dilthey W ([1931] 1968) Weltanschauungslehre. Abhandlungen zur philosophie der philosophie. Gesammelte schriften, vol. VIII. Teubner, Stuttgart
Dilthey W (1990) Logik und system der philosophischen wissenschaften. Vorlesungen zur erkenntnistheoretischen logik und methodologie (1864–1903). Gesammelte schriften, vol. XX. Vandenhoeck & Ruprecht, Göttingen
Eliaeson S (2002) Max Weber's Methodologies. Interpretation and Critique, Polity Press, Cambridge
Hahn A (1999) Die Systemtheorie Wilhelm Diltheys. Für Niklas Luhmann. Berl. Journ. f. Soz. 9:5–24
Kern I (1964) Husserl und Kant. Eine Untersuchung über Husserls Verhältnis zu Kant und zum Neukantianismus, Nijhoff, Den Haag
Kettler D, Meja V, Stehr N (1989) Politisches wissen. Studien zu karl mannheim. Suhrkamp, Frankfurt am Main
Köhnke KC ([1986] 1991) The rise of neo-Kantianism. German academic philosophy between idealism and positivism. Translated by R.J. Hollingdale. Originally published as entstehung und aufstieg des neukantinismus. Die deutsche Universitätsphilosophie zwischen Idealismus und Positivismus. (Suhrkamp, Frankfurt am Main) Cambridge University Press, Cambridge
Köhnke KC (1989) Die Wechselwirkung zwischen Diltheys Soziologiekritik und Simmels soziologischer Methodik. Dilthey-Jahrb.f.Phil.u.Gesch.d.Geisteswiss. 6:303–326
Köhnke KC (1996) Der junge simmel in theoriebeziehungen und sozialen bewegungen. Suhrkamp, Frankfurt am Main
Krijnen C (2001) Nachmetaphysischer sinn. Eine problemgeschichtliche und systematische studie zu den prinzipien der wertphilosophie heinrich rickerts. Königshausen & Neumann, Würzburg
Lepenies W (1981) Einleitung. Studien zur kognitiven, sozialen und historischen Identität der Soziologie. In: Geschichte der soziologie. Studien zur kognitiven, sozialen und historischen identität einer disziplin, vol. 1. Suhrkamp, Frankfurt am Main, pp. I–XXXV
Loader C (1985) The intellectual Development of Karl Mannheim: Culture, Politics, and Planning. Cambridge University Press, Cambridge
Mannheim K (1922) Die strukturanalyse der erkenntnistheorie. von Reuther & Reichard, Berlin
Mannheim K ([1929] 1968) Ideology and Utopia. An Introduction to the Sociology of Knowledge. Translated by Louis Wirth and Edward Shils. Originally published as Ideologie und Utopie. Cohen, Bonn. Harcourt, Brace & World, New York
Mannheim K (1969) Essays on Sociology and Social Psychology. Routledge, London
Mannheim K (1970) Wissenssoziologie. Auswahl aus dem werk. Luchterhand, Berlin
Mannheim K ([1980] 1982) Structures of Thinking. Translated by Jeremy J. Shapiro and Shierry Weber Nicholson. Originally published as Strukturen des Denkens. (Suhrkamp, Frankfurt am Main) Routledge, London
Oakes G (1990) Die Grenzen kulturwissenschaftlicher begriffsbildung. Heidelberger Max Weber – Vorlesungen 1982. Suhrkamp, Frankfurt am Main
Rickert H ([1899] 1986a) Kulturwissenschaft und naturwissenschaft. Originally published by Mohr, Tübingen. Reclam, Stuttgart
Rickert H (1913) Die Grenzen der naturwissenschaftlichen begriffsbildung. Eine logische Einleitung in die historischen wissenschaften, 2 revised ednth edn. Mohr, Tübingen
Rickert H ([1926] 1986b) The Limits of Concept Formation in Natural Science: A logical introduction to the historical sciences. Translated by Guy Oakes. Originally published as Die Grenzen der naturwissenschaftlichen Begriffsbildung. Eine logische Einleitung in die historischen Wissenschaften. (Mohr, Tübingen) Cambridge University Press, Cambridge
Rickert H (1929) Die grenzen der naturwissenschaftlichen begriffsbildung. Eine logische einleitung in die historischen wissenschaften, 5th (amended) edn. Mohr, Tübingen
Rickert H (1999) Philosophische aufsätze. In: Bast RA (ed.). Mohr, Tübingen
Ringer FK (1969) The Decline of the German Mandarins. The German Academic Community, 1890–1933. Harvard University Press, Cambridge

Simmel G ([1892] 1989a) Die Probleme der Geschichtsphilosophie. Eine erkenntnistheoretische Studie, 1st edn. in Georg-simmel-gesamtausgabe, vol. 2. Originally published by Duncker & Humblot, Leipzig, Suhrkamp, Frankfurt am Main, pp 297–421
Simmel G ([1892] 1989b) Einleitung in die Moralwissenschaft. Eine Kritik ethischer Grundbegriffe, vol. 1. Georg-simmel-gesamtausgabe, vol. 3. Originally published by Hertz, Berlin, Suhrkamp, Frankfurt am Main
Simmel G ([1900] 1990) The philosophy of money. Translated by Tom Bottomore and David Frisby. Originally published as Philosophie des Geldes. Duncker & Humblot, Leipzig, Routledge, London
Simmel G ([1905] 1977) The problems of the philosophy of history. An epistemological essay. Translated by Guy Oakes. Originally published as Die probleme der geschichtsphilosophie. eine erkenntnistheoretische studie, 2nd and revised edn. Duncker & Humblot, Leipzig, Free Press, New York
Simmel G ([1905] 1997) Kant. Die Probleme der Geschichtsphilosophie (Zweite Fassung 1905/07). Georg-simmel-gesamtausgabe, vol. 9. Originally published by Duncker & Humblot, Leipzig, Suhrkamp, Frankfurt am Main
Simmel G (1958) Aus einer unvollendeten Selbstdarstellung. In: Gassen K, Landmann M (eds) Buch des dankes an georg simmel. Briefe, Erinnerungen, Bibliographie. Duncker & Humblot, Berlin, pp 9, 10
Simmel G (1980) Essays on Interpretation in Social Science. Translated by Guy Oakes. Totowa, Rowman and Littlefield
Simmel G (2000) Aufsätze und Abhandlungen 1909–1918. Georg-simmel-gesamtausgabe, vol. 13. Suhrkamp, Frankfurt am Main
Soeffner H-G (2004) Prämissen einer Sozialwissenschaftlichen Hermeneutik. In: Soeffner H-G Auslegung des alltags – der alltag der auslegung. Universitätsverlag Konstanz, Konstanz, pp 78–113
Šuber D (2007) Zur soziologischen Kritik der philosophischen vernunft. Zum Verhältnis von Soziologie und Philosophie um 1900. Transcript: Bielefeld
Wanstrat R (1950) Das sozialwissenschaftliche Verstehen bei Dilthey und Max Weber. Schmollers Jahrbuch 70:19–44
Weber M ([1903] 1975) Roscher and Knies and the logical problems of historical economics. Translated by Guy Oakes. Originally published as Roscher und Knies und die logischen Probleme der historischen Nationalökonomie. In: Jahrbuch für Gesetzgebung, Verwaltung und Volkswirtschaft im Deutschen Reich 27(3), 1181–1221. Free Press, New York
Weber M ([1904] 2004) Objectivity in social science and social policy. In Whimster S (ed.) The Essential Weber. A reader. Routledge, London, pp 359–403
Weber M ([1921] 2004) Basic sociological concepts. In: Whimster S (ed.) The essential Weber. A reader. Originally published as Soziologische Grundbegriffe. In: Weber M, Grundriss der Sozialökonomik III. Abteilung: Wirtschaft und Gesellschaft. Erster Teil, Mohr, Tübingen, 1921, §§ 1–7; Routledge, London, pp 311–358
Weber M ([1922] 1980) Wirtschaft und Gesellschaft. Mohr, Tübingen
Weber M (1949) Max Weber on the methodology of the social sciences. Shils EA, Finch HA (eds). Translated by Edward A. Shils and Henry A. Finch. Originally published as three different essays (1904) 1905, and 1917. The Free Press, Glencoe, IL
Weber M (1968) Gesammelte aufsätze zur wissenschaftslehre. Mohr, Tübingen
Weber M (1994) Sociological writings. Heydebrand W (ed). Continuum, New York
Weber M (2004) The essential Weber. A reader. Whimster S (ed). Routledge, London
Weiß J (1992) Das verstehen des lebens und die verstehende Soziologie (Dilthey und Weber). Annali di Sociologia 8:353–368
Windelband W ([1894] 1998) History and natural science. Theory Psychol 8(1). Originally published as Geschichte und naturwissenschaft. Heitz & Mündel, Straßburg, pp 5–22
Wolff KH (1964) Karl Mannheim in seinen Abhandlungen bis 1933. In: Wolff KH (ed.) Karl Mannheim. Wissenssoziologie. Luchterhand, Berlin, pp 12–65
Zocher R (1939) Die philosophische grundlehre. Eine studie zur kritik der ontologie. Mohr, Tübingen

# Chapter 15
# Opposition to *Verstehen* in Orthodox Logical Empiricism

**Thomas Uebel**

## 15.1 Orthodox Logical Empiricism and *Verstehen*: An Overview

Let's begin with an unexciting commonplace about logical empiricism in order to raise a question about what follows from it. In trying to answer it we shall find that matters are not as cut and dried as is often supposed – even before we get to heterodox representatives of the movement like Otto Neurath.

Logical empiricism always opposed the separation of *Naturwissenschaften*, the natural sciences, from *Geisteswissenschaften*, the human sciences, and promoted a (variously understood) model of unified science. Does this mean that logical empiricism *therefore* also opposed all interpretive procedures in the social sciences? If it does not mean that and logical empiricism did allow some such procedures to play a role, in *what form* and *what capacity* did it allow them to do so?

We shall find that it not only depends on whether we are looking at orthodox or heterodox logical empiricism, but also on what developmental phases of the orthodoxy are at issue – and on which wing of the orthodoxy. That said, for present purposes it suffices to distinguish three phases. The first covers the later 1920s and most of the 1930s, the second covers the 1940s and the third covers the 1950s and 1960s. (It is during the third phase that we can distinguish two varieties of the orthodoxy.) Each of the phases can be interrogated for its characteristic stance vis-à-vis *Verstehen* and for its characteristic argumentative strategy in support of that stance.

Phase one finds logical empiricism still in Central Europe, with Vienna and Berlin its main centers. Orthodox representatives with regard to our topic are Rudolf Carnap with his small book *Scheinprobleme in der Philosophie* (*Pseudoproblems in Philosophy*) (Carnap [1928a] 2003) and the essay "Psychologie in physikalischer Sprache" ("Psychology in Physicalist Language", Carnap [1932] 1959) and Carl Gustav Hempel with his essay "Analyse logique de la psychologie" ("The Logical Analysis of Psychology", Hempel [1935] 1949).

T. Uebel (✉)
Philosophy, School of Social Sciences, University of Manchester,
Manchester M13 9PL, U.K
e-mail: thomas.e.uebel@manchester.ac.uk

U. Feest (ed.), *Historical Perspectives on Erklären and Verstehen*, Archimedes, Vol. 21,
DOI 10.1007/978-90-481-3540-0_15,  

Phase two finds logical empiricism in the process of enculturation in its North American exile. Hempel continues as its representative for our topic with two papers, "The Function of General Laws in History" (Hempel [1942] 1965) and "Studies in the Logic Explanation" (Hempel and Oppenheim [1948] 1965).

Phase three sees logical empiricism as the dominant force in North American philosophy of science, now with indigenous support. Representatives are Theodore Abel with his much-anthologized essay "The Operation Called *Verstehen*" (1948), Richard Rudner with his monograph *Philosophy of Social Science* (1966) and again Hempel with the essays "Typological Methods in the Natural and Human Sciences" (Hempel [1952] 1965), "Rational Action" (Hempel [1962] 2001) and "Aspects of Scientific Explanation" (Hempel 1965). As we shall see, it is Abel and Rudner who present positions close to (and in one case even coarser than) the orthodoxy of the previous phase and Hempel who moved on to a new understanding of the issues under discussion.

The positions expressed vis-à-vis *Verstehen* are of two sorts. During phase one it is held that *Verstehen* can at best be of heuristic interest in formulating hypotheses but cannot play a role in their validation. This position is upheld in phase two and by the ultra-orthodox even during phase three. Yet phase three also see the development of a more sophisticated view which accords a validational role to *Verstehen*.

The argumentative strategies adopted are the following. During phase one, the argument for the merely heuristic role of *Verstehen* is based on the general doctrine of logical behaviourism (or a liberalised form thereof). During phase two and the ultra-orthodox version of phase three it is based on the argument that all explanation follows the deductive-nomological model. The sophisticated stage three position that also grants a validational role to *Verstehen* meanwhile is based on a conception of psychological explanation as incorporating ideal rational types and thereby satisfying the deductive-nomological model.

Across the three stages of its development, we can see changes in the logical empiricists' argumentative strategy against the separatism of *Geisteswissenschaften* and also in what the logical empiricists actually objected to. Once the reductive doctrine of logical behaviourism had run into trouble, the argument based itself on the deductive-nomological model of explanation. And once intentional psychological terminology was admitted as *bona fide* theoretical talk, all that remained objectionable about *Verstehen* was the traditional claim that it pertained to a non-physical and separate domain of mental or spiritual agency.

The overall conclusion I will come to is the following. That logical empiricism always opposed the separation of the *Natur-* from the *Geisteswissenschaften* does *not* mean that it also opposed all interpretive procedures in the social sciences. Even when only orthodox logical empiricism is considered, we find that all of its representatives allowed *Verstehen* as heuristics and did so during all phases of its development. However, the later Hempel also came to allow *Verstehen* a validational role concerning the hypotheses it suggested and so opened a door to "post-positivist"developments.

## 15.2 The Rejection of *Verstehen* by Carnap 1928–1932

Beginning with early logical empiricism before it was forced into exile, we can turn first to the Vienna Circle's "manifesto", signed by Carnap, Neurath and Hans Hahn.

Given the centrality of the doctrine of unified science there, it is not without interest to note that doctrines of empathetic knowledge and their relation to the separatism of the social sciences do not merit a specific discussion of their own (i.e., rejection) in this manifesto. Needless to say, it was stressed that "the goal ahead is unified science" (Carnap et al. [1929] 1973, 306) and claims about the power of intuitive knowledge generally were dismissed in the following paragraph:

> Intuition which is especially emphasised by metaphysicians as a source of knowledge, is not rejected as such by the scientific world-conception. However, rational justification has to pursue all intuitive knowledge step by step. The seeker is allowed any method; but what has been found must stand up to testing. The view which attributes to intuition a superior and more penetrating power of knowing, capable of leading beyond the contents of sense experience and not to be confined by the shackles of conceptual thought – this view is to be rejected (Ibid, 308f).

Clearly, this rejection also holds for the empathetic intuition in social science. But the point that was particularly stressed in the manifesto's section "Foundations of the Social Sciences" was not this but only a related one concerned with methodological individualism (understood methodologically).

> It is not too difficult to drop concepts like 'folk spirit' (*Volksgeist*) and instead to choose, as our object, groups of individuals of a certain kind. Scholars from the most divers trends, such as Quesnay, Adam Smith, Ricardo, Comte, Marx, Menger, Walras, Müller-Lyer, have worked in the sense of the empiricist, anti-metaphysical attitude. The object of history and economics are people, things and their arrangement (Ibid, 315).

Remarkably, the manifesto also asserted that while the "purification" of the social sciences from metaphysics had "not yet reached the same degree as in physics", it was "less urgent perhaps" (Ibid). One can only speculate who of the manifesto's authors was responsible for this particular slip. Most important, however, is that the overall shape of the implied argument against *Verstehen* in the manifesto is also representative for orthodox logical empiricism. The rejection of *Verstehen* follows on from the general rejection of knowledge claims based on intuition which was part of early logical empiricism's general strategy to reject all claims to knowledge that resisted the demand for discursive, logical justification. *Verstehen* was rejected in so far as it fell under that rubric, not because it postulated the human sciences to be different from the natural sciences in some more particular methodological aspects.

What also is worth noting, therefore, is that the extensive annotated bibliography of members and sympathisers of the Vienna Circle, which was attached to the Circle's manifesto, contains a reference to a little-known paper by Viktor Kraft on intuitive *Verstehen* in historiography (Kraft 1929). The annotation makes clear its central relevance to the unified science problematic and provides an argument that backs up the manifesto's overall stance.

> The *Geistewissenschaften* as well do not find in *Verstehen* a special justification of their knowledge claims. For intuition cannot serve as an independent legitimation base, as it is

subjectively conditioned and does not allow for a decision when conflicting results are reached. It can only serve in a heuristic function and still must be verified by a logical proof (Carnap et al. [1929, 39]; not translated in 1973 edition).

In its stripped down form one might call this the proto-argument upon which all other logical empiricist rejections of *Verstehen* as an autonomous methodology build. It instantiates the general anti-intuition argument given in the text of the manifesto and so provides a very schematic blueprint for the orthodox logical empiricist response as a whole. This blueprint invokes a distinction that Reichenbach was to codify as that between the contexts of discovery and justification in his *Experience and Prediction* (1938) but had already been invoked by Carnap as one between the realistic description and the rational reconstruction of cognition in his *Aufbau* (Carnap [1928b] 2003). Rational reconstructions were there said to be purged of the sociological and psychological considerations that are pertinent to descriptions of processes of discovery and instead concerned purely rational connections between hypotheses and evidence. The context of discovery concerns how findings are as a matter of historical fact hit upon, whereas the context of justification concerns the systematic validation of knowledge claims. Paradigmatically, Kraft stressed the need for independent validation of any knowledge claims deriving from *Verstehen*.

In a still more radical fashion than Kraft, Carnap had dealt with the problem of the cognition of other minds – he called it "knowledge of the heteropsychological" – already one year earlier in his *Scheinprobleme in der Philosophie*. Knowledge of other minds, he stated, is gained either by reports issued by such an other ($E_1$), by observation of expressive motions or acts ($E_2$), or, occasionally, by knowledge of the other's being in certain external conditions ($E_3$). "There is no other way to gain knowledge of the heteropsychological." Most notably, "in each of the cases, $E_1$, $E_2$, $E_3$, the cognition of the heteropsychological is connected with the perception of physical facts." (Carnap [1928a] 2003, 317) Carnap then gave the following "epistemological analysis", the opening sentence of which makes clear that it was intended to replace the empathy account:

If a psychologist is to justify or defend against doubt the assertion that certain psychological events have taken place within subject A, then no one will be satisfied if he claims that he has simply experienced or clearly felt them. Rather, one demands of him that he should state in which if the three ways, $E_1$, $E_2$, $E_3$, his knowledge was obtained. … in case $E_1$ … he must at least be able to report that he has heard or read some words which were of such a nature that from them the particular psychological events of A can be inferred. Similarly in case $E_2$: the most satisfactory justification consists in describing observed expressive motions or other acts of A, and it is indispensable for any justification that that acts of A can be indicated from which the particular psychological events of A can be inferred. Finally, in case $E_3$ the justification is accomplished through a description of the perceived outward circumstances of A and his already known character (Ibid, 319f.).

Carnap allowed for the possibility of error and lying in reports and the need for prior knowledge of the meaning of the words used in the report, for instance. Nonetheless, he claimed to have "demonstrated that in all cases where heteropsychological occurrences are recognized, the epistemological nucleus of the experience in which the recognition takes place contains nothing but perceptions of physical events" (Ibid, 320f.). Those physical events, in case of $E_1$ and $E_2$, are behavioural events.

Against this it can be argued that Carnap's analysis of our cognition of other minds clearly depends, not only on linguistic understanding, but also on understanding the context in which the other takes herself to be in. The greatest vulnerability of Carnap's argument, however, lies in its reductionism. This reductionism looks like behaviourism, to be precise, the doctrine of logical behaviourism that mental states can be reduced to behavioural expressions, for that is one of the forms taken by Carnap's reductionist attitude towards the psychological. But it was not the only one: in the *Aufbau* (Carnap [1928b] 2003, §57, 92) and "Psychology in Physical Language" (Carnap [1932] 1959, 175f.) Carnap also allowed for the reduction of the psychological to brain processes. Given either behaviourist or physiologist reductionism, reasoning along the lines indicated by Carnap would provide the "logical proof" that Kraft suggested was required to validate empathetic understanding. (Since Carnap admitted that, due to the state of physiology, so far only behaviourist reductions could be provided, we may speak of his view as a form of logical behaviourism if the required qualification is remembered.)

However, even for Carnap this reductionist claim was not uncontentious. How was logical behaviourism to be justified? In the *Aufbau*, he drew the conclusion – twice over! – that "it is in principle possible to reduce all psychological objects to physical objects", but on both occasions did so on somewhat speculative scientific grounds, without either providing concrete instances of the reduction of properties of the psychological processes to properties of the brain process or of the reduction of statements about psychological objects to statements about their behavioural indicators (Carnap [1928b] 2003, §57, 92f.). Carnap begged off providing a proof for his assertion, not wholly convincingly, as not being pertinent to the project of the *Aufbau*; instead, he set out to provide it in *Scheinprobleme* along the lines surveyed above. Behind his epistemological analysis there – again he claimed only to have provided a rational reconstruction of knowledge of other minds – lay a verificationist principle of meaning: to be meaningful – to have "factual content" – a statement must be testable:

> if a statement is testable, then it has always factual content, but the converse does not generally hold. If it is impossible, not only for the moment but in principle, to find an experience which will support a given statement then that statement does not have factual content (Carnap [1928a] 2003, 327).

Note that this was still a fairly liberal verificationism: not only was not the actuality but only the conceivability of verification involved, but also not the conclusive verification of a statement but only its fallible confirmation or disconfirmation. (Wittgenstein's "The meaning of a statement is its method of verification" is considerably stricter in this latter respect.) Even so, we are still owed an argument to the effect that the direct perception of other minds is impossible. Why should this be impossible?

It took Carnap a while before coming up with an answer he himself felt was satisfactory. To be noted here, first, is that for Carnap, the conceivability that the verificationist criterion of meaning turns on was not logical conceivability as such, but conceivability within the framework established by the laws of empirical science: any putative verification that depended on travel at speeds faster than light or communication from beyond the grave was thus ruled out. Importantly, Carnap's meaning criterion presupposed the probity of scientific procedures and did not seek to validate them (despite its reductionist tendencies even his *Aufbau* was not an epistemologically foundationalist account). What Carnap required to recognise in addition was the fact that the enterprise of science presupposes the availability of a physicalistic, non-phenomenalist language – a language speaking of physical objects rather than of how one is "appeared to" – in order to sustain the intersubjective intelligibility of its discourse. With that insight in place, he could argue that the physical language is the "universal language of science" and that therefore, on pain of becoming meaningless otherwise, "psychological sentences ... are always translatable onto physical, language" (Carnap [1932] 1959, 197).

Logical behaviourism (whether liberalised to allow for physiological reductions or not) is mistaken, however. That mental states do not map one-to-one onto behaviour was what Hempel and Carnap freely admitted from the 1950s onwards. Given the inferential nature of our knowledge of other minds, our attributions can be mistaken even if, say, the reporting subject did not lie or err. Yet in 1928 or 1932 Carnap did not allow for this. This raises the question of what the point of logical behaviourism was. Unlike Watson's behaviourism, logical behaviourism was not conceived with eliminative intentions, but rather in order to render psychology respectable as a science that dealt with spatio-temporal processes like all other sciences. Carnap's and Hempel's later retreat to considering terms for mental states as theoretical terms simply amounted to an admission that, while psychological talk requires behavioural criteria for its terms in order to be meaningful, it cannot be definitionally reduced to them.

What prompted the change away from logical behaviourism in the first place, however, were not reflections about psychology but Carnap's investigations into the logic of dispositional concepts generally. As he argued first in 1935 and put it canonically in "Testability and Meaning", if we understand by "disposition concepts" "predicates which enunciate the disposition of a point or body for reacting in such and such a way to such and such conditions", then "disposition-terms cannot be defined by means of the terms by which these conditions and reactions are described" (Carnap 1936–1937, 440). (Carnap's so-called "reduction-sentences" were unable to define but merely served to introduce disposition terms by reference to typical instances of application.) With this then new non-reductive interpretation of dispositions, logical behaviourism lost whatever verificationist bite it once had. Hempel's nomologically based arguments from the 1940s, which incorporate this insight, may therefore be regarded as developing the demand for (fallible) logical proof for attributions of mental states by the formal deductive-nomological model of explanation in place of Carnap's and at the time Hempel's own – we need not consider his version here (Hempel [1935] 1949) except to note that his was an orthodox version – substantive commitment to logical behaviourism.

## 15.3 Hempel, the Deductive-Nomological Model of Scientific Explanation and Empathetic Understanding

The paper commonly held to be the *locus classicus* of the logical empiricist campaign against the separation of the human from the natural sciences in the English-speaking world is C.G. Hempel's "The Function of General Laws in the History" of

1942. In previous assertions of the ideal of the unity of science in English language publications, the anti-separatist consequence tended to be clearly stated but not to be expressly argued for. In one respect, it was just as well that the writings just discussed were unavailable in English: it allowed for a fresh beginning of the anti-*Verstehen* campaign, unencumbered by the need to retract claims made earlier on the basis of logical behaviourism. In another respect, however, continuity was preserved: once again the argument proceeded at a height of lofty abstraction, with no mention at all of who the *Verstehen* theorists were whose claims were being refuted. Again, anti-*Verstehen* animus sprang from very general considerations: as Carnap had put it concerning physics three years earlier, the "understanding which alone is essential in the field of knowledge and science" is "knowing how to use the symbol ... in the calculus in order to derive predictions which we can test by observations" (Carnap 1939, 69). What Hempel set out to show was that history was no different in this respect: as in physics, there was no need for "intuitive understanding" (Ibid, 69).

In his 1942 paper Hempel presented the model of deductive-nomological (DN) model of explanation as a model that covers all forms of scientific explanation in both the natural and the social sciences (together with its variant, the inductive-statistical model). Like Reichenbach who did not invent but canonise the distinction between the contexts of discovery and justification, so Hempel did not invent but canonise the concept of explanation that naturally goes with the by then long-standing idea of hypothetico-deductivism as representative of "the" scientific method. The basic idea of the DN model is that a phenomenon is explained by its subsumption under a general law (thus also the expression "covering law explanation"). The explanation thus contains (as *explanans*) a statement of initial conditions and of the relevant law(s) such that the phenomenon to be explained (the *explanandum*) follows deductively from it. The thought behind this model was that the once contested notion of scientific explanation – it had been thought by theorists like Mach or Duhem to carry objectionable metaphysical baggage – was rendered legitimate by rendering it as a formal argument.

The specific point of Hempel's paper lay in arguing that even an archetypically "individualising" form of inquiry like history depended for its explanations on such general laws. Hempel freely admitted that "most explanations offered in history or sociology, however, fail to include an explicit statement of the regularities they presuppose" (Hempel [1942] 1965, 236). But he argued that this was so either because "the universal hypotheses in question frequently relate to individual or social psychology, which somehow is supposed to be familiar to everybody through his everyday experience; thus they are tacitly taken for granted" or that "it would often be very difficult to formulate the underlying assumption explicitly with sufficient precision and at the same time in such a way that they are in agreement with all the relevant experience available" (Ibid). All this meant, Hempel continued, that such explanations as proffered were really just "explanation sketches" that needed "'filling out' in order to turn into a fully fledged explanation" (Ibid, 238). Those explanation sketches are at least initially acceptable that "indicate, at least roughly, what kind of evidence would be relevant in testing them, and what findings would tend to confirm them" (Ibid).

Hempel noted that the covering law model was in conflict "with the familiar view that genuine explanation in history is obtained by a method which characteristically distinguishes the social from the natural sciences, namely, the method of empathetic understanding". This method he characterised as employing "imaginary self-identification" to arrive "at an understanding and thus at an adequate explanation" of the actions of the historical actors under investigation (Ibid, 239). Against this Hempel argued that such empathetic understanding

> does not in itself constitute an explanation; it rather is essentially a heuristic device; its function is to suggest psychological hypotheses which might serve as explanatory principles in the case under consideration. Stated in crude terms, the idea underlying this function is the following: The historian tries to realize how he himself would act under the given conditions, and under the particular motivations of his heroes; he tentatively generalises his findings into a general rule and uses the latter as an explanatory principle in accounting for the actions of the persons involved. Now, this procedure may sometimes prove heuristically helpful; but it does not guarantee the soundness of the historical explanation to which it leads. The latter rather depends upon the factual correctness of the generalizations which the method of understanding may have suggested. Nor is the use of the method indispensible for historical explanation. A historian may, for example, be incapable of feeling himself into the role of a paranoic personality, and yet he may well be able to explain certain of his actions by reference to the principles of abnormal psychology (Ibid, 239f.).

In short, as Hempel put it some years later, "the existence of empathy on the part of the scientist is neither a necessary nor a sufficient condition for the explanation, or the scientific understanding, of any human action" (Hempel and Oppenheim [1948] 1965, 258). It is not sufficient, since empathetic understanding stands in need of further confirmation (we could have empathetic understanding without explanation), and it is not necessary, since explanation of action is possible without recourse to empathy (we could have explanation without empathetic understanding). Note that Hempel's argument presupposes the distinction between the so-called contexts of discovery and justification: it was in the context of justification that empathy had no role to play.

Defenders of *Verstehen* had several ways of responding open to them. Either, to declare that *Verstehen* did not work like that, or to deny that *Verstehen* incurred the problems alleged. Under the first heading it could be argued that empathetic understanding was merely one aspect of *Verstehen* which also included the comprehension, not only of matters of cultural significance, but also the understanding of other people's utterances and of texts generally. Clearly, however, Hempel was concerned only with the former, as were other logical empiricist critics following him. (I will return to this criticism below.) Another response would have been to challenge the context distinction itself, but that was rarely taken in the *Verstehen* debate.

Accepting the limitations of Hempel's argument, it could be argued that empathetic understanding does not recur to general laws or lawlike generalisations. This incurred the obligation to explain how *Verstehen* did work and hand-waving towards something like a holistic perception of the psychic *Gestalt* at issue was met with the counter that such claims especially stood in need of independent validation. The fall-back position here was similar to the position of the *Verstehen* camp that simply denied, under the second heading, that independent validation was needed.

Independent validation was not possible, but neither was it needed, it was argued, for to claim that it was to foist the straightjacket of natural scientific explanation upon a cognitive enterprise of an entirely different nature. Such defenders of *Verstehen* thus argued that Hempel's argument begged the question: it presupposed, and so could not, establish, the unity of science. This response left open whether with such in-principle lack of validation defenders of *Verstehen* should drop the claim to 'science' for their enterprise (as logical empiricists would argue).

Another stance taken under the second heading took an entirely different path and conceded the need for validation of empathetic understanding, yet also claimed to be able to provide for this. This position was adopted by Weber ([1921] 1978, 9–12) and we shall return to it below; here we merely note that Hempel's arguments in 1942 and 1948 did not take account of it and for that reason deserve criticism (even if one accepts the limited range of his argument): not all *Verstehen* theorists held it to constitute an autonomous methodology. Yet Hempel's argument raised difficulties not only for some *Verstehen* theorists, but also to supporters of unified science: how were they to understand the everyday practice of intentional explanation? Was it reasonable to think of them, when true, as substandard versions of scientific explanations (explanation sketches) and so to demand that the status of the psychological generalisations implicitly relied upon be that of laws?

## 15.4 Abel on the Operation called *Verstehen*

Theodore Abel's influential critique of "The Operation Called *Verstehen*", published in *The American Journal of Sociology* in 1948, is closely associated with the emerging "positivist" dominance of Anglo-American social science. Like Hempel, Abel was concerned with the view that the methodology of *Verstehen* accounts for the dichotomy between the physical and the social sciences. (In fact, Abel's paper can be viewed as an elaboration and amplification of Hempel's 1942 argument.)

Abel provided examples of behavioural analysis that dealt with a single case, a generalisation and a statistical regularity. In all three cases Abel found that "the operation of *Verstehen* involves three steps: (1) internalizing the [observed] stimulus, (2) internalizing the [observed] response, and (3) applying behaviour maxims." (Abel 1948, 215) As regards the internalisation of the observed stimulus, it was held to depend largely upon the use of imagination and "our ability to describe a situation or event by categorizing it and evoking a personal experience which fits into that category." As regards the internalisation of the response, it too depended largely upon the use of the imagination in inferring the motive behind the observed behaviour and selecting one such that according to it that behaviour "fits" as a "solution" to the perceived "problem". As regards the behaviour maxims, finally, Abel declared these to express functional dependencies between "feeling-states" associated with human behaviours: "the functional dependence consists of the fact that the feeling-state we ascribe to a given human action is directed by the feeling-state we presume is evoked by an impinging situation or event. Anxiety directs caution;

a feeling of cold, the seeking of warmth; a feeling of insecurity, a desire for something that will provide reassurance" (Ibid, 215f.).

Abel concluded, first, that the operation called *Verstehen* was based upon the application of personal experience to observed behaviour: "since the operation consists of the application of knowledge we already possess, it cannot serve as a means of discovery. At best it can only confirm what we already know" (Ibid, 216). Second, that "owing to the relative inaccessibility of emotional experiences, most interpretations will remain mere expressions of opinion, subject only to the 'test' of plausibility" (Ibid) Third, that the operation "is not a method of verification": "From the point of view of *Verstehen* alone, any connection that is possible is equally certain." (Ibid) "These limitations", Abel stated, "virtually preclude the use of the operation of *Verstehen* as a scientific tool of analysis". He noted, however, "one positive function which the operation can perform in scientific investigations". It was this: "It can serve as a aid in preliminary explorations of a subject", particularly "in setting up hypotheses, even though it cannot be used to test them" (Ibid, 217).

Abel's analysis appears tied too closely to feeling-states we have personally experienced: most practitioners would want to be able to draw on the entire stock of human emotions that are intelligible to an interpreter. Beyond that, the first criticism is utterly mistaken, whatever we make of the other two. To charge that *Verstehen* "cannot serve as a means of discovery" appears false in several respects. It is false, first, once we observe the distinction between the contexts of discovery and justification. For then we can claim that while it cannot provide confirmation of a hypothesis, *Verstehen* at least provides hypotheses for confirmation. (As Abel put his first criticism, it would even rule out the one "positive function" which he allows *Verstehen* to perform.) Second, to claim, as Abel did, that *Verstehen* does not produce new knowledge is to overlook that knowledge does not only exist at the level of generalisations but also at the level of understanding individual events or actions. Even given the analysis of *Verstehen* as presented, supposing that the interpretation is true, it is clear that new knowledge is gained, namely, of why this individual on this occasion acted as she did. To respond in this second way to Abel's first charge, however, would be to overlook the force of the second and third criticisms: by "supposing that the interpretation is true" the issue is begged against them. For these criticisms say, in effect, that proceeding by *Verstehen* alone we will never be in a position to claim justification for our interpretation beyond some initial subjective plausibility: it is the possession of justification that turns mere true opinion into knowledge.

One issue here is just how strictly we want to understand Abel's talk of "verification". Granting fallibilism for claims to scientific knowledge, one cannot demand certainty for our belief that a given interpretation is correct. What more then than plausibility is required? First of all, that plausibility must be intersubjectively intelligible: it must be plausible to a wide range of interpreters of a given behaviour. Second, there must be some method by which to assess at least rough degrees of plausibility. As described by Abel, however, the method of interpretation does not include such additional resources. So readers of Abel may conclude either that his description of the operation called *Verstehen* is deficient in that it neglects a constituent part of the methodology, or that the operation called *Verstehen* is deficient in that it overlooks what is an essential part of any scientific methodology. Against the latter diagnosis it would not help greatly to claim that the very lack in principle of such "positivist" legitimation distinguishes the social sciences as a class; as noted above, this line is not without grave difficulties.

Defenders of *Verstehen* may be better advised to opt for the former diagnosis. For besides then-contemporary American champions of *Verstehen* in sociology (Cooley, Znaniecki, Sorokin, MacIver) and older philosophical antecendents (Vico, Comte), Abel also cited a long list of German authors dealing with *Verstehen*, among them Dilthey, Jaspers, Rickert, Rothacker, Simmel, Spranger and Max Weber. Just the latter author, as noted already, spent considerable effort on the problem of validating an interpretation in the opening chapter to his monumental *Wirtschaft und Gesellschaft* of 1921. Weber spelled out the difference between an interpretation "adequate on the level of meaning" – one that captures a possible meaning nexus – and one that is "causally adequate" – one that correctly explains the individual behaviour under discussion – and then set out the different conditions of adequacy. Importantly, Weber insisted that however plausible an attribution of a motive for an action to an agent may be on the level of meaning, such a hypothesis will not be considered causally adequate unless there is at least some plausibility that the agent in question did indeed act so for that reason (Weber [1921] 1978, 12); for discussion see Ringer (1997, Chs. 3, 4) and Heidelberger (2009, this volume). Independently of whether Weber's hermeneutic antecedents overlooked the dimension of scientific confirmation or not, Weber himself was not guilty of it. This is a notable weakness in Abel's argument, especially since Weber's considerations figure prominently in the work of his that Abel cited.

## 15.5 Rudner on *Verstehen* and the Unity of Science

For our last example of orthodox logical empiricism, consider a prominent philosopher of science on the same topic barely 20 years later. With Rudner we move into the third phase of the development of logical empiricism. Like Hempel and Abel, Richard Rudner equated *Verstehen* and empathetic understanding from the start. Moreover, he ascribed to Max Weber the view that "to understand Christian martyrdom, to validate hypotheses about its socio-cultural character, to accomplish the requisite task of capturing the 'meaning' martyrdom had for the martyr, we imagine ourselves in the place of, or 'recreate' the psychological states of, the martyr – and thus, presumably, come to the requisite understanding and effect the requisite validation" (Runder 1966, 72). Rudner, it will be seen, counts as ultra-orthodox according to our introductory taxonomy.

Rudner divided the issue into two parts, first, "is *Verstehen* a validational method?"; second, whether it is an indispensable method for social science (Ibid). Yet he presented but one argument to settle both issues. To start with, Rudner pointed out that we need a check on whether such an empathetic state is veridical.

With regard to this, he found the outlook bleak. On the one hand, he asked, "how can we establish independently the reliability of such an empathetic act without having had previous knowledge ... of the very psychological state that is the object of empathy?" On the other hand, he noted, "if we do have this presupposed knowledge ... what more could be methodologically required?" Moreover, were such evidence available, that would only show that other means, besides empathy, are available for acquiring the knowledge sought by empathy (Ibid, 73). By the same argument that establishes when empathetic understanding is trustworthy, it is also established that if it is trustworthy, then it does not represent an indispensible method in social science.

Even more than Hempel and Abel, we must criticise this critic of *Verstehen* of selective reading. On what basis did Rudner "presume" that, for Weber, the validation of an interpretation was given with its mere adoption on part of an interpreter? Readers are not told but instead treated to a discussion of the Weberian concept of cultural significance (albeit not under its original name) and of Winch's use of the Wittgensteinian concept of rule-following (Winch 1958, Chs. 1–2). Rudner held Weberian appeals to the concept of cultural significance responsible for barring the use of normal scientific methods of validation in social science, but rejected such appeals because the question whether a value judgement has been made is not beyond the pale of ordinary fallible methods (again undercutting the claim that *Verstehen* is the only way in). Rudner rejected Winch's conception because the claim that to understand the meanings of actions we need to assume a participant's perspective allegedly turned on the reproductive fallacy: the only understanding appropriate to social science would accordingly be one that consists of a reproduction of the conditions or states of affairs being studied. Both of these arguments have a point in stressing that a certain understanding of valuations being undertaken and a certain understanding of intentional behaviour is possible without *Verstehen*. But the question is, of course, whether we can get, in any other way than by some kind of *Verstehen*, at the specific content of the valuation or intended action. So Rudner's arguments against these further points of Weber's and Winch's beg the central question at issue. Moreover, Rudner's arguments about cultural significance do nothing to ameliorate his failing to even only note, never mind take due account of, Weber's arguments concerning what it takes to render "causally adequate" an interpretation that so far only "adequate on the level of meaning" (Weber [1921] 1978, 11f.).

## 15.6 Discussion of the Orthodox Logical Empiricist Critique and Some Counter-Critiques

We saw that the orthodox logical empiricist position considered so far consisted not just in the rejection of the separation of the natural from the social sciences that was deemed to spring from their allegedly separate methodologies, but also in the rejection of that very *Verstehen* as a legitimate autonomous methodology (in the sense that it not only made for a useful heuristic but also provided the validation of knowledge claims about other minds). Let's now ask whether missing out on Weber's defence was really all that Hempel, Abel and Rudner were mistaken about concerning interpretation in social science. What about their limiting *Verstehen* to empathetic understanding?

In their influential reader on interpretive social science Fred Dallmayr and Thomas McCarthy noted about Charles Taylor's "Interpretation and the Sciences of Man" (1971) – of which it can be said that it put hermeneutics squarely on the map of analytical philosophy – that it rejected the basic presuppositions of the orthodox logical empiricist discussions of *Verstehen*.

> *Verstehen* has to do not with 'inner-organic', 'psychological' states, but with 'intersubjective meanings' which are constitutive of social life. These are not grasped through empathy, but through procedures not unlike those employed in the hermeneutic interpretation of texts. And the logic of such interpretation is markedly different from that obtaining in the nonsocial phenomena... interpretative understanding 'cannot meet the requirements of intersubjective, non-arbitrary verification' that the neopositivists consider essential to science (Dallmayr and McCarthy 1977, 79f.).

Taylor's radical anti-critique charged that the neopositivist critics got the basic facts of interpretation wrong – importantly, not only that they presupposed and so did not establish the unity of science.

Since any discussion of Taylor or, indeed, Dallmayr and McCarthy's reading of Taylor would move us into self-consciously post-positivist territory, I must leave it here (being concerned with excavations on positivist territory) but for two comments. First, that there arises the question of sources, namely whether Taylor's own challenges to the positivist understanding of *Verstehen* (especially its intersubjectivisation tactic) derive from older sources in the *Verstehen* tradition (and if so which ones) or whether they build essentially on the Wittgensteinian conception of meaning and understanding that became available only in the 1950s. Second, we may note that Dallmayr and McCarthy's implicit suggestion that Weber's concern with verification in *Economy and Society* amounts breaking faith with Rickert's Neokantian founding of "cultural science" in a separate ontological domain (a realm which constitutes *Geltung* as opposed to *Sein*) finds its mirror image in more recent criticisms of Taylor's conception of interpretive science – namely, that Taylor overlooks and discounts at his peril the strong causal element that is subserved by interpretation and its narratives (see Martin 1994) – for it is precisely the unconcern with causal explanation that Dallmayr and McCarthy seem to laud as distinctive of and pathbreaking by the Kantian tradition (Dallmayr and McCarthy 1977, 19–21).

But back to our focus on neo-positivism. Dallmayr and McCarthy concede:

> The neopositivist analysis of *Verstehen* as a heuristic device based on empathetic identification is certainly relevant to some of the earlier versions in which the theory of understanding was historically propounded – for instance, to the psychologically oriented conceptions of Schleiermacher and the early Dilthey.

So Dallmayr and McCarthy admit that against certain proponents of *Verstehen* the orthodox logical empricist critique works. Yet they continue:

> Less obvious is its relevance to other versions – for instance, to the neo-Kantian approaches derived from a transcendental conception of culture as constituted by certain *Wertbeziehungen* or value relations (Rickert); to approaches incorporating Hegel's notion of the objective spirit (later Dilthey); or to approaches based on a hermeneutics of language (Heidegger, Gadamer) (Dallmayr and McCarthy 1977, 137).

What has to be considered here, however, is whether against these proponents it may not be possible to field arguments other than those used by Hempel, Abel and Rudner which turn on the alleged impossibility of validating subjective attributions of psychological content. All three of these approaches (Rickert, Dilthey, Heidegger), after all, willfully seek to transcend the limits of mere empirical science. In so far as the possibility of empirical social science is concerned then, these approaches do not appear to be particularly helpful. The same has to be said for hermeneutic critiques based on the work of the later Wittgenstein, whose anti-scientism and abhorrence of all theory undermines the very ambition of social inquiry as a form of science – a state of affairs discerned correctly by its early proponent Winch (1958) and since by Taylor.

Yet does this excuse the limitation of *Verstehen* to empathetic understanding in the orthodox logical empiricist critiques? It is of interest here to note that once the limitation of his argument was pointed out to him (by McCarthy 1973), Abel accepted that "no motivational understanding of an action is possible unless one is familiar with the cultural milieu in which the action takes place" (Abel 1975, 100). Abel, in other words, accepted the idea that there is a *Verstehen* that is wider than empathetic understanding and that the latter may presuppose. Yet Abel also reaffirmed his "commitment to the points" he made about empathetic understanding and set forth the hypothesis "that we grasp the meaning of a cultural phenomenon by conceiving it as an instance of a cultural insight, i.e., a generalization derived from a personal experience in the process of enculturation" (Ibid, 99, 101).

Abel did not say whether he thereby also meant to raise the validational bar for cultural and linguistic understanding. Nevertheless that is one interpretation of his brief return to the topic and it calls our attention to the dispute within the *Verstehen* camp between objectivist (Betti 1962; Hirsch 1967) and transcendental hermeneutics (Gadamer [1960] 1975): can there be objective truth in interpretation through capturing intended meanings? Transcendentalists would answer that it makes no sense to ask for validation of cultural insight as its basis, the process of enculturation, must be regarded as fixing the very horizon of experience and interpretation. Like Taylor, such hermeneuticists grant primacy to intersubjective cultural meanings and resist their reduction to the contents of mental states of individuals – and so intend to challenge the methodological individualism that underlies all logical empiricist social science. Short of abandoning their paradigm then, logical empiricists could only embrace a *Verstehen* that is wider than empathetic understanding and that the latter may presuppose, if even that *Verstehen* were subject to some empirical control (which the *Verstehen* of the transcendentalist hermeneuticists does not seem to be).

This returns us to our earlier question which, differently put, asked whether – short of de-psychologising interpretation, inter-subjectivising meaning and de-scientising social studies – *Verstehen* theorists can escape the neopositivist charges other than by the Weberian route. Here we must note that Hempel's DN-model of scientific explanation did not remain unchallenged in its own domain, natural science, either. Hempel's model itself was charged with putting conditions on scientific explanations that are neither necessary nor sufficient. Thus it was argued, first, that some scientific explanations do not cite general laws and, second, that the deduction of a state of affairs from a set of laws and initial conditions does not yet make for an acceptable explanation of that state of affairs obtaining (for an overview of these discussions; see Salmon [1989]). With philosophers of natural science therefore settling for a menu of different types of explanations (causal explanations, explanations by symmetry, etc.) instead of pressing all of them into one over-arching formalist schema, methodologists of social science tended to feel relieved of the pressure of the positivist-inspired critique that their supposed explanations of individual events fall short of basic standards.

Such relief, however, threatens to be short-lived for the issue that this critique has unearthed does not simply go away. Typically, causal explanations do not abandon the covering law conception altogether but require that the causal mechanisms they invoke be backed by appropriate causal laws. Thus, if *Verstehen* is to lead to causal explanation – especially if a more realist conception of causation than a Humean regularity theory is adopted – something like the positivist challenge remains to be answered. But even if one were to adopt a still more robust conception of causation and adopted an Aristotelian theory of causal powers and capacities, that question would not go away: just what are the causal powers and capacities that an interpretation attributes to agents? So either interpretations need independent confirmation of the presupposed law or generalisation or they need independent confirmation of the presupposed powers and capacities. *Verstehen* will still not work on its own as an explanatory strategy. Weber's proposals remain central – if the ambition to provide causal explanations is to be upheld.

## 15.7 Hempel on Ideal Types and Rational Explanation in Social Science

In light of these anti-positivist criticisms it is of interest to note that, still before most of these were voiced, Hempel himself had moved on. Unlike Rudner and still before him, Hempel developed a detailed and sophisticated and, as it were, anticipatory response during phase three of logical empiricism.

By the time he returned to the topic in 1952, Hempel actually had read his Weber. His argument was no longer that all claims to arrive at knowledge by the method of empathetic interpretation were deficient. Instead, Hempel remarked concerning Weber's attempts to derive criteria for the causal adequacy of interpretations – in particular by the method of "'imaginary experiment' which consists in thinking away certain elements of a chain of motivation and working out the course of action which would then probably ensue, thus arriving at a causal judgement"

– that "Weber's fascinating illustration of the proposed method by reference to interpretative problems of historiography ... shows how well he was aware of the close connection between contrary-to-fact conditionals and general laws" (Hempel [1952] 1965, 162). Now Hempel's worries lay elsewhere. His aim was to show that the alleged differences between the explanatory use of ideal types and the methods of explanation in natural science were "spurious" (Ibid).

Hempel read Weber as conceding that any knowledge gained from *Verstehen* depends on nomological knowledge. Hempel took Weber to have responded independently to the concerns that his own earlier and Abel's critiques made explicit. (Weber accepted the need for independent validation of his explanatory hypotheses.) However, Hempel rejected Weber's "limitation of the explanatory principles of sociology to 'meaningful' rules of intelligible behavior" as "untenable" for it "bars from the field of sociology any theory of behavior which foregoes the use of 'subjectively meaningful' motivational concepts" (Ibid, 163f.). (Like Martin [1994] against Taylor [1971], Hempel argued against Weber that this limitation amounted to an ill-grounded restriction of the objects and methods of social science.) Hempel then went on to compare Weber's use of ideal types in thought experiments to the use of theoretical models in natural science. He noted that such constructs are not like simple concepts at all but require being introduced by reference to "a set of characteristics (such as pressure, temperature, and volumes in the case of an ideal gas) *and* a set of general hypotheses concerning these characteristics: these ideal constructs are like theoretical systems" (Ibid, 168). Needless to say, Hempel also stressed this difference between ideal types in physics and social science: the latter tended to be more intuitive and less quantitative and the class of phenomena for which they are meant to be explanatory was "not always clearly specified" (Ibid, 170). Still, Hempel concluded that the parallel way in which ideal types operate in natural social science, shows "an important logical and methodological similarity between divers branches of empirical science" (Ibid, 171).

Unlike Rudner, then, Hempel did not reject Weber's ideal type method – which relies centrally on *Verstehen* – but instead gave a reconstruction of it that moved it rather closely to certain procedures in the natural sciences. Rather, Hempel concentrated his fire on those who see a fundamental difference between two types of science in the use of the method of *Verstehen*. It is notable too that in doing so he moved beyond the excessively individualistic psychologisation of *Verstehen* that Dallmayr and McCarthy, following Taylor, identified as typical of neopositivistic responses to the problem (that Hempel's explication is not intersubjective enough for them is another matter). In sum, Hempel now was ready to accept hermeneutic methods under a certain description, while upholding the ideal of the unity of science. Just whether hermeneutical practice, however, is happily understood in parallel to explanations by physical idealisations is another question. (Something like that, incidentally, was the reaction of a leading empirical social scientist to Hempel's paper: for Paul Lazarsfeld it exemplified the problematic relation of the schematism of the so-called logic of science to the complexities of actual empirical social science itself (Lazarsfeld 1962 ).

In the early 1960s Hempel returned to the issue of explanation by reasons, paying special attention to what his previous discussions had left out: their double-sidedness in being both normative and explanatory, in allowing critical appraisal as well providing an empirical hypothesis. (This reflects an important line of the partly independent Wittgensteinian critique of talk of reasons as causes in which some analytical defenders of *Verstehen* like Dray [1957] were also involved.) Noting that the "rationality of an action will be understood in a strictly relative sense, as its suitability, judged by the given information, for achieving the specified objective" (Hempel 1965, 465), Hempel suggested that the theory of decision under risk and uncertainty with its principle of maximising expected utility may serve as a broad clarification of the normative aspect of reason explanations even in this relative sense, but he held that that this normative aspect must nevertheless be distinguished from the explanatory office of giving reasons for actions.

Importantly, by this time Hempel had given up any behaviourist ambitions, however residual, and accepted that rational explanations involved recourse to definitionally irreducible mental states. Hempel realised that the concept of a rational agent cannot be explicated "simply by dispositions to respond to specifiable stimuli with certain characteristic modes of behaviour" and that what is required are "higher-order dispositions". The reason was that the relevant situations in which the agent can be expected to behave in a certain way cannot be described in terms of external conditions and stimuli alone but need to invoke "the agent's objectives and beliefs" (Ibid, 473). Since "belief attributions and goal attributions are epistemically interdependent" (Ibid, 475), however, this only raised the question whether the agent was indeed rational at the time such that overall consistency, etc., can be assumed. Moreover, one may wonder whether this is not an indispensible but forever untestable assumption – which would be problematic finding for a supposedly empirical type of explanation. Hempel demurred and outlined circumstances that would allow retention of assumptions about the agent's objectives and beliefs yet would discard the assumption of his rationality: such circumstances arise when the agent makes a variety of mistakes in reasoning. This means that "the agent's beliefs should be represented by some set of sentences that is not closed under logical derivability" (Ibid, 478), such that they may not be taken to believe every logical consequence of what they do believe.

More generally, Hempel believed that the explanatory model concept of consciously rational action tended to be limited to cases where the decision problem of the agent "is clearly structured and permits of a relatively simple solution" and the agent is "sufficiently intelligent" and "circumstances permit careful deliberation free from disturbing influences" (Ibid, 482), though occasionally it can be stretched to non-deliberative actions, just like people can sometimes be described as acting rationally in the utility-maximising sense of decision theory even though they are not aware of calculating the relevant probabilities and utilities, as effecting "a type of conscious decision which is unconsciously rational with quantitative precision" (Ibid, 483). Notably nothing, then, in all of this – neither the recognition of irreducible intentionality nor the recognition of what later writers would call the "holism of the mental" – swayed Hempel's belief that accepting rational explanations as in principle legitimate did not occasion a radical separation between the natural and the social sciences.

Readers will notice that Hempel's explication leads straight to questions indicative of two problems that are central to Donald Davidson's philosophy of mind and, indeed, to rational choice theory. These questions concern, on the one hand, the extent to which the normative element in the model of rational action in fact influences the attribution of beliefs and objectives. Given the epistemic interdependence of their attribution, it would seem that instances of irrationality quite different from those which Hempel was able to discern will remain indiscernible: incompatibilities of beliefs and objectives that get "straightened out" in our attributions without our even noticing. The normative aspect of reason explanation may not be as easily separated from the causal aspect as Hempel supposed. On the other hand, the questions to which Hempel's explication gives rise concern the issue whether the "general hypotheses" concerning the characteristics invoked by the ideal types in rational explanations are or are not laws. If they are not (say they're not even *ceteris paribus* laws), the further question arises just where their explanatory force derives from. Connect this to the issue of singular causal explanation (noted at the end of the previous section) and the entire Wittgensteinian problematic about whether reasons can be causes is rekindled. Again, I cannot pursue these contemporary and decidedly post-positivist issues here. But I also cannot help remarking that Hempel's sophisticated logical empiricist response to the hermeneutic problematic lies remarkably close to current central problems in the philosophy of social science.

## References


Abel T (1948) The operation called *Verstehen*. Am J Sociol 54:211–218

Abel T (1975) Verstehen I and Verstehen II. Theor Decis 6:99–102

Betti E (1962) Die hermeneutik als allgemeine methodik der geisteswissenschaften. Mohr, Tübingen

Carnap R ([1928a] 2003) Pseudoproblems in philosophy. In: The logical structure of the world. pseudoproblems in philosophy. Originally published as Scheinprobleme in der philosophie: das fremdpsychische und der realismusstreit. Weltkreisverlag, Berlin, Open Court, Chicago, pp 301–343

Carnap R ([1928b] 2003) The logical structure of the world. In: The logical structure of the world. Pseudoproblems in philosophy. Originally published as Der logische aufbau der welt. Weltkreisverlag, Berlin, Open Court, Chicago, v–300

Carnap R ([1932] 1959) Psychology in physicalist language. In: Ayer AJ (ed) Logical positivism. Originally published as "Psychologie in physikalischer Sprache", Erkenntnis 3. The Free Press, New York, pp 165–198

Carnap R (1936–1937) Testability and meaning. Philos Sci 3:419–471, 4:1–40

Carnap R (1939) Foundations of Logic and Mathematics. University of Chicago Press, Chicago, IL

Carnap R, Hahn H, Neurath O ([1929] 1973) The scientific conception of the world. The Vienna circle. In: Neurath O (author), Neurath M, Cohen RS (eds) Empiricsm and sociology. Originally published as Wissenschaftliche Weltauffassung. Der Wiener Kreis. Wolf, Wien, pp 299–318

Dallmayr F, McCarthy T (eds) (1977) Understanding and Social Inquiry. University of Notre Dame Press, Notre Dame

Dray W (1957) Laws and Explanation in History. Oxford University Press, Oxford

Gadamer H-G ([1960] 1975) Truth and method. Originally published as Wahrheit und methode. Translation is of the 2nd German ed. 1965. Mohr, Tübingen

Heidelberger M (2009) *Erklären* und *Verstehen* bei Max Weber, Unter Rückgriff auf Johannes von Kries, this volume
Hempel CG ([1935] 1949) The logical analysis of psychology. In: Feigl H, Sellars W (eds) Readings in philosophical analysis. Originally published as Analyse logique de la psychologie, revue de synthése 10. Appleton-Century-Crofts, New York, pp 373–384
Hempel, CG ([1942] 1965) The function of general laws in the history. In: Aspects of Scientific Explanation and Other Essays. Originally published in Philosophy of Science 15. The Free Press, Glencoe, IL, pp 231–243
Hempel CG ([1952] 1965) Typological methods in the natural and social sciences. In: Aspects of scientific explanation and other essays. Originally published by American Philosophical Association (ed). Science, Language and Human Rights. University of Pennsylvania Press, Philadelphia, PA, pp 155–171
Hempel CG ([1962] 2001) Rational action. In: Fetzer JH (ed) The philosophy of Carl G. Hempel. Originally published in Proceedings of the American Philosophical Association 35. Oxford University Press, Oxford, pp 311–328
Hempel CG (1965) Aspects of scientific explanation. In: Aspects of scientific explanation and other essays. The Free Press, Glencoe, IL, pp 331–496
Hempel CG, Oppenheim P ([1948] 1965) Studies in the logic of explanation. In: Hempel CG Aspects of Scientific Explanation and Other Essays. Originally published in Philosophy of Science 15. The Free Press, Glencoe, IL, pp 245–295
Hirsch ED Jr (1967) Validity in Interpretation. Yale University Press, New Haven, CT
Kraft V (1929) Intuitives Verstehen in der Geschichtswissenschaft, Mitteilungen des Instituts für österreichische Geschichte Ergänzungs-Band 11
Lazarsfeld P (1962) Philosophy of science and empirical social research. In: Nagel E, Suppes P, Tarski A (eds) Logic, Methodology and Philosophy of Science. Stanford University Press, Stanford, pp 463–473
Martin M (1994) Taylor on interpretation and the sciences of man. In: Martin M, MacIntyre L (eds) Readings in the Philosophy of Social Science. MIT Press, Cambridge, MA, pp 259–280
McCarthy T (1973) On misunderstanding understanding. Theor Decis 3:351–370
Ringer F (1997) Max Weber's Methodology. The Unification of the Cultural and Social Sciences. Harvard University Press, Cambridge, MA
Rudner R (1966) Philosophy of Social Science. Prentice-Hall, Englewood Cliffs, NJ
Salmon WC (1989) Four Decades of Scientific Explanation. University of Minnesota Press, Minneapolis
Taylor C (1971) Interpretation and the sciences of man. Rev Metaphys 25:3–51
Weber M ([1921] 1978) Economy and society. Originally published as Wirtschaft und gesellschaft, 5th edn, 1972. Mohr, Tübingen.
Winch P (1958) The Idea of Social Science. Routledge, London

# Index

**I**

MIX
Papier aus verantwortungsvollen Quellen
Paper from responsible sources
FSC® C105338

If you have any concerns about our products,
you can contact us on
**ProductSafety@springernature.com**

In case Publisher is established outside the EU,
the EU authorized representative is:
**Springer Nature Customer Service Center GmbH**
**Europaplatz 3, 69115 Heidelberg, Germany**

Printed by Libri Plureos GmbH
in Hamburg, Germany